Quaternion Orders, Quadratic Forms, and Shimura Curves

Volume 22

CRM MONOGRAPH SERIES

Centre de Recherches Mathématiques
Université de Montréal

Quaternion Orders, Quadratic Forms, and Shimura Curves

Montserrat Alsina
Pilar Bayer

The Centre de Recherches Mathématiques (CRM) of the Université de Montréal was created in 1968 to promote research in pure and applied mathematics and related disciplines. Among its activities are special theme years, summer schools, workshops, postdoctoral programs, and publishing. The CRM is supported by the Université de Montréal, the Province of Québec (FCAR), and the Natural Sciences and Engineering Research Council of Canada. It is affiliated with the Institut des Sciences Mathématiques (ISM) of Montréal, whose constituent members are Concordia University, McGill University, the Université de Montréal, the Université du Québec à Montréal, and the Ecole Polytechnique. The CRM may be reached on the Web at www.crm.umontreal.ca.

American Mathematical Society
Providence, Rhode Island USA

Both authors were partially supported by MCYT BFM2000-0627. The first author has also been supported by 2000BEAI400232 and PR2001-0116.

The production of this volume was supported in part by the Fonds pour la Formation de Chercheurs et l'Aide à la Recherche (Fonds FCAR) and the Natural Sciences and Engineering Research Council of Canada (NSERC).

2000 *Mathematics Subject Classification.* Primary 11R52, 11S45, 11F06, 11G18, 11E16, 11E20; Secondary 30F35, 51M10.

For additional information and updates on this book, visit
www.ams.org/bookpages/crmm-22

Library of Congress Cataloging-in-Publication Data

Alsina, Montserrat, 1966–
Quaternion orders, quadratic forms, and Shimura curves / Montserrat Alsina, Pilar Bayer.
p. cm. — (CRM monograph series, ISSN 1065-8599; v. 22)
Includes bibliographical references and index.
ISBN 0-8218-3359-6 (acid-free paper)
1. Quaternions. 2. Forms, Quadratic. 3. Shimura varieties. I. Bayer i Isant, Pilar. II. Title. III. Series.
QA196.A47 2004
512′.5–dc22 2004040995

AMS softcover ISBN: 978-0-8218-9019-6

∞ The paper used in this book is acid-free and falls within the guidelines
established to ensure permanence and durability.
This volume was submitted to the American Mathematical Society
in camera ready form by the Centre de Recherches Mathématiques.
Visit the AMS home page at http://www.ams.org/

10 9 8 7 6 5 4 3 2 1 17 16 15 14 13 12

Contents

List of Tables

List of Figures

Introduction

The purpose of this monograph is to provide an introduction to Shimura curves from a theoretical and algorithmic perspective. Shimura curves lie at the crossroads of many areas, including complex analysis, p-adic analysis, arithmetic, Diophantine geometry, algebraic geometry, algebra, and noncommutative algebra. Our approach to them has two objectives: to construct fundamental domains in the Poincaré half-plane, and to determine their complex multiplication points. Our presentation is based on a previous study of quadratic forms attached to orders in quaternion algebras. The algorithms needed for the computations have been compiled in a package, named `Poincare`, which has been implemented in *Maple* V.

Let H denote a quaternion algebra defined over a totally real number field K of degree d, and let $\mathcal{O}$ be an order in H. We shall deal exclusively with indefinite algebras and, moreover, we suppose that $\mathbb{R} \otimes_{\mathbb{Q}} H = M(2, \mathbb{R}) \times \mathbb{H}^{d-1}$, where $\mathbb{H}$ denotes the skew field of the Hamilton quaternions. By considering an embedding of H into the matrix algebra $M(2, \mathbb{R})$, the group of the units of $\mathcal{O}$ of positive norm provides an arithmetic Fuchsian group $\Gamma \subseteq \mathrm{SL}(2, \mathbb{R})$. The group Γ acts on the Poincaré half-plane $\mathcal{H}$ and the quotient $\Gamma \backslash \mathcal{H}$ yields a Riemann surface which is compact unless $H = M(2, K)$. The projective nonsingular curve attached to this Riemann surface will be denoted by $X(\Gamma)$. From 1960 onward, Shimura outlined a theory for these curves, which came to be known as *Shimura curves*. If $K = \mathbb{Q}$ and Γ is commensurable with a subgroup of the modular group $\mathrm{SL}(2, \mathbb{Z})$, then the curves $X(\Gamma)$ are the classical modular curves, whose study goes back to Klein and Fricke, in the nineteenth century.

One of Shimura's main contributions in the topic that concerns us is the theory of the canonical models. To prove their existence, Shimura describes the curves $X(\Gamma)$ as moduli spaces of principally polarized abelian surfaces with quaternion multiplication and level structure. To prove their uniqueness, Shimura considers what today are called the complex multiplication points of the curves, which correspond to those abelian surfaces which are also endowed with complex multiplication. This modular point of view generalizes the well-known interpretation of the classical modular curves as moduli spaces of elliptic curves with level structure and sheds new light on the theory of complex multiplication.

Our study will focus mainly on the compact case and $K = \mathbb{Q}$. The algorithmic approach to Shimura curves which are nonmodular differs considerably from the modular case, mainly due to the absence of cusps and the lack of numerical information on abelian surfaces, in contrast to the abundant numerical data available for elliptic curves.

In order to develop an effective approach to Shimura curves, quadratic forms are important tools. An interplay mentioned by Poincaré between Fuchsian groups and indefinite ternary quadratic forms has been especially valuable. Indeed, Poincaré used Fuchsian groups in his studies of indefinite ternary quadratic forms:

> *Envisageons une forme quadratique indéfinie F à coefficients entiers [...]. Considérons le groupe principal de F formé de toutes les substitutions à coefficients entiers qui n'altèrent pas cette forme. [...] au groupe principal de F correspondra un groupe fuchsien G, qui sera le groupe fuchsien principal de F.*
>
> H. Poincaré [**Poi87**]

In one of his early articles, Shimura also mentioned Poincaré's point of view:

> [...] *part of the paper is devoted to the theory of a certain type of automorphic functions of one variable known in the literature as functions belonging to indefinite ternary quadratic forms* [**Poi87, FK97**]. *They occur as moduli of abelian varieties of dimension 2 whose endomorphism rings are isomorphic to an order of an indefinite quaternion algebra.*
>
> G. Shimura [**Shi59**]

The study of complex multiplication points in Shimura curves leads to the study of families of binary quadratic forms with algebraic coefficients and to their classification by arithmetic Fuchsian groups. Once a fundamental domain is determined, each Shimura curve provides a reduction theory for a family of binary quadratic forms attached to it. In this regard, the theory parallels Gauss' theory on the classification of binary quadratic forms with integral coefficients by the action of the modular group $\mathrm{SL}(2, \mathbb{Z})$.

Our presentation draws on a variety of papers and books on hyperbolic geometry, quadratic forms, and quaternion orders.

For the study of the action of Fuchsian groups on the Poincaré half-plane, we should mention the works of Ford [**For51**], Lehner [**Leh64**] and Siegel [**Sie71**], which contain results for isometric circles and their role in the construction of fundamental domains.

The main sources for the arithmetic of quaternion orders are still Eichler's articles [**Eic37, Eic38, Eic55b, Eic55a**]. Some of Eichler's results can be found in the text of Vignéras [**Vig80**]. For more general results on orders, we also quote Deuring [**Deu68**] and Reiner [**Rei75**]. We note the works of Hijikata [**Hij74**], in which local Eichler orders are taken into account, and also quote the works of Pizer [**Piz73, Piz76a, Piz76b, Piz80**], where connections between quadratic forms and the arithmetic of Eichler orders are considered, in the context of definite quaternion algebras.

Binary quadratic forms were studied by Gauss in *Disquisitiones Arithmeticae* [**Gau01**]. For higher numbers of variables, we have Ogg [**Ogg69**], Jones [**Jon67**], Serre [**Ser73**], and Lehman [**Leh92**], among many others. Mostly, the forms considered in these works have rational, or rational integer, coefficients. We also note that in the usual computational packages the commands dealing with quadratic forms are restricted to integral quadratic forms.

On relations between quadratic forms and quaternion algebras, we mention the classical work of Latimer [**Lat37**], where a partial correspondence was stated, and those of Brandt [**Bra24, Bra28, Bra43**], where a complete correspondence was presented. The works of Brzezinski ([**Brz80, Brz82, Brz83, Brz90, Brz95**]) focus on the algebraic study of orders; he uses a relationship between quadratic forms and orders and generalizes some results known for Eichler orders.

Next we describe the contents of the monograph, summarizing the contributions contained in each chapter.

Chapter 1 is devoted to general facts on quaternion algebras and their orders. We give the simplest examples of them. To establish notation, the concepts are reviewed. We define what is meant by *nonramified* or by *small ramified* quaternion $\mathbb{Q}$-algebras and prove a classification theorem for them. We introduce the concepts of small ramified algebras of *type A* and of *type B*, in accordance with the classification theorem. These algebras will play a special role throughout the whole monograph, since they will provide the context in which our results will be made explicit. Section 1.2 is devoted to the arithmetic theory of quaternion orders. We derive some basic properties of Eichler orders $\mathcal{O}(D,N)$, of level N in a quaternion $\mathbb{Q}$-algebra of discriminant D. We provide tables of quaternion algebras and tables of Eichler orders for small ramified algebras of type A or B.

In Chapter 2, Shimura curves $X(D,N)$ are formally introduced. Section 2.1 contains a brief description of concepts relative to hyperbolic geometry in the Poincaré half-plane which are needed in the sequel. In Section 2.3, the Fuchsian groups $\Gamma(D,N)$ are introduced. We make them explicit for the small ramified algebras of type A or B. The tables contain constants for Shimura curves $X(D,N)$. For sake of completeness, we also provide tables of known equations of nonmodular Shimura curves, and the list of all hyperelliptic nonmodular Shimura curves, which was obtained by Ogg [**Ogg83**].

Chapter 3 presents a basic approach to ternary and quaternary quadratic forms attached to quaternion algebras or to their orders. The quaternary and ternary normic forms $n_{H,4}$ and $n_{H,3}$ attached to H turned out to be K-forms, a concept which goes back to Brandt. Section 3.6 is devoted to the study of quadratic forms associated to quaternion orders $\mathcal{O}$, in particular the quaternary and ternary normic forms, denoted by $n_{\mathcal{O},4}$ and $n_{\mathcal{O},3}$. We specify relationships between invariants associated to the order and to the quadratic forms, and characterize those normic forms which are K-forms. Explicit criteria are provided in the case of Eichler orders.

In Chapter 4, we reformulate Eichler's theory of optimal embeddings of quadratic orders into quaternion orders in terms of quadratic forms. This point of view has been particularly important for performing effective computations. In Section 4.4, we define the set $\mathcal{H}(\mathcal{O})$ of binary forms associated to an order and give explicit results. In Section 4.5, we obtain bijective mappings between the set of embeddings of orders into orders and families of binary quadratic forms. These forms turn out to have semi-integer quadratic coefficients. Specifically, we make explicit the binary forms which correspond to the nonramified case, and to small ramified cases of type A or B.

In Chapter 5, we construct hyperbolic polygons that are fundamental domains for Shimura curves in the nonramified and small ramified cases. Section 5.1 describes certain properties of fundamental domains for some modular curves of prime level. In Section 5.2 , we characterize quaternion transformations using results for embeddings and quadratic forms from Chapter 4. In particular, in Section 5.3 we study the hyperbolic transformations fixing infinity and some symmetries, analogous to those of the modular case. We show the existence of a principal homothety which replaces the usual translation of the nonramified case. In Section 5.5, we show the graphic representations of explicit fundamental domains for the Shimura curves $X(6,1)$, $X(10,1)$, and $X(15,1)$. We include tables which contain cycles and presentations of the Fuchsian groups involved.

In Chapter 6, we study complex multiplication points of the Shimura curves $X(D,N)$ by using results on the hyperbolic uniformization and the interplay between embeddings and quadratic forms from previous chapters. We compile the graphic results in Sections 6.4 and 6.5. The commands implemented allow us to classify quadratic orders for which a given curve $X(D,N)$ has special complex multiplication points, to obtain the quadratic orders involved, and to compute the complex multiplication points for a fixed quadratic order.

In Chapter 7, we present the instructions for the `Poincare` package. We describe the technical characteristics and we provide a full index of the implemented instructions.

The `Poincare` package has been constructed in conjunction with the results presented throughout this work. It has contributed to the evolution of the results, giving support to the computations, and it has been improved constantly by the theoretical results. It is a good tool for dealing with the arithmetic of the quaternion orders and with the computation of hyperbolic domains for Shimura curves. It allows us to handle quaternions, quaternion orders, quadratic forms, embeddings of orders, points of Shimura curves, among others.

Tables in this monograp are collected in Appendix A. In Appendix B we present a survey of further results on Shimura curves. In Appendix C we also include a list of applications of these curves to the study of outstanding problems of number theory.

In general, at the beginning of each chapter or section, we give references for the known results for which the proofs are omitted.

Acknowledgments. The authors thank Bruce W. Jordan, Carlos Moreno, and Noriko Yui for valuable comments on earlier versions of this monograph, and Rafal Ablamowicz for his collaboration in the first stage of the `Poincare` package. They also thank the hospitality of the Graduate Center and University School of CUNY provided to the first author during the academic year 2000–2001.

CHAPTER 1

Quaternion Algebras and Quaternion Orders

In this chapter we review the prerequisites for quaternion algebras and their orders. The first section contains some notations and background needed in our study. We consider quaternion algebras over a field K of $\mathrm{char}(K) \neq 2$, with special emphasis on the case $K = \mathbb{Q}$. The necessary background can be found in [**Alb34, Ser73, Rei75, Gér79, Vig80**].

Our presentation includes the concept of quaternion $\mathbb{Q}$-algebra with small ramification, tailored to our needs. A classification theorem for these algebras is proved in Theorem 1.22. Besides the matrix algebra, $\mathrm{M}(2, \mathbb{Q})$, which is nonramified, two types of small ramified quaternion algebras will be relevant for our purposes: those of type A, $H_{\mathrm{A}}(p)$, and those of type B, $H_{\mathrm{B}}(p, q)$, for $p \neq q$ rational primes. Their discriminant equals 1, $2p$ and pq, respectively. Most of our calculations in the subsequent chapters will be focused on these cases.

We devote the second section to the arithmetic theory of orders in quaternion algebras, with special emphasis on maximal orders and Eichler orders. We give explicit results for orders in small ramified algebras. The references for the results due to Eichler are principally [**Eic37, Eic38**]. Most of Eichler's results can be found in [**Vig80**]. For Eichler orders in local matrix algebras, see also [**Hij74**].

1.1. Introduction to quaternion algebras

We review the concept of Eichler order in a quaternion algebra and calculate explicit bases for some of the three cases just mentioned.

A K-algebra A (associative and with unity) is a vector space over K with ring structure and with unity, 1_A, in such a way that the inner product in the ring and the product by an scalar are related by $k(uv) = (ku)v = u(kv)$, $u, v \in A$, $k \in K$. The elements $\{k \cdot 1_A : k \in K\}$ are central; that is, $K \subseteq Z(A)$, where $Z(A)$ denotes the center of A. A K-algebra A is called central if $Z(A) = K$. It is called simple if it has no nontrivial bilateral ideals. An homomorphism of K-algebras $\varphi\colon A \to B$ is a K-linear homomorphism of rings.

As usual, we denote by $\mathrm{M}(n, R)$ the ring of $n \times n$ matrices with entries in a ring R; by $\mathrm{GL}(n, R)$ the linear group containing the invertible elements in $\mathrm{M}(n, R)$; and by $\mathrm{SL}(n, R)$ the special linear subgroup of those elements in $\mathrm{GL}(n, R)$ with determinant equal to 1.

1.1.1. Definitions and results.

Definition 1.1. A quaternion K-algebra H is a central simple K-algebra of dimension 4 over K. We denote by H^* its group of units, $H^* = \{u \in H : \exists v \in H, uv = vu = 1\}$.

Over a field K of characteristic different from 2, every quaternion algebra H has a K-basis $\{1, i, j, ij\}$ satisfying the relations $i^2 = a$, $j^2 = b$, and $ij = -ji$, for some $a, b \in K^*$. Conversely, a K-basis and relations such as the previous ones, plus the associative property, define a quaternion K-algebra. In this case, we denote by $\left(\frac{a,b}{K}\right)$ the quaternion algebra H and the basis $\{1, i, j, ij\}$ is called the canonical basis. Of course different couples may lead to isomorphic quaternion K-algebras (cf. Remark 1.11).

DEFINITION 1.2. A quaternion $\omega = x + yi + zj + tij$ in H is called pure if $x = 0$. We denote by H_0 the K-vector space of pure quaternions.

The notion of pure quaternion is independent of the choice of the canonical basis $\{1, i, j, ij\}$, as is shown in the following result.

PROPOSITION 1.3. *Let H be a quaternion K-algebra and consider $\omega \in H$, $\omega \neq 0$. Then $\omega \in H_0$ if and only if $\omega \notin K$ and $\omega^2 \in K$.*

By the Skolem–Noether theorem, the K-automorphisms of a quaternion K-algebra H are the inner automorphisms (that is, the conjugations $\omega \mapsto \sigma^{-1}\omega\sigma$, where $\sigma \in H^*$). Therefore the group $\mathrm{Aut}_K(H)$ of K-automorphisms of H is isomorphic to the quotient group H^*/K^*. As a consequence, for every separable quadratic K-algebra $L \subseteq H$, there exist elements $\theta \in K^*$ and $\omega \in H$ such that $H = L + L\omega$, with $\omega^2 = \theta$ and $\omega m = m'\omega$ for all $m \in L$, where $m \mapsto m'$ denotes the nontrivial K-automorphism of L. This gives another interpretation of quaternion K-algebras.

A quaternion K-algebra is either a skew field or an algebra isomorphic to the matrix algebra $\mathrm{M}(2, K)$; in the first case it is called a division K-algebra, and in the second one, a matrix K-algebra. If K is algebraically closed, we only obtain matrix algebras. If K is a local field ($\neq \mathbb{C}$), there exists a unique division quaternion K-algebra up to isomorphism. If $K = \mathbb{R}$, the unique quaternion division algebra $\mathbb{H}$ is the Hamilton quaternion algebra.

DEFINITION 1.4. Every quaternion K-algebra $H = \left(\frac{a,b}{K}\right)$ is provided with a K-endomorphism which is an involutive antiautomorphism called conjugation; it is denoted by $\omega \mapsto \overline{\omega}$. If $\omega = x + yi + zj + tij$, with $x, y, z, t \in K$, then $\overline{\omega} = x - yi - zj - tij$. The reduced trace and the reduced norm are defined by $\mathrm{tr}(\omega) = \omega + \overline{\omega}$ and $\mathrm{n}(\omega) = \omega\overline{\omega}$, respectively. Thus, $\mathrm{tr}(\omega) = 2x$ and $\mathrm{n}(\omega) = x^2 - ay^2 - bz^2 + abt^2$. Note $\omega \in H_0$ if and only if $\overline{\omega} = -\omega$; in fact, H_0 is the set of quaternions of reduced trace equal to 0. The elements in H^* are the elements of nonzero reduced norm.

If $H = \mathrm{M}(2, K)$, then for $\omega = \left(\begin{smallmatrix} c & d \\ e & f \end{smallmatrix}\right) \in H$, we have $\overline{\omega} = \left(\begin{smallmatrix} f & -d \\ -e & c \end{smallmatrix}\right)$. The reduced trace and reduced norm of an element $\omega \in \mathrm{M}(2, K)$ coincide with the trace and the determinant of ω as a matrix.

PROPOSITION 1.5. *Let H be a quaternion K-algebra. Then, the reduced trace map leads to a nondegenerate symmetric bilinear K-form $\tau\colon H \times H \to K$ defined by $\tau(\alpha, \beta) = \mathrm{tr}(\alpha\beta)$, for $\alpha, \beta \in H$. Moreover, the form τ is associative; that is, $\tau(\alpha\beta, \gamma) = \tau(\alpha, \beta\gamma)$, for $\alpha, \beta, \gamma \in H$.*

LEMMA 1.6. *Let $\psi\colon H \to H'$ be an isomorphism of quaternion K-algebras. Consider $H = \left(\frac{a,b}{K}\right)$ endowed with the canonical basis $\{1, i, j, ij\}$. The following properties hold:*

(i) $\psi(i)^2 = a$, $\psi(j)^2 = b$.
(ii) $\psi(H_0) = H'_0$.
(iii) $\psi(\overline{\omega}) = \overline{\psi(\omega)}$, *for* $\omega \in H$.

(iv) $\mathrm{n}(\psi(\omega)) = \mathrm{n}(\omega)$, $\mathrm{tr}(\psi(\omega)) = \mathrm{tr}(\omega)$, *for* $\omega \in H$.

From now on, K will denote a number field and for each place v of K, K_v will denote the corresponding local field. We denote by R and R_v their rings of integers.

DEFINITION 1.7. Let H be a quaternion K-algebra. For each place v of K, $H_v := K_v \otimes H$ is a quaternion K_v-algebra. If H_v is a division algebra, we say that H is ramified at v; otherwise, we say that H is nonramified at v.

Consider $H = \left(\frac{a,b}{K}\right)$ and v a place of K. The Hasse invariant at v is defined as

$$\varepsilon\left(\frac{a,b}{K}\right)_v = \begin{cases} 1, & \text{if } v \text{ does not ramify in } H, \\ -1, & \text{if } v \text{ ramifies in } H. \end{cases}$$

Recall that, if $K = \mathbb{Q}$, the Hasse invariant at p of the quaternion algebra $H = \left(\frac{a,b}{K}\right)$ coincides with the Hilbert symbol $(a,b)_p$.

The following classification theorem is well known.

THEOREM 1.8. (i) *A quaternion K-algebra H is ramified at a finite even number of places.*

(ii) *Two quaternion K-algebras are isomorphic if and only if they are ramified at the same places.*

(iii) *Given an even number of noncomplex places of K, there exists a quaternion K-algebra that ramifies exactly at these places.*

DEFINITION 1.9. The reduced discriminant D_H of a quaternion K-algebra H is the integral ideal of R equal to the product of prime ideals of R that ramify in H.

COROLLARY 1.10. *Two quaternion K-algebras are isomorphic if and only if they have the same reduced discriminant. In particular, a quaternion K-algebra H is a matrix K-algebra if and only if $D_H = R$.*

If R is a principal ideal domain, we may identify the ideals of R with their generators, up to units. Thus, in a quaternion $\mathbb{Q}$-algebra H, D_H is an integer. The isomorphism class of the matrix $\mathbb{Q}$-algebra $\mathrm{M}(2,\mathbb{Q})$ is characterized by $D_H = 1$.

REMARK 1.11. By using general properties of the Hilbert symbol, the following isomorphisms are obtained:

$$\mathrm{M}(2,\mathbb{Q}) \simeq \left(\frac{1,-1}{\mathbb{Q}}\right) \simeq \left(\frac{1,b}{\mathbb{Q}}\right) \simeq \left(\frac{a,-a}{\mathbb{Q}}\right) \simeq \left(\frac{a,1-a}{\mathbb{Q}}\right), \quad a,b \in \mathbb{Q}^*, a \neq 1.$$

We describe explicitly an isomorphism ψ from $\mathrm{M}(2,\mathbb{Q})$ to $\left(\frac{1,b}{\mathbb{Q}}\right)$, which will be used specially for $b=-1$:

$$\mathrm{M}(2,\mathbb{Q}) \xrightarrow{\psi} \left(\frac{1,b}{\mathbb{Q}}\right)$$

$$\begin{pmatrix} x+y & z+t \\ b(z-t) & x-y \end{pmatrix} \leftarrow\!\shortmid x + yi + zj + tij,$$

$$\begin{pmatrix} \alpha & \beta \\ \gamma & \delta \end{pmatrix} \mapsto \tfrac{1}{2}\big((\alpha+\delta) + (\alpha-\delta)i + (\beta + b^{-1}\gamma)j + (\beta - b^{-1}\gamma)ij\big).$$

The canonical basis of $\left(\frac{1,b}{\mathbb{Q}}\right)$ is mapped to the following basis of $\mathrm{M}(2,\mathbb{Q})$:

$$\left\{ \begin{pmatrix} 1 & 0 \\ 0 & 1 \end{pmatrix}, \begin{pmatrix} 1 & 0 \\ 0 & -1 \end{pmatrix}, \begin{pmatrix} 0 & 1 \\ -1 & 0 \end{pmatrix}, \begin{pmatrix} 0 & 1 \\ 1 & 0 \end{pmatrix} \right\}.$$

DEFINITION 1.12. Let H be a quaternion K-algebra and F a field extension of K. The field F splits H if $H_F := F \otimes_K H \simeq \mathrm{M}(2, F)$.

Given a quaternion K-algebra H, there always exists a field $F \supseteq K$ that splits H. In fact, if $\overline{K}$ is an algebraic closure of K, then $\overline{K}$ splits H. It is clear that if F splits H, then every field $L \supseteq F$ also splits H. Let F be a field splitting H and fix an isomorphism $\varphi\colon F \otimes H \xrightarrow{\sim} \mathrm{M}(2, F)$. Then, the determinant and the trace of the matrix $\varphi(1 \otimes \omega)$ coincide with the norm and the reduced trace of $\omega \in H$, respectively. In particular, the determinant and the trace of the matrix $\varphi(1 \otimes \omega)$ do not depend on the isomorphism φ nor on the field F splitting H.

We will now review the conditions under which a quadratic field over K splits a quaternion K-algebra.

PROPOSITION 1.13. *A finite extension $F \mid K$ splits a quaternion K-algebra H if and only if F_w splits H_v for every place $w \mid v$ of F.*

PROPOSITION 1.14. *Let H be a quaternion K-algebra and F a quadratic field over K. Then the following conditions are equivalent:*

(a) *F splits H.*
(b) *F is K-isomorphic to a maximal subfield of H containing K.*
(c) *There exists a K-embedding $F \hookrightarrow H$.*
(d) *Every place v in K that ramifies in H is not totally split in F.*

The equivalences between conditions (a), (b), and (d) follow from theorems due to Hasse, [**Has33**]. An easy and direct proof for the case $K = \mathbb{Q}$ can be found in [**Lat36**]. Condition (d) follows easily from (c). Namely, if $v \mid D_H$, then H_v is a division K_v-algebra; thus, for the existence of an embedding of F into H, we need an embedding of F_v into H_v, and F_v must be a field. This implies that v does not split in F.

REMARK 1.15. In the case of matrix algebras, it is clear that every quadratic field splits. If $F = K(\sqrt{\delta})$, with $\delta \in K$, (b) and (c) are satisfied considering the following K-embedding in which the norm and the trace are obviously preserved.

$$\begin{aligned} \psi\colon\ K(\sqrt{\delta}) &\hookrightarrow \mathrm{M}(2, K) \\ x + y\sqrt{\delta} &\mapsto \begin{pmatrix} x & y \\ \delta y & x \end{pmatrix}. \end{aligned}$$

REMARK 1.16. Let $F = K(\alpha)$ be a quadratic field that splits a quaternion K-algebra H. An embedding $\psi\colon F \hookrightarrow H$ is characterized by an element $\psi(\alpha) \in H$ such that $\mathrm{n}\big(\psi(\alpha)\big) = \mathrm{n}(\alpha)$ and $\mathrm{tr}\big(\psi(\alpha)\big) = \mathrm{tr}(\alpha)$.

If $F \subseteq H$, then the set of embeddings of F into H is in bijective mapping with the conjugation class of the element $\psi(\alpha)$ in H^*, $\{\sigma^{-1}\psi(\alpha)\sigma \mid \sigma \in H^*\} = \{\omega \in H \mid \mathrm{n}(\omega) = \mathrm{n}(\alpha), \mathrm{tr}(\omega) = \mathrm{tr}(\alpha)\}$.

DEFINITION 1.17. Let K be a totally real number field of degree $[K{:}\mathbb{Q}] = m$. Let H be a quaternion K-algebra. Fix an isomorphism $\mathbb{R} \otimes_{\mathbb{Q}} H \simeq \mathrm{M}(2, \mathbb{R})^r \times \mathbb{H}^{m-r}$. If $r = 0$, H is said to be definite. If $r \geq 1$, H is said to be indefinite.

By the norm theorem, we have $\mathrm{n}(H) = K_H$, where K_H denotes the set of elements in K that are positive in the real infinite places of K that ramify in H. Moreover, suppose K is a totally real field with ring of integers R and consider H

an indefinite quaternion K-algebra. Then, for all $\delta \in R \cap H^*$, there exists $\omega \in H^*$ such that $n(\omega) = \delta$ and $\operatorname{tr}(\omega) \in R$.

The definite or indefinite character of a quaternion $\mathbb{Q}$-algebra H can be read off from the discriminant D_H. An odd number of factors of D_H corresponds to the definite case, and an even number to the indefinite case. The real field $\mathbb{R}$ splits indefinite quaternion $\mathbb{Q}$-algebras. Proposition 1.19 gives an explicit embedding in this case.

REMARK 1.18. If $F = \mathbb{Q}(\sqrt{d})$ splits H, then there exists $b \in \mathbb{Q}^*$ such that $H \simeq \left(\frac{d,b}{\mathbb{Q}}\right)$. In particular, if H is an indefinite quaternion algebra, since $\mathbb{Q}(\sqrt{D_H}) \subseteq H$, we may identify $H \simeq \left(\frac{D_H,b}{\mathbb{Q}}\right)$.

PROPOSITION 1.19. *Let $H = \left(\frac{a,b}{\mathbb{Q}}\right)$ be an indefinite quaternion algebra with $a > 0$. An embedding $\Phi : H \hookrightarrow \mathrm{M}(2,\mathbb{R})$ is obtained by:*

$$\Phi(x + yi + zj + tij) = \begin{pmatrix} x + y\sqrt{a} & z + t\sqrt{a} \\ b(z - t\sqrt{a}) & x - y\sqrt{a} \end{pmatrix}.$$

In fact, Φ is an embedding of H into the matrix algebra $\mathrm{M}(2,\mathbb{Q}(\sqrt{a}))$.

1.1.2. Nonramified and small ramified quaternion $\mathbb{Q}$-algebras. Let p, q be rational primes. Consider now the indefinite quaternion $\mathbb{Q}$-algebras $H = \left(\frac{p,q}{\mathbb{Q}}\right)$.

Denote by $\left(\frac{\cdot}{\cdot}\right)$ the multiplicative symbol of quadratic residues in a wide sense; that is, for an odd prime p:

$$\left(\frac{d}{p}\right) = \begin{cases} 1 & \text{if } d \text{ is a nonzero square module } p, \\ 0 & \text{if } p \mid d, \\ -1 & \text{if } d \text{ is not a square module } p, \end{cases}$$

and for $p = 2$:

$$\left(\frac{d}{2}\right) = \begin{cases} 1 & \text{if } d \equiv \pm 1 \bmod 8, \\ 0 & \text{if } 2 \mid d, \\ -1 & \text{if } d \equiv \pm 5 \bmod 8. \end{cases}$$

In the following lemmas, for a quaternion $\mathbb{Q}$-algebra $H = \left(\frac{p,q}{\mathbb{Q}}\right)$, we characterize the case of division $\mathbb{Q}$-algebras and we calculate the discriminants, depending on p and q. Proofs are straightforward using properties of Hilbert symbol, and are left to the reader.

LEMMA 1.20. *Let $H = \left(\frac{p,q}{\mathbb{Q}}\right)$, p, q primes. Then, there always exists an embedding $H \hookrightarrow \mathrm{M}(2,\mathbb{R})$. Moreover, H is a matrix algebra if and only if one of the following conditions are satisfied: $p = q = 2$; $p = q \equiv 1 \bmod 4$; $q = 2$ and $p \equiv \pm 1 \bmod 8$; $p \neq q$, $p \neq 2$, $q \neq 2$, $\left(\frac{q}{p}\right) = 1$, and either p or q is congruent to $1 \bmod 4$.*

LEMMA 1.21. *Let $H = \left(\frac{p,q}{\mathbb{Q}}\right)$, p, q primes. If $p \equiv q \equiv 3 \bmod 4$ and $\left(\frac{q}{p}\right) \neq 1$, then $D_H = 2p$. If $q = 2$, $p \equiv 3 \bmod 8$, then $D_H = pq = 2p$. If $p \neq q$, p or $q \equiv 1 \bmod 4$ and $\left(\frac{p}{q}\right) = -1$, then $D_H = pq$.*

Using both lemmas, we state the following proposition classifying quaternion algebras $H = \left(\frac{p,q}{\mathbb{Q}}\right)$.

THEOREM 1.22. *Let $H = \left(\frac{a,b}{\mathbb{Q}}\right)$ be a quaternion algebra.*

(i) *If* $D_H = 1$, *then* $H \simeq \mathrm{M}(2,\mathbb{Q}) \simeq \left(\frac{1,-1}{\mathbb{Q}}\right)$.
(ii) *If* $D_H = 2p$, *p prime and* $p \equiv 3 \bmod 4$, *then* $H \simeq \left(\frac{p,-1}{\mathbb{Q}}\right)$.
(iii) *If* $D_H = pq$, *p, q primes,* $q \equiv 1 \bmod 4$ *and* $\left(\frac{p}{q}\right) = -1$, *then* $H \simeq \left(\frac{p,q}{\mathbb{Q}}\right)$.

If a and b are prime numbers, the algebra H satisfies one, and only one, of the three previous statements.

NOTATION 1.23. We denote by $H_{\mathrm{A}}(p)$ the quaternion algebra $\left(\frac{p,-1}{\mathbb{Q}}\right)$, with p prime and $p \equiv 3 \bmod 4$; and we denote by $H_{\mathrm{B}}(p,q)$ the quaternion algebra $\left(\frac{p,q}{\mathbb{Q}}\right)$, with p, q primes, $q \equiv 1 \bmod 4$ and $\left(\frac{p}{q}\right) = -1$.

DEFINITION 1.24. According to Definition 1.7, we call the quaternion $\mathbb{Q}$-algebras having discriminant 1 *nonramified $\mathbb{Q}$-algebras*; that is, the algebras isomorphic to $\mathrm{M}(2,\mathbb{Q})$. We call the quaternion $\mathbb{Q}$-algebras having discriminant greater that 1 *ramified $\mathbb{Q}$-algebras.* We call the quaternion $\mathbb{Q}$-algebras having discriminant equal to the product of two distinct prime numbers *small ramified $\mathbb{Q}$-algebras.* We say that a small ramified $\mathbb{Q}$-algebra is of type A if it is isomorphic to $H_{\mathrm{A}}(p)$, for some prime p; we say that it is of type B if it is isomorphic to $H_{\mathrm{B}}(p,q)$, for some primes p, q.

We complete Theorem 1.22 by giving a representative for each isomorphism class of small ramified $\mathbb{Q}$-algebras.

PROPOSITION 1.25. *Given p and q two different prime numbers, let H be a quaternion $\mathbb{Q}$-algebra of discriminant $D_H = pq$.*

(i) *If* $p \equiv 3 \bmod 4$ *and* $q = 2$, *then* $H \simeq \left(\frac{p,-1}{\mathbb{Q}}\right)$.
(ii) *If* $p \equiv 5 \bmod 8$ *and* $q = 2$, *then* $H \simeq \left(\frac{p,2}{\mathbb{Q}}\right)$.
(iii) *If* $p \equiv 1 \bmod 8$ *and* $q = 2$, *then* $H \simeq \left(\frac{2p,-r}{\mathbb{Q}}\right)$, *where r is a prime number such that* $\left(\frac{r}{p}\right) = \left(\frac{r}{2}\right) = -1$.
(iv) *If p or* $q \equiv 1 \bmod 4$ *and* $\left(\frac{q}{p}\right) \neq 1$, *then* $H \simeq \left(\frac{p,q}{\mathbb{Q}}\right)$.
(v) *If p or* $q \equiv 1 \bmod 4$ *and* $\left(\frac{q}{p}\right) = 1$, *then* $H \simeq \left(\frac{pq,-r}{\mathbb{Q}}\right)$, *where r is a prime number such that* $\left(\frac{r}{s}\right) = \pm 1$ *according to* $s \equiv \mp 1 \bmod 4$, *respectively, for* $s = p, q$; *moreover, if p or* $q \equiv 3 \bmod 4$, *necessarily* $r \equiv 3 \bmod 4$.
(vi) *If* $p \equiv q \equiv 3 \bmod 4$, *then* $H \simeq \left(\frac{pq,-1}{\mathbb{Q}}\right)$.

PROOF. The conditions over the primes p and q cover all the possibilities. Thus, it is only necessary to check the discriminant in each case.

Statements in (i) and (iv) come from previous lemmas.

Note that $(2,p)_p = (2,p)_2 = -1$ if and only if $p \equiv \pm 5 \bmod 8$, and the case $p \equiv 3 \bmod 8$ is included in (i). This proves (ii).

For (iii), let us see that there always exists at least one prime r verifying those conditions. The condition $\left(\frac{r}{2}\right) = -1$ is equivalent to $r \equiv \pm 5 \bmod 8$; fix for example the positive sign. To satisfy the other condition, fix $b \in \mathbb{Z}$ such that $\left(\frac{b}{p}\right) = -1$. We want to look for primes r satisfying $r \equiv b \bmod p$. Since $(8,p) = 1$, by Chinese remainder theorem, the system of congruences $x \equiv 5 \bmod 8$, $x \equiv b \bmod p$ has a solution: $x \equiv x_0 \bmod 8p$. The solutions in $\mathbb{Z}$ form the arithmetic progression $\{x_0, x_0 + 8p, x_0 + 2 \cdot 8p, \dots\}$, and by Dirichlet theorem, assuming $(x_0, 8p) = 1$, there are infinitely many primes r. Actually, we have $2 \nmid x_0$ since $x_0 \equiv 5 \bmod 8$, and $p \nmid x_0$ since $x_0 \equiv b \bmod p$ and $b \neq 0$. It just remains to check that the discriminant of the quaternion algebra in (iii) is $pq = 2p$. Computations with the Hilbert symbol give $\varepsilon_v := (2p, -r)_v = (2, -1)_v(2, r)_v(p, -1)_v(p, r)_v = (2, r)_v(p, r)_v$. In particular, $\varepsilon_p = \left(\frac{r}{p}\right)$, $\varepsilon_2 = \left(\frac{r}{2}\right)$ and $\varepsilon_p = -\left(\frac{p}{r}\right)$. The conditions in (iii) for r are equivalent to asking $\varepsilon_p = \varepsilon_2 = -1$. Now the quadratic reciprocity law gives $\varepsilon_r = 1$.

Analogous arguments work for (v). In this case, to find the discriminant, we have $\varepsilon_v = (pq, -r)_v = (p, -1)_v(p, r)_v(q, -1)_v(q, r)_v$. Assume $p \equiv 1 \bmod 4$; otherwise, replace p with q. Then, $\varepsilon_p = (p, -1)_p(p, r)_p = \left(\frac{r}{p}\right) = -1$. If $q \equiv 1 \bmod 4$, then $\left(\frac{r}{q}\right) = -1$; so, $\varepsilon_q = (q, -1)_q(q, r)_q = -1$ and $\varepsilon_r = (p, r)_r(q, r)_r = 1$; thus the discriminant is pq. If $q \equiv 3 \bmod 4$, we have $\left(\frac{r}{q}\right) = 1$; so, $\varepsilon_q = (q, -1)_q(q, r)_q = -1$, and $\varepsilon_r = (p, r)_r(q, r)_r = 1$, since $r \equiv 3 \bmod 4$; thus, the discriminant is also pq. The condition over $\left(\frac{r}{q}\right)$ is equivalent to $\varepsilon_q = -1$. Note, that for $\varepsilon_q = -\varepsilon_r$ to be satisfied, it is necessary to exclude the case $q \equiv 3 \bmod 4$, $r \equiv 1 \bmod 4$.

Finally, it is clear that the quaternion algebra in (vi) has discriminant pq. □

REMARK 1.26. The algebras in the statement (i) are the small ramified algebras of type A, and those in statements (ii) and (iv) are the small ramified algebras of type B. In fact, note (ii) is a particular case of (iv), and statement (iii) can be also included in (v), removing $q \neq 2$.

Some results about the classification of $\mathbb{Q}$-quaternion algebras given by a pair of primes are shown in Table A.1, in Appendix A, for the primes p, $q < 55$. Table A.2 gives a representative of the isomorphism classes of the small ramified quaternion algebras of discriminant $D < 240$. In Table A.3, we give the quaternion algebras H of discriminant given by product of four primes, $D_H = p_1p_2p_3p_4 \leq 1000$. Enlarging Table A.2 up to discriminant 1000, we would have a representative of all the isomorphism classes of indefinite quaternion algebras of discriminant smaller than 1000.

1.2. Introduction to quaternion orders

Let R denote a Dedekind ring and K its field of fractions. Let H be a quaternion K-algebra.

1.2.1. Orders and ideals.

DEFINITION 1.27. An element $\alpha \in H$ is said to be integral over K if $\mathrm{n}(\alpha)$ and $\mathrm{tr}(\alpha)$ are in R. In general, the set of integral elements in a quaternion algebra is not a ring.

DEFINITION 1.28. An R-lattice Λ of H is a finitely generated R-torsion-free R-module contained in H. An R-ideal I of H is an R-lattice such that $K \otimes_R \Lambda \simeq H$. The inverse of an ideal I is the R-ideal $I^{-1} = \{h \in H \mid IhI \subseteq I\}$. An R-ideal is said to be integral if all its elements are integral.

Every quaternion K-algebra has nontrivial ideals. For example, by taking a K-basis $\{v_1, v_2, v_3, v_4\}$ of H, then $I = R[v_1, v_2, v_3, v_4]$ is an R-ideal, and I_v is an R_v-ideal for any place v of R.

LEMMA 1.29. *Let I and I' be R-ideals of H. Then there exists $\lambda \in R$, $\lambda \neq 0$, such that $\lambda I \subseteq I'$.*

PROPOSITION 1.30. *Let $\mathcal{O}$ be a subset of H. The following conditions are equivalent:*

(i) *$\mathcal{O}$ is a ring whose elements are integral, it contains R and $K \otimes \mathcal{O} = H$.*

(ii) $\mathcal{O}$ *is an R-ideal that is a ring.*

A subset $\mathcal{O}$ *which satisfies these conditions is called an R-order of H.*

The left and right orders associated with an R-ideal I in H are

$$\mathcal{O}_{\mathrm{r}}(I)=\{h\in H \mid Ih\subseteq I\} \quad \text{and} \quad \mathcal{O}_{\mathrm{l}}(I)=\{h\in H\mid hI\subseteq I\}.$$

An R-ideal I is said to be integral if it is contained in the associated orders. An R-ideal can be right, left or bilateral R-ideal in these orders. The reduced norm $\mathrm{n}(I)$ of an R-ideal I is defined as the fractional ideal in R generated by the reduced norms of its elements.

DEFINITION 1.31. The different $\mathcal{D}_{\mathcal{O}}$ of an R-order $\mathcal{O}$ is the bilateral R-ideal of $\mathcal{O}$ computed as the inverse of the dual of $\mathcal{O}$ by the bilinear form given by the reduced trace; that is, $\alpha\in\mathcal{D}_{\mathcal{O}}^{-1}$ if and only if $\mathrm{tr}(\alpha\mathcal{O})\subseteq R$. The reduced discriminant $D_{\mathcal{O}}$ of an R-order $\mathcal{O}$ is the reduced norm of $\mathcal{D}(\mathcal{O})$.

PROPOSITION 1.32. *Let* $\mathcal{O}$ *be an R-order in H. The reduced discriminant* $D_{\mathcal{O}}$ *has the following properties:*

(i) $D_{\mathcal{O}}^2$ *is the ideal of R generated by* $\{\det(\mathrm{tr}(\omega_i\omega_j)) : 1\leq i,j\leq 4, \omega_i,\omega_j\in\mathcal{O}\}$.
(ii) *If* $\{v_1,\dots,v_4\}$ *is an R-basis of the order* $\mathcal{O}$*, then* $D_{\mathcal{O}}^2=R\det(\mathrm{tr}(v_iv_j))$.
(iii) *Let* $\mathcal{O}\subseteq\mathcal{O}'$ *be R-orders in H. Then,* $D_{\mathcal{O}'}$ *divides* $D_{\mathcal{O}}$ *in R and the two are equal if and only if* $\mathcal{O}=\mathcal{O}'$.

COROLLARY 1.33. *Let* $\mathcal{O}\subseteq\left(\frac{a,b}{\mathbb{Q}}\right)$ *be an order. Fix a* $\mathbb{Z}$*-basis* $\mathcal{B}$ *of* $\mathcal{O}$*. Let P be the matrix of the change of basis from the* $\mathcal{B}$*-coordinates to the canonical coordinates. Then,*

$$D_{\mathcal{O}}=|4ab\det P|.$$

PROOF. By the above proposition, $D_{\mathcal{O}}^2=|\det(\mathrm{tr}(v_iv_j))|$ for $\mathcal{B}=\{v_1,v_2,v_3,v_4\}$. We refer by M to the matrix of the bilinear form defined by the trace with respect to the canonical basis. By properties of change of basis, we have $D_{\mathcal{O}}^2=|\det M|(\det P)^2=16a^2b^2(\det P)^2$. □

In any quaternion K-algebra each R-order is contained in a maximal R-order. Next we will discuss more orders derived from maximal orders. They are relevant for the construction of Shimura curves.

DEFINITION 1.34. An Eichler R-order in a quaternion algebra H is the intersection of two maximal R-orders of H.

LEMMA 1.35. *Let* $\psi\colon H\to H'$ *be a K-isomorphism between two K-quaternion algebras. The following properties hold:*

(i) $\alpha\in H$ *is an integral element if and only if* $\psi(\alpha)\in H'$ *is an integral element.*
(ii) $\mathcal{O}\subseteq H$ *is an R-order if and only if* $\psi(\mathcal{O})\in H'$ *is an R-order. Thus,* $\mathcal{O}$ *is a maximal order if, and only if,* $\psi(\mathcal{O})$ *is a maximal order.*
(iii) $\mathcal{O}\in H$ *is an Eichler R-order if and only if* $\psi(\mathcal{O})\in H'$ *is an Eichler R-order.*
(iv) *If* $\mathcal{O}\in H$ *is an R-order, then the equality of discriminants* $D_{\psi(\mathcal{O})}=D_{\mathcal{O}}$ *holds. In particular, if* $\mathcal{O}$ *and* $\mathcal{O}'$ *are conjugate R-orders of H, then they have the same discriminant.*

In what follows, we consider the case $K = \mathbb{Q}$ and determine bases with special conditions.

DEFINITION 1.36. Let $\mathcal{B} = \{v_1, v_2, v_3, v_4\}$ be a $\mathbb{Q}$-basis of H. We define the character $\chi(\mathcal{B})$ of the basis $\mathcal{B}$ by the 1×4 matrix over $\mathbb{Q}$ associated with the trace linear form: $\chi(\mathcal{B}) = \big(\mathrm{tr}(v_1), \mathrm{tr}(v_2), \mathrm{tr}(v_3), \mathrm{tr}(v_4)\big)$.

DEFINITION 1.37. Let $\mathcal{B} = \{v_1, v_2, v_3, v_4\}$ be a $\mathbb{Q}$-basis of H. We say that $\mathcal{B}$ is a normalized basis if $v_1 = 1$ and its character is $\chi(\mathcal{B}) = (2,0,0,1)$ or $(2,0,0,0)$; that is, $v_1 = 1$, $v_2, v_3 \in H_0$ and $\mathrm{tr}(v_4) \in \{0,1\}$. We say that $\mathcal{B}$ is an even normalized basis in the first case, and an odd normalized basis in the second one.

LEMMA 1.38. *Let $\mathcal{O}$ be an order in a quaternion $\mathbb{Q}$-algebra H.*

(i) *There exist normalized $\mathbb{Z}$-bases in $\mathcal{O}$. Its character is an invariant of the order, denoted by $\chi(\mathcal{O})$.*
(ii) *Let $Q \in \mathrm{GL}(4,\mathbb{Z})$. Then Q is the matrix of a change of basis between two normalized $\mathbb{Z}$-bases in $\mathcal{O}$ if and only if $\chi(\mathcal{O})$ is a left eigenvector corresponding to the eigenvalue 1 of Q.*

PROOF. Let us construct a normalized $\mathbb{Z}$-basis of $\mathcal{O}$. The order $\mathcal{O}$ admits a $\mathbb{Z}$-basis of the form $\{1, u_2, u_3, u_4\}$, cf. [**God87**]. Put $u_i' = u_i - [\mathrm{tr}(u_i)/2]$, for $i = 2$, 3, 4. If $u_i' \in H_0$, for every i, we are done. Otherwise, up to permutation, we can assume $\mathrm{tr}(u_4') = 1$. Then, by putting $v_4 = u_4'$ and $v_i = u_i' - \mathrm{tr}(u_i')u_4'$ for $i = 2$, 3, we obtain a normalized basis $\{1, v_2, v_3, v_4\}$. Note that if $\mathrm{tr}(v_4) = 0$, then every element in $\mathcal{O}$ has even trace. The converse is also true. Hence, $\mathrm{tr}(v_4) = 1$ if and only if there exists an element in $\mathcal{O}$ with odd trace. Thus, any normalized basis in $\mathcal{O}$ has the same character.

The condition in (ii) simply means $\chi(\mathcal{O}) = \chi(\mathcal{O})Q$, which follows from (i) by the change of basis relation. □

Using this result we give the following definitions.

DEFINITION 1.39. Let $\mathcal{O}$ be a $\mathbb{Z}$-order in a quaternion $\mathbb{Q}$-algebra H and $\mathcal{B}$ any normalized basis in $\mathcal{O}$. The character of the order $\mathcal{O}$, denoted by $\chi(\mathcal{O})$, is the character of the basis $\mathcal{B}$. Thus, we say that an order is odd if $\chi(\mathcal{O}) = (2,0,0,1)$ or, equivalently, if there exists $\omega \in \mathcal{O}$ with $\mathrm{tr}(\omega) = 1$. Analogously, we say that an order is even if $\chi(\mathcal{O}) = (2,0,0,0)$ or, equivalently, if there exists no $\omega \in \mathcal{O}$ with $\mathrm{tr}(\omega) = 1$; that is, when all its elements have even trace.

LEMMA 1.40. *Let $\mathcal{O}$ be an order in a quaternion $\mathbb{Q}$-algebra. We put $k = 0$ or $k = 1$, depending on whether $\mathcal{O}$ is even or odd. A matrix $Q = (q_{ij}) \in \mathrm{GL}(4,\mathbb{Z})$ yields a change of basis between two normalized basis if and only if $q_{11} = 1$ and $q_{i1} = 0$, for $i = 2$, 3, 4; $2q_{1i} = -kq_{4i}$, for $i = 2$, 3; and $2q_{14} = k(1 - q_{44})$.*

LEMMA 1.41. *Let $\mathcal{O}$ be an order in a quaternion $\mathbb{Q}$-algebra H and $\mathcal{B}$ a normalized basis of $\mathcal{O}$. Then for every $\sigma \in H^*$, $\sigma^{-1}\mathcal{B}\sigma$ is a normalized basis of $\sigma^{-1}\mathcal{O}\sigma$. In particular, we have $\chi(\sigma^{-1}\mathcal{O}\sigma) = \chi(\mathcal{O})$.*

Next, we introduce the concept of denominator of a $\mathbb{Z}$-order.

DEFINITION 1.42. Let $\mathcal{O}$ be an order in a quaternion $\mathbb{Q}$-algebra H. We define the denominator $m_{\mathcal{O}}$ of $\mathcal{O}$ as the minimal positive integer such that $m_{\mathcal{O}} \cdot \mathcal{O} \subseteq \mathbb{Z}[1,i,j,ij]$. Then the ideal $(m_{\mathcal{O}})$ is the conductor of $\mathcal{O}$ in $\mathbb{Z}[1,i,j,ij]$.

LEMMA 1.43. *Let $\mathcal{O}$ be an order in a quaternion $\mathbb{Q}$-algebra H. Fix a basis $\mathcal{B}$ of $\mathcal{O}$ and the canonical basis $\mathcal{C} = \{1, i, j, ij\}$ of H. Let P be the matrix of change of basis from the $\mathcal{B}$-coordinates to the $\mathcal{C}$-coordinates. Then $m_{\mathcal{O}}$ is the minimal positive integer such that the matrix $m_{\mathcal{O}}P$ has entries in $\mathbb{Z}$. Moreover, if $\mathcal{O} \subseteq \mathcal{O}'$, then $m_{\mathcal{O}} \mid m'_{\mathcal{O}}$.*

REMARK 1.44. Consider the order $\mathrm{M}(2,\mathbb{Z})$ in the quaternion algebra $H = \mathrm{M}(2,\mathbb{Q})$. Consider the following bases of H:

$$\mathcal{C} = \left\{ \begin{pmatrix} 1 & 0 \\ 0 & 1 \end{pmatrix}, \begin{pmatrix} 1 & 0 \\ 0 & -1 \end{pmatrix}, \begin{pmatrix} 0 & 1 \\ -1 & 0 \end{pmatrix}, \begin{pmatrix} 0 & 1 \\ 1 & 0 \end{pmatrix} \right\},$$
$$\mathcal{B} = \left\{ \begin{pmatrix} 1 & 0 \\ 0 & 0 \end{pmatrix}, \begin{pmatrix} 0 & 1 \\ 0 & 0 \end{pmatrix}, \begin{pmatrix} 0 & 0 \\ 1 & 0 \end{pmatrix}, \begin{pmatrix} 0 & 0 \\ 0 & 1 \end{pmatrix} \right\},$$
$$\mathcal{B}' = \left\{ \begin{pmatrix} 1 & 0 \\ 0 & 1 \end{pmatrix}, \begin{pmatrix} 0 & 1 \\ 0 & 0 \end{pmatrix}, \begin{pmatrix} 0 & 0 \\ 1 & 0 \end{pmatrix}, \begin{pmatrix} 0 & 0 \\ 0 & 1 \end{pmatrix} \right\}.$$

The canonical basis $\mathcal{C}$ of H, given by the isomorphism in Remark 1.11, is not a $\mathbb{Z}$-basis of the order $\mathrm{M}(2,\mathbb{Z})$, in spite of having integral entries in all the matrices. Both $\mathcal{B}$ and $\mathcal{B}'$ are bases of $\mathrm{M}(2,\mathbb{Z})$. Although $\mathcal{B}$ is the usual one, only $\mathcal{B}'$ is normalized.

The denominator of $\mathrm{M}(2,\mathbb{Z})$ is 2.

1.2.2. Local Eichler orders. Fix a finite place v of a number field K and consider the local field K_v, and the quaternion algebra $H_v = K_v \otimes H$. We will study Eichler orders in this setting. Let R_v denote the ring of integers of K_v and let π be an uniformizer of R_v.

We recall that H_v is either a division algebra or a matrix algebra. We can restrict ourselves to the study of the orders in these cases since, by Lemma 1.35, the property of being an Eichler order is preserved by isomorphisms. The following results describe maximal and Eichler R_v-orders.

Assume that H_v is a division algebra. Let v be a discrete valuation of K_v. Then for $h \in H_v$, $\tilde{v}(h) := v(\mathrm{n}(h))$ defines a discrete valuation in H_v.

The set $\mathcal{O}_v = \{h \in H_v : \mathrm{n}(h) \in R_v\} \subseteq H_v$ is a ring, called the valuation ring of $\tilde{v}$, cf. [**Vig80**]. For every local field F_v, with $K_v \subseteq F_v \subseteq H_v$, the restriction of $\tilde{v}$ to F_v is also a discrete valuation having valuation ring $\mathcal{O}_v \cap F_v$ equal to the ring of integers in F_v. Hence all the elements in $\mathcal{O}_v$ are integral over R_v and $\mathcal{O}_v$ is an order. Since it contains all the integral elements of H_v, it is a maximal order, the unique one. Therefore we obtain the following result.

LEMMA 1.45. *Let H_v be a local division K_v-algebra. Then H_v contains a unique maximal R_v-order, $\mathcal{O}_v = \{h \in H_v : \mathrm{n}(h) \in R_v\}$. Hence, $\mathcal{O}_v$ is the unique Eichler R_v-order in H_v. Moreover the ideal πR_v ramifies: $\pi\mathcal{O}_v = \mathfrak{p}^2$, where $\mathfrak{p}$ is the unique ideal maximal of $\mathcal{O}_v$.*

Now, assume H_v is the matrix K_v-algebra $\mathrm{M}(2, K_v)$. Then we have the following results.

LEMMA 1.46. *Let $H_v = \mathrm{M}(2, K_v)$. Then the maximal R_v-orders in H_v are the $\mathrm{GL}(2, K_v)$-conjugate orders of $\mathcal{O}_v = \mathrm{M}(2, R_v)$.*

PROPOSITION 1.47. *Let $\mathcal{O}_v \subseteq \mathrm{M}(2, K_v)$ be an R_v-order. The following conditions are equivalent:*

(a) *$\mathcal{O}_v$ is an Eichler order.*
(b) *There exists a unique pair $\{\mathcal{O}_1, \mathcal{O}_2\}$ of maximal orders of $\mathrm{M}(2, K_v)$ such that $\mathcal{O}_v = \mathcal{O}_1 \cap \mathcal{O}_2$.*
(c) *There exists a unique $n \in \mathbb{N} \cup \{0\}$ such that the order $\mathcal{O}_v$ is conjugate to the order*

$$\mathcal{O}_{\underline{n}} := \begin{pmatrix} R_v & R_v \\ \pi^n R_v & R_v \end{pmatrix} = \mathrm{M}(2, R_v) \cap \begin{pmatrix} R_v & \pi^{-n} R_v \\ \pi^n R_v & R_v \end{pmatrix},$$

which is an Eichler R_v-order, called the canonical Eichler order of level $\pi^n R_v$.
(d) *The order $\mathcal{O}_v$ contains a subring conjugate to*

$$\begin{pmatrix} R_v & 0 \\ 0 & R_v \end{pmatrix} := \left\{ \begin{pmatrix} r_1 & 0 \\ 0 & r_2 \end{pmatrix} : r_1, r_2 \in R_v \right\}.$$

The ideal $N_{\mathcal{O}_v} := \pi^n R_v$, determined in statement (c), *is called the level of the local Eichler order $\mathcal{O}_v \subseteq \mathrm{M}(2, K_v)$.*

Condition (c) tells us that all the local Eichler orders of the same level are conjugate. Actually the local rings of integers R_v are principal. When the global ring of integers R is also principal and the quaternion algebra is indefinite, then the global Eichler orders are also conjugate, by the Eichler result (cf. Theorem 1.59).

The following definition makes precise the concept of level for any local Eichler order $\mathcal{O}_v$.

DEFINITION 1.48. Let $\mathcal{O}_v$ be an Eichler order in a quaternion K_v-algebra H_v. The level of $\mathcal{O}_v$ is the ideal

$$N_{\mathcal{O}_v} = \begin{cases} R_v & \text{if } H_v \text{ is a division algebra,} \\ N_{\varphi(\mathcal{O}_v)} & \text{where } \varphi\colon H_v \to \mathrm{M}(2, K_v) \text{ is an isomorphism.} \end{cases}$$

REMARK 1.49. For the canonical Eichler order $\mathcal{O}_{\underline{n}} \in \mathrm{M}(2, K_v)$, it is easy to check that $D_{\mathcal{O}_{\underline{n}}} = \pi^n R_v$. Since the discriminant is preserved by isomorphisms, for any Eichler order $\mathcal{O}_v \subseteq H_v \simeq \mathrm{M}(2, K_v)$ the discriminant and the level coincide: $D_{\mathcal{O}_v} = N_{\mathcal{O}_v}$

1.2.3. Global Eichler orders. Suppose that H is a quaternion algebra over a number field K with ring of integers R.

For an R-order $\mathcal{O}$, put $\mathcal{O}_v := R_v \otimes \mathcal{O}$. If v is a finite place, $\mathcal{O}_v$ is a local R_v-order; if v is an infinite place, consider $R_v = K_v$ and $\mathcal{O}_v = H_v$. Then $\mathcal{O} = H \cap (\prod_v \mathcal{O}_v)$. It is easy to prove that $(D_{\mathcal{O}})_v = D_{\mathcal{O}_v}$.

PROPOSITION 1.50. *Let $\mathcal{O}$ be an order of H. Then, $\mathcal{O}$ is a maximal order if and only if $\mathcal{O}_v$ is an R_v-maximal order for every finite place v. Moreover, $\mathcal{O}$ is a maximal order if and only if $D_{\mathcal{O}} = D_H$. In particular, all maximal orders have the same discriminant.*

This result is useful for recognizing maximal orders. For example, $\mathrm{M}(2, R)$ is a maximal R-order of $\mathrm{M}(2, K)$, since it has reduced discriminant equal to R.

PROPOSITION 1.51. *Let $\mathcal{O}$ be an order in a quaternion K-algebra H. Then, $\mathcal{O}$ is an Eichler order if and only if $\mathcal{O}_v$ is an Eichler R_v-order for every finite place v.*

DEFINITION 1.52. The level $N_{\mathcal{O}}$ of a global Eichler order $\mathcal{O}$ is the unique integral ideal N in R such that N_v is the level of each $\mathcal{O}_v$ at each finite place v of K. Thus, $N_{\mathcal{O}} := \prod_v N_{\mathcal{O}_v}$.

The level of an Eichler order $\mathcal{O}$ is well defined. Namely, let $\mathcal{O} = \mathcal{O}' \cap \mathcal{O}'' \subseteq H$ be a global Eichler order, with $\mathcal{O}'$ and $\mathcal{O}''$ maximal orders. By Lemma 1.29, there exist $a, b \in K^*$ such that $a\mathcal{O}' \subseteq \mathcal{O}'' \subseteq b\mathcal{O}'$. But a_v and b_v are units for almost all v. Thus $\mathcal{O}'_v = \mathcal{O}''_v$ and hence $\mathcal{O}_v$ is maximal, of level $N_{\mathcal{O}_v} = R_v$, for almost all v. Hence, there exists a unique ideal N such that $N_v = N_{\mathcal{O}_v}$ for all v.

We will denote by $\mathcal{O}(D, N)$ an Eichler order of level N in a quaternion algebra of discriminant D.

The following proposition, stated in the case $K = \mathbb{Q}$, gives characterizations of Eichler orders.

PROPOSITION 1.53. *Let $\mathcal{O}$ be an order in a quaternion $\mathbb{Q}$-algebra H of discriminant D_H. Let $N \subset \mathbb{Z}$ be an ideal coprime to D_H. Then, the following conditions are equivalent:*

(a) *$\mathcal{O}$ is an Eichler order of level N.*
(b) *For every prime $p \in \mathbb{Z}$, $\mathcal{O}$ satisfies: if $p \nmid N$, the local $\mathbb{Z}_p$-order $\mathcal{O}_p$ is maximal, and if $p \mid N$, $\mathcal{O}_p$ is isomorphic to the order $\begin{pmatrix} \mathbb{Z}_p & \mathbb{Z}_p \\ N\mathbb{Z}_p & \mathbb{Z}_p \end{pmatrix}$.*
(c) *For every prime $p \in \mathbb{Z}$, $\mathcal{O}$ satisfies: if $p \mid D_H$, the local $\mathbb{Z}_p$-order $\mathcal{O}_p$ is maximal, and if $p \nmid D_H$, $\mathcal{O}_p$ is isomorphic to the order $\begin{pmatrix} \mathbb{Z}_p & \mathbb{Z}_p \\ N\mathbb{Z}_p & \mathbb{Z}_p \end{pmatrix}$.*

PROOF. By Proposition 1.51, we can reduce ourselves to proving the equivalences for any finite place p, since this is a local property.

If $p \nmid D_H \cdot N$, then on the one hand we have $H_p \simeq \mathrm{M}(2, \mathbb{Q}_p)$, and on the other hand $\begin{pmatrix} \mathbb{Z}_p & \mathbb{Z}_p \\ N\mathbb{Z}_p & \mathbb{Z}_p \end{pmatrix} = \mathrm{M}(2, \mathbb{Z}_p)$, because N_p is a unit in $\mathbb{Z}_p$. Hence, the statement in (a) means $\mathcal{O}_p$ is an Eichler order of level 1, that is, a maximal order, which coincides directly with (b). In this same case, (c) claims $\mathcal{O}_p$ is isomorphic to $\mathrm{M}(2, \mathbb{Z}_p)$, which is equivalent to being maximal, because all maximal orders are conjugate to this one, by Lemma 1.45.

If $p \mid D_H$, then H_p is a division algebra, hence there is a unique maximal order. This is the condition stated in (c) and in (b), because N and D_H are coprime. In this case, condition (a) means that $\mathcal{O}_p$ is an Eichler order of level 1, also equivalent to being maximal.

Finally, in the case $p \mid N$, H_p is the local matrix algebra. Applying Lemma 1.35 and Proposition 1.47, condition (a) is equivalent to the fact that $\mathcal{O}_p$ is a conjugate of $\mathcal{O}_{\underline{n}}$, up to isomorphism, which agrees with (b) and (c). □

Unlike the case of maximal orders (cf. Proposition 1.50), there is no explicit characterization of Eichler orders in terms of their discriminant, but the following properties allow us to determine some Eichler orders.

PROPOSITION 1.54. *Let $\mathcal{O}$ be an order in a quaternion $\mathbb{Q}$-algebra H of discriminant D_H.*

(i) *If $\mathcal{O}$ is an Eichler order, then $D_{\mathcal{O}} = D_H N_{\mathcal{O}}$ and $\gcd(D_H, N_{\mathcal{O}}) = 1$.*
(ii) *If $D_{\mathcal{O}} = D_H N$ is a square-free integer, then $\mathcal{O}$ is an Eichler order of level N.*
(iii) *Let $\mathcal{O}$ and $\mathcal{O}'$ be conjugate $\mathbb{Z}$-orders in H. Then, $\mathcal{O}$ is an Eichler order of level N if and only if $\mathcal{O}'$ is an Eichler order of level N.*

PROOF. Since the concepts of maximality and level are local it is only necessary to check (i) locally. For the ramified places p, apply Lemma 1.45. For the nonramified places, as we noted in the above remark, the level and the discriminant coincide. Now, assume they are not coprime; that is, let $p \mid D_H$, $p \mid N$. Then, $\mathcal{O}_p$ would be an Eichler order in H_p of level $N_p \neq 1$; but since H_p is a field, it has a unique maximal order, and therefore a unique Eichler order of level 1.

Statement (ii) is proved in [**Eic55b**, Satz3].

Finally, statement (iii) is obtained by applying statement (i) above together with Lemma 1.35. □

PROPOSITION 1.55. *Let H be a quaternion $\mathbb{Q}$-algebra. Let $\mathcal{O}(D,N) \subseteq \mathcal{O}(D,1)$ be an Eichler order of level N and a maximal order in H, respectively. Denote by $[\mathcal{O}(D,1){:}\mathcal{O}(D,N)]$ the index as $\mathbb{Z}$-module. Then, $[\mathcal{O}(D,1){:}\mathcal{O}(D,N)] = N$.*

PROOF. As a $\mathbb{Z}$-module, we have $\mathcal{O}(D,1)/\mathcal{O}(D,N) \simeq \bigoplus_p \mathcal{O}(D,1)_p/\mathcal{O}(D,N)_p$, and $[\mathcal{O}(D,1){:}\mathcal{O}(D,N)] = \prod_p [\mathcal{O}(D,1)_p{:}\mathcal{O}(D,N)_p]$. By Proposition 1.53, for the primes $p \nmid N$, we have $\mathcal{O}(D,1)_p = \mathcal{O}(D,N)_p$; hence $[\mathcal{O}(D,1)_p{:}\mathcal{O}(D,N)_p] = 1$. For the primes $p \mid N$, put $N = p^r N'$, with $p \nmid N'$. It is clear that $[\mathcal{O}(D,1)_p{:}\mathcal{O}(D,N)_p] = p^r$. Thus, $[\mathcal{O}(D,1){:}\mathcal{O}(D,N)] = N$. □

REMARK 1.56. Let $H = \left(\frac{a,b}{\mathbb{Q}}\right)$ be a quaternion algebra of discriminant D and canonical basis $\mathcal{C}$. Let $\mathcal{O}(D,N)$ be an Eichler order in H of level N and $\mathcal{B}$ a $\mathbb{Z}$-basis of $\mathcal{O}(D,N)$. Let P be the matrix of the base change from the $\mathcal{B}$-coordinates to the $\mathcal{C}$-coordinates. Since $D_{\mathcal{O}(D,N)} = DN$, we have $|\det P| = DN/|4ab|$, by Corollary 1.33. Thus, for any basis $\mathcal{B}$ the determinant $|\det P|$ only depends on parameters of the algebra: namely a, b and D, and on the level N. Moreover, if N is square-free and $\det P = r/s \in \mathbb{Q}$ is a reduced fraction, then $2 \mid s$. We also have $2 \mid m_{\mathcal{O}}$.

The following result is well known (cf. [**Vig80**]).

PROPOSITION 1.57. *Let Λ be a lattice in a quaternion K-algebra H. For each finite place v in K, let us consider a local lattice L_v in the local quaternion algebra H_v. Assume that $L_v = \Lambda_v$ for almost all v. Then, there exists a lattice Λ' in H such that $\Lambda'_v = L_v$ for any finite place v.*

In particular, from this proposition we deduce the following corollary.

COROLLARY 1.58. *Let H be a quaternion $\mathbb{Q}$-algebra of discriminant D. Then, for each integer N such that $\gcd(D,N) = 1$, there exist Eichler orders of level N.*

PROOF. It is only necessary to apply the above proposition using the local description of the Eichler orders given by Proposition 1.53. Thus, we obtain a lattice Λ' with the desired local characteristics. Note that Λ'_v is a $\mathbb{Z}_v$-order for all place v. Therefore, $\Lambda' = \bigcap_v \Lambda'_v$ is a $\mathbb{Z}$-order in H which is by construction an Eichler order of the desired level N. □

In Appendix A we give tables with explicit bases for Eichler orders in small ramified algebras.

As we have seen, Eichler orders of the same level are locally conjugate. The following result proved by Eichler states a global conjugation. In particular, it ensures that Eichler orders with the same level in indefinite quaternion $\mathbb{Q}$-algebras are also globally conjugate.

THEOREM 1.59 (cf. [**Vig80**]). *Let K be a totally real number field, and let H be an indefinite quaternion K-algebra. If the ideal class number of K is odd, there is only one conjugacy class of Eichler orders having the same level.*

Given an order $\mathcal{O}$ in H, its normalizer is defined by

$$\mathrm{Nor}(\mathcal{O}) := \{\sigma \in H^* : \sigma\mathcal{O}\sigma^{-1} = \mathcal{O}\}.$$

It is clear that conjugate orders have the same normalizer. Moreover $\mathrm{Nor}(\mathcal{O}) = \{h \in H^* : h \in \mathrm{Nor}(\mathcal{O}_v), \text{for any finite place } v \text{ of } K\}$, hence the normalizer can be studied locally.

Assume that $\mathcal{O}$ is an Eichler order. If $v \mid D_H$, then $\mathcal{O}_v$ is the unique maximal order in the division algebra H_v; thus $\mathrm{Nor}(\mathcal{O}_v) = H_v^*$. If $v \nmid D_H$ and $\mathcal{O}_v$ is a maximal order in $\mathrm{M}(2, K_v)$, then $\mathrm{Nor}(\mathcal{O}_v) = K^*\mathcal{O}_v^*$. If $v \nmid D_H$ and $\mathcal{O}_v$ is the canonical Eichler order $\mathcal{O}_{\underline{n}}$ in $\mathrm{M}(2, K_v)$, then $\mathrm{Nor}(\mathcal{O}_{\underline{n}})$ is generated by $K^*\mathcal{O}_v^*$ and $\left(\begin{smallmatrix} 0 & 1 \\ \pi^n & 0 \end{smallmatrix}\right)$.

1.2.4. Eichler orders in nonramified and small ramified $\mathbb{Q}$-algebras. In the following proposition we provide explicit Eichler orders in nonramified and small ramified quaternion $\mathbb{Q}$-algebras. In the particular case $N = 1$, they are maximal orders.

PROPOSITION 1.60. *Let $N \geq 1$, and p, q different primes.*

(i) *$\mathcal{O}_0(1, N) := \left\{\left(\begin{smallmatrix} a & b \\ cN & d \end{smallmatrix}\right) : a, b, c, d \in \mathbb{Z}\right\}$ is an Eichler order of level N in the matrix algebra $\mathrm{M}(2, \mathbb{Q})$.*
$\mathcal{O}_M(1, N) := \mathbb{Z}[1, (j + ij)/2, N(-j + ij)/2, (1 - i)/2]$ is an Eichler order of level N in the matrix algebra $\left(\frac{1,-1}{\mathbb{Q}}\right)$.

(ii) *$\mathcal{O}_{\mathrm{A}}(2p, N) := \mathbb{Z}[1, i, Nj, (1 + i + j + ij)/2]$ is an Eichler order of level N in the $\mathbb{Q}$-algebra $H_{\mathrm{A}}(p)$, for $N \mid (p - 1)/2$, N square-free.*

(iii) *$\mathcal{O}_{\mathrm{B}}(pq, N) := \mathbb{Z}[1, Ni, (1 + j)/2, (i + ij)/2]$ is an Eichler order of level N, in the $\mathbb{Q}$-algebra $H_{\mathrm{B}}(p, q)$, for $N \mid (q - 1)/4$, $\gcd(N, p) = 1$, N square-free.*

(iv) *$\mathbb{Z}[1, i, j, (1+i+j+ij)/2]$ is an Eichler order of H of level q in the $\mathbb{Q}$-algebra $H = \left(\frac{p,q}{\mathbb{Q}}\right)$ of type A, with $q \equiv 3 \bmod 4$ and $D_H = 2p$.*

PROOF. The case (i) is clear.

Firstly, we need to ensure that the given $\mathbb{Z}$-module $\mathcal{O}$ is an order in the corresponding quaternion algebra. Secondly, we compute the discriminant $D_{\mathcal{O}}$ of the order and use the fact that $D_{\mathcal{O}} = D_H N$ is an square-free integer. Then $\mathcal{O}$ will be an Eichler order of level N.

Let us prove (ii). The condition $N \mid (p-1)/2$ ensures us that $\mathcal{O}_{\mathrm{A}}(2p, N)$ is a $\mathbb{Z}$-order in $H_{\mathrm{A}}(p)$. We compute $D_{\mathcal{O}_{\mathrm{A}}(2p,N)} = 2pN$ and compare with $D_{H_{\mathrm{A}}(p)} = 2p$, cf. Lemma 1.21. Note that, putting together the conditions over N and $p \equiv 3 \bmod 4$, N and $2p$ are coprime. Then, $\mathcal{O}_{\mathrm{A}}(2p, N)$ is an Eichler order of level N.

Let us prove (iii). The condition $N \mid (q-1)/4$ ensures us that $\mathcal{O}_{\mathrm{B}}(pq, N)$ is a $\mathbb{Z}$-order in $H_{\mathrm{B}}(p, q)$. Its discriminant is $D_{\mathcal{O}_{\mathrm{B}}(pq,N)} = pqN$. In this case, $D_{H_{\mathrm{B}}(p,q)} = pq$, and automatically $\gcd(N, q) = 1$. Limiting ourselves to the case N and p coprime, we see that $\mathcal{O}_{\mathrm{B}}(pq, N)$ is an Eichler order of level N.

In case (iv), the condition on the discriminant implies $p \equiv 3 \bmod 4$. Therefore, it is easy to check that $\mathbb{Z}[1, i, j, (1 + i + j + ij)/2]$ is a $\mathbb{Z}$-order of discriminant $2pq$. We thus obtain that the given order is an Eichler order of level q. ☐

We also are able to construct Eichler orders for other levels. In Tables A.4–A.7 in Appendix A, we compute explicit $\mathbb{Z}$-bases for representative orders of the conjugacy classes of Eichler orders of consecutive levels N in the small ramified algebras $H_{\mathrm{A}}(3)$, $H_{\mathrm{B}}(2,5)$, $H_{\mathrm{A}}(7)$, and $H_{\mathrm{B}}(3,5)$.

CHAPTER 2

Introduction to Shimura Curves

The goal of this chapter is to introduce Shimura curves $X(D,N)$ attached to Fuchsian groups defined from Eichler orders $\mathcal{O}(D,N)$ in quaternion $\mathbb{Q}$-algebras. First, we describe the hyperbolic structure of the Poincaré half-plane and we give some results related to its automorphism group. Next, we define the arithmetic Fuchsian groups $\Gamma(D,N)$ and calculate them in the nonramified and small ramified cases of type A or B. The moduli point of view of Shimura and the main facts concerning the canonical models are summarized in Section 2.4. A compilation of known results allows us to implement instructions in the `Poincare` package in order to compute the constants associated to the Shimura curves $X(D,N)$ and to present them in tables collected in Appendix A.

2.1. The Poincaré half-plane

We denote by $\iota \in \mathbb{C}$ an imaginary complex number such that $\iota^2 = -1$ and by $\mathrm{Re}(z)$ and $\mathrm{Im}(z)$ the real and the imaginary part of a complex number $z = \mathrm{Re}(z) + \mathrm{Im}(z)\iota$. The Poincaré half-plane is the complex upper half-plane $\mathcal{H} = \{z \in \mathbb{C} : \mathrm{Im}(z) > 0\}$ endowed with the structure given by the hyperbolic metric.

We use the usual notion of point and angle that $\mathcal{H}$ inherits from $\mathbb{C}$. The hyperbolic lines are the semilines that are orthogonal to the real axis $\mathbb{R}$ and the semicircles centered on real points.

The hyperbolic distance $\delta(z_1, z_2)$ between two points $z_1, z_2 \in \mathcal{H}$ is defined by

$$\delta(z_1,z_2) := \left|\operatorname{arc\,cosh}\left(1 + \frac{|z_1 - z_2|^2}{2\,\mathrm{Im}(z_1)\,\mathrm{Im}(z_2)}\right)\right|.$$

The hyperbolic distance can be computed in different ways. If two points $z_i = x_i + \iota y_i$ have the same real part, then $\delta(z_1,z_2) = \big|\ln|y_2/y_1|\big|$. If the hyperbolic line that connects the two points is an arc of circle, then $\delta(z_1,z_2) = \big|\ln|\tan(\frac{1}{2}\theta_1)/\tan(\frac{1}{2}\theta_2)|\big|$, where θ_1 and θ_2 are the angles determined by the points z_1 and z_2. The hyperbolic measure of the angles coincides with the Euclidean one.

The hyperbolic distance is additive over the hyperbolic lines and satisfies the triangular inequality. The geodesics are the hyperbolic lines.

The hyperbolic circles, that is, the geometric locus of the points in $\mathcal{H}$ with hyperbolic distance from a fixed point z_0 equal to r, coincide with the Euclidean circles contained in $\mathcal{H}$.

A hyperbolic polygon of $\mathcal{H}$ is a subset of $\mathcal{H}$ bounded by a simple curve, formed by a finite number of geodesic segments. The hyperbolic volume of a hyperbolic polygon can be computed with the measure

$$d\omega = \frac{4}{(1-|z|^2)^2}.$$

PROPOSITION 2.1. *Let n be the number of vertices of a hyperbolic polygon P and let θ_i be the interior angle in the vertex v_i, $1 \leq i \leq n$. Then the hyperbolic volume of the polygon is*

$$V_h(P) = (n-2)\pi - (\theta_1 + \cdots + \theta_n).$$

2.2. Linear fractional transformations

In this section, we describe the transformations acting on the Poincaré half-plane. We classify them by their Jordan matrix and by their associated binary quadratic form. General definitions and known results can be found in [**Poi87, Shi71, Sie71**].

A linear fractional transformation (or homographic transformation) $\gamma\colon \mathcal{H} \to \mathcal{H}$ is defined by

$$\gamma(z) = \frac{az+b}{cz+d}, \quad \text{where } a,b,c,d \in \mathbb{R}, ad-bc=1.$$

The set of these transformations of $\mathcal{H}$ is a group isomorphic to $\mathrm{PSL}(2,\mathbb{R}) = \mathrm{SL}(2,\mathbb{R})/\pm\mathrm{Id}$. Usually, we deal with groups of matrices $\Gamma \subseteq \mathrm{SL}(2,\mathbb{R})$. As a group of transformations, it is necessary to consider its image $\overline{\Gamma}$ in $\mathrm{PSL}(2,\mathbb{R})$.

Linear fractional transformations can also be considered as mappings of the complex plane $\mathbb{C}$ onto itself. Below, we describe another important mapping of the complex plane onto itself, namely, the inversions. Although they are not conformal mappings, they are useful for the geometric interpretation of linear fractional transformations.

DEFINITION 2.2. Let $C = C(o,r)$ denote the circle with center $o \in \mathbb{C}$ and radius r. The inversion with respect to the circle C is the mapping $f\colon \mathbb{C}\cup\{\infty\} \to \mathbb{C}\cup\{\infty\}$ interchanging the points o and ∞ and such that for each point $z \in \mathbb{C}-\{o\}$, $f(z) = w$ lies in the line determined by z and o, in such a way that the scalar product of $\overrightarrow{oz}\cdot\overrightarrow{ow}$ (as vectors in $\mathbb{R}^2$) equals r^2. The circle C can be recovered as the fixed point set for f.

The inversion with respect to a circle is the composition of the complex conjugation and a conformal mapping. Thus this inversion changes the orientation of the angles, but preserves their magnitude. Therefore, it is an anticonformal mapping. Every linear fractional transformation of $\mathrm{PSL}(2,\mathbb{R})$ can be expressed as a composition of two inversions with respect to suitable circles. In the following lemma we give analytic expressions for the inversion with respect to a circle C.

LEMMA 2.3. *Let C be a circle in the complex plane and $f\colon \mathbb{C}\cup\{\infty\} \to \mathbb{C}\cup\{\infty\}$ the inversion with respect to C.*

(i) *If the equation of the circle C is $az\bar{z} + bz + \bar{b}\bar{z} + c = 0$, then $f(z) = (-\bar{b}\bar{z} - c)/(a\bar{z} + b)$.*
(ii) *If $C = C(o,r)$ is the circle of center $o \in \mathbb{C}$ and radius r, then $f(z) = o + r^2/(\bar{z} - \bar{o})$.*

REMARK 2.4. Expression (i) includes the case $a = 0$, where C is a vertical line; in this case the inversion is exactly the symmetry with respect to C. Using expression (ii), if C is centered at the origin, then $f(z) = r^2/\bar{z}$. In particular, for the unit circle we obtain the standard expression $f(z) = 1/\bar{z}$. If C is a hyperbolic line in the half-plane $\mathcal{H}$, then the inversion with respect to C is called a hyperbolic symmetry with respect to this line.

DEFINITION 2.5. Given $\gamma = \left(\begin{smallmatrix} a & b \\ c & d \end{smallmatrix}\right) \in \mathrm{SL}(2,\mathbb{R})$ such that $c \neq 0$, the circle $C_\gamma := \{z \in \mathbb{C} : |cz + d| = 1\}$ is called isometric circle of γ. We denote by r_γ and o_γ its radius and center, respectively.

LEMMA 2.6. *For each $\gamma \in \mathrm{SL}(2,\mathbb{R})$ there exists a line, denoted by L_γ, with equation $x = t$, for some $t \in \mathbb{R}$, such that γ is equivalent to the inversion with respect to the circle C_γ followed by the reflection with respect to the line L_γ.*

REMARK 2.7. Using the above notations, the radius and the center of C_γ are the real numbers $o_\gamma = -d/c$ and $r_\gamma = 1/|c|$. Thus, isometric circles are hyperbolic lines of $\mathcal{H}$. This allows us to reformulate the former lemma, to saying that the homography γ is the composition of the two hyperbolic symmetries with respect to the associated isometric circle C_γ and line L_γ.

To compute the fixed points of $\gamma \in \mathrm{SL}(2,\mathbb{R})$ acting on $\mathbb{C} \cup \{\infty\}$, it is necessary to solve the quadratic equation $cz^2 + (d-a)z - b = 0$. This leads to the following well-known equivalent definitions.

DEFINITION 2.8. Let $\gamma = \left(\begin{smallmatrix} a & b \\ c & d \end{smallmatrix}\right) \in \mathrm{SL}(2,\mathbb{R})$. Assume that γ defines a transformation of $\mathbb{C}$ different from $\pm\,\mathrm{Id}$.

(a) We call γ an hyperbolic transformation if it has two different fixed points in $\mathbb{R} \cup \{\infty\}$; equivalently, if $(a+d)^2 > 4$; or if $|\operatorname{tr}(\gamma)| > 2$.
(b) We call γ an elliptic transformation if it has a fixed point $z \in \mathcal{H}$ and the other fixed point is $\bar{z}$; equivalently, if $(a+d)^2 < 4$; or if $|\operatorname{tr}(\gamma)| < 2$.
(c) We call γ a parabolic transformation if it has a unique fixed point in $\mathbb{R} \cup \{\infty\}$; equivalently, if $(a+d)^2 = 4$; or if $|\operatorname{tr}(\gamma)| = 2$.

Each matrix $\gamma \in \mathrm{SL}(2,\mathbb{R})$, $\gamma \neq \pm\,\mathrm{Id}$, is conjugate over $\mathbb{C}$ to one of the following canonical Jordan forms:

$$\begin{pmatrix} \lambda & 0 \\ 0 & \lambda^{-1} \end{pmatrix}, \text{ with } \lambda \in \mathbb{C}^*, \lambda \neq \pm 1, \quad \text{or} \quad \begin{pmatrix} \pm 1 & 0 \\ 1 & \pm 1 \end{pmatrix}.$$

The parabolic, hyperbolic or elliptic condition of a transformation can be read from the eigenvalues of its matrix, as we can see in the following lemma.

LEMMA 2.9. *Given a transformation $\gamma \in \mathrm{SL}(2,\mathbb{R})$, let λ, λ^{-1} denote its complex eigenvalues and put $\mu := \lambda/\lambda^{-1} = \lambda^2$. Then,*

(i) *γ is hyperbolic if and only if $\mu \in \mathbb{R}^+$ and $\mu \neq 1$. In particular, if γ is hyperbolic, then the two eigenvalues are different and γ is diagonalizable.*
(ii) *γ is elliptic if and only if $\mu = e^{\iota\theta}$ and $0 < \theta < 2\pi$; in this case $\lambda + \lambda^{-1} = 2\cos\theta$. In particular, if γ is elliptic, then the two eigenvalues are different and γ is diagonalizable.*
(iii) *γ is parabolic if and only if $\mu = 1$. In particular, if γ is parabolic, it only has one eigenvalue but the eigenspace has dimension 1, thus γ is not diagonalizable.*

In the nonparabolic cases, the value μ is called the multiplicator of γ. It has the following geometric interpretation. Consider the transformation γ given by some matrix M, and consider the change of variables transforming M in the corresponding Jordan matrix. If γ is hyperbolic or elliptic, this change of variables maps the two fixed points to 0 and infinity, respectively. Then the transformation is geometrically an homothety of ratio μ with center in the origin in the hyperbolic case, and a rotation of angle θ equal to the argument of μ around the origin in the elliptic case. In the case of a parabolic transformation, the change of variables maps its unique fixed point to infinity and geometrically it is a translation.

Elliptic transformations such that $\theta = r\pi$, with $r \in \mathbb{Q}^*$, have finite order. Parabolic transformations have infinite order.

LEMMA 2.10. *Let $\gamma \in \Gamma \subseteq \mathrm{SL}(2,\mathbb{R})$ define an elliptic transformation in the group $\mathrm{PSL}(2,\mathbb{R})$. If $\mathrm{tr}(\gamma) = 0$, then γ has order 2 or 4 in $\mathrm{PSL}(2,\mathbb{R})$, depending on $-\mathrm{Id} \in \Gamma$ or $-\mathrm{Id} \notin \Gamma$, respectively. If $\mathrm{tr}(\gamma) = 1$, then γ has order 3 or 6 in $\mathrm{PSL}(2,\mathbb{R})$, depending on $-\mathrm{Id} \in \Gamma$ or $-\mathrm{Id} \notin \Gamma$, respectively. These are the two only possibilities for the elliptic transformations with integral traces.*

The classification of the transformations and their fixed points can be also interpreted in terms of suitable binary quadratic forms and their associated points. General definitions about quadratic forms can be found in Chapter 3.

DEFINITION 2.11. Given a matrix $\gamma = \left(\begin{smallmatrix} a & b \\ c & d \end{smallmatrix}\right) \in \mathrm{M}(2,\mathbb{R})$, the binary quadratic form associated with it is $f_\gamma(X,Y) := cX^2 + (d-a)XY - bY^2$.

For a binary quadratic form f of real coefficients, $f(X,Y) = AX^2 + BXY + CY^2$, put $\mathcal{P}(f)$ the set of complex points with non negative imaginary part which are roots of the quadratic equation $AX^2 + BX + C = 0$. That is,

$$\mathcal{P}(f) = \{z : Az^2 + Bz + C = 0, \mathrm{Im}(z) \geq 0\}.$$

If $\mathcal{P}(f) \cap \mathcal{H} \neq \varnothing$, then $\mathcal{P}(f)$ contains only one point, denoted by $\tau(f)$.

LEMMA 2.12. *Let $\gamma \in \mathrm{M}(2,\mathbb{R})$.*

(i) *For all $\lambda, \mu \in \mathbb{Q}$, we have $f_{\lambda\gamma} = \lambda f_\gamma$ and $f_{\gamma+\mu\,\mathrm{Id}} = f_\gamma$; in particular, $\mathcal{P}(f_{\lambda\gamma+\mu\,\mathrm{Id}}) = \mathcal{P}(f_\gamma)$.*
(ii) *Let $z \in \mathcal{H} \cup \mathbb{R}$. Then z is a fixed point of γ if and only if $z \in \mathcal{P}(f_\gamma)$.*
(iii) *Let $P \in \mathrm{GL}(2,\mathbb{R})$. Then $f_{P^{-1}\gamma P} = (\det P^{-1}) P^t f_\gamma P$.*

The geometric interpretation given in the following proposition justifies the names of the transformations.

PROPOSITION 2.13. *Let $\gamma \in \mathrm{SL}(2,\mathbb{R})$ and consider the associated binary form f_γ. Let $\kappa \in \mathbb{R}$, $\kappa \neq 0$. Then:*

(i) *γ is hyperbolic if and only if the conic $f_\gamma = \kappa$ is a hyperbola.*
(ii) *γ is elliptic if and only if the conic $f_\gamma = \kappa$ is an ellipse.*
(iii) *γ is parabolic if and only if the conic $f_\gamma = \kappa$ is a parabola.*

PROOF. Consider the binary quadratic form f_γ attached to γ. On the one hand, we have $\det_1(f_\gamma) = -bc - \frac{1}{4}(d-a)^2) = 1 - \frac{1}{4}(a+d)^2$, since $\det\gamma = 1$. Hence, according to its sign, the quadratic form f_γ is indefinite, definite or degenerate according to whether $(a+d)^2$ is greater, lesser or equal to 4, respectively. This condition is precisely equivalent to the fact that the transformation is hyperbolic, elliptic or parabolic, respectively. On the other hand, it is clear that the indefinite, definite or degenerate condition of f_γ determine the fact that the conic $f_\gamma = \kappa$ is a hyperbola, an ellipse or a parabola, respectively. □

The following lemma contains a technical condition that we shall need in later chapters.

LEMMA 2.14. *Let $\gamma \in \mathrm{GL}(2,\mathbb{R})$ with $\det(\gamma) > 0$. Then $\sigma^{-1}\gamma\sigma \neq -\gamma$, for any $\sigma \in \mathrm{GL}(2,\mathbb{R})$.*

PROOF. If $\operatorname{tr}(\gamma) \neq 0$, the statement is true, since the trace is preserved by conjugation and $\operatorname{tr}(-\gamma) = -\operatorname{tr}(\gamma)$. Assume $\operatorname{tr}(\gamma) = 0$ and consider the associated quadratic forms. If $\det(\gamma) > 0$, then f_γ and $f_{-\gamma}$ are definite binary quadratic forms, one positive definite and the other negative definite. The signature is an invariant of the equivalence class of quadratic forms over $\mathbb{R}$; hence, they cannot be equivalent. $\square$

REMARK 2.15. The above result does not extend to the case $\det(\gamma) < 0$. For example, we have $\sigma^{-1}\gamma\sigma = -\gamma$ for $\sigma = \left(\begin{smallmatrix} 0 & 1 \\ 1 & 0 \end{smallmatrix}\right)$ and $\gamma = \left(\begin{smallmatrix} 1 & 0 \\ 0 & -1 \end{smallmatrix}\right)$.

In particular, if z is a real fixed point of γ, then f_γ is isotropic over $\mathbb{R}$, since $f_\gamma(z,1) = 0$. The elliptic case corresponds to the quadratic form f_γ being anisotropic over $\mathbb{R}$.

DEFINITION 2.16. Let $\Gamma \subseteq \mathrm{SL}(2,\mathbb{R})$ be a subgroup acting on the Poincaré half-plane $\mathcal{H}$. The action of Γ is proper and discontinuous if there exist a point z_0 and a real number $\varepsilon > 0$ such that, for all $\gamma \in \Gamma$, $\gamma \neq \pm\mathrm{Id}$, the condition $|\gamma(z_0) - z_0| > \varepsilon$ is satisfied. In this case, z_0 is called a standard point with respect to Γ.

The definition of proper and discontinuous action is equivalent to the fact that Γ is a discrete subgroup of $\mathrm{SL}(2,\mathbb{R})$.

DEFINITION 2.17. The action of Γ in $\mathcal{H}$ gives an equivalence relation between the points: two points $z, z' \in \mathcal{H}$ are called equivalent with respect to Γ if and only if $z' = \gamma(z)$ for some $\gamma \in \Gamma$.

DEFINITION 2.18. A point $x \in \mathbb{R}\cup\{\infty\}$ is called parabolic (or hyperbolic) with respect to Γ if there exists a transformation $\gamma \in \Gamma$ which is parabolic (or hyperbolic) and such that $\gamma(x) = x$. A point $z \in \mathcal{H}$ is called elliptic with respect to Γ if there exists an elliptic transformation $\gamma \in \Gamma$, $\gamma \neq \pm\mathrm{Id}$, such that $\gamma(z) = z$. The isotropy group of a point z with respect to Γ is the group $\Gamma_z = \{\gamma \in \Gamma \mid \gamma(z) = z\}$.

If Γ is a discrete subgroup of $\mathrm{SL}(2,\mathbb{R})$, the isotropy group of an elliptic point is finite and cyclic. The matrices of order 2 lead to the involutions and they have multiplicator $\mu = -1$.

DEFINITION 2.19. The order of an elliptic point $z \in \mathcal{H}$ with respect to Γ is the order of its isotropy group with respect to $\overline{\Gamma}$ in $\mathrm{PSL}(2,\mathbb{R})$. That is, the order of an elliptic point z is $\sharp\Gamma_z$ if $-\mathrm{Id} \notin \Gamma$, or $\frac{1}{2}\sharp\Gamma_z$ if $-\mathrm{Id} \in \Gamma$.

If z is an elliptic point with respect to Γ, then $\gamma(z)$ is also an elliptic point with respect to Γ, for any $\gamma \in \Gamma$. Moreover, equivalent elliptic points have the same order, since their isotropy groups are conjugate: $\Gamma_{\gamma(z)} = \gamma\Gamma_z\gamma^{-1}$.

LEMMA 2.20. *Let $\Gamma \subseteq \mathrm{SL}(2,\mathbb{R})$ be a subgroup such that for each matrix $\gamma = \left(\begin{smallmatrix} a & b \\ c & d \end{smallmatrix}\right)$ in Γ, the matrix $\gamma' := \left(\begin{smallmatrix} -a & b \\ c & -d \end{smallmatrix}\right)$ lies in Γ. Then two points in $\mathcal{H}$ are Γ-equivalent if and only if their symmetric points with respect to the imaginary axis are also Γ-equivalent. Moreover, a point in $\mathcal{H}$ is elliptic with respect to Γ if and only if its symmetric point with respect to the imaginary axis is also elliptic. In this case they have the same order.*

DEFINITION 2.21. A connected closed polygon $\mathcal{D} \subseteq \mathcal{H} \cup \mathbb{R} \cup \{\infty\}$ is a fundamental domain for the action of Γ in $\mathcal{H}$ if any two points in the interior of $\mathcal{D}$ are not Γ-equivalent and if each point in $\mathcal{H}$ is Γ-equivalent to some point in $\mathcal{D}$. Obviously the fundamental domain for a group Γ is not unique.

DEFINITION 2.22. A cycle of a fundamental domain is an orbit of vertices under the Γ-action. A cycle is called elliptic of order k if it consists of elliptic vertices of order k. A cycle is called parabolic if it is formed by parabolic vertices; then by convention its order is $k = \infty$. The number of vertices in a cycle is called the length of the cycle.

2.3. Groups of quaternion transformations

Let $D, N \geq 1$ be natural numbers such that $\gcd(D, N) = 1$.

Consider a quaternion $\mathbb{Q}$-algebra H of discriminant D, which is determined up to isomorphism. Let $\mathcal{O}(D, N)$ be an Eichler order in H of level N. Consider the group of quaternion units with positive norm equal to 1:

$$\mathcal{O}(D,N)_+^* := \{\alpha \in \mathcal{O}(D,N)^* : n(\alpha) = 1\}.$$

As a consequence of Eichler's results [**Eic38**], the group of quaternion units $\mathcal{O}(D,N)_+^*$ has the following properties.

THEOREM 2.23. *Let $\mathcal{O}(D, N)$ be an Eichler order in a quaternion $\mathbb{Q}$-algebra H. Then*:

(i) *If H is indefinite, then $\mathcal{O}(D, N)$ has units of reduced norm -1. Therefore, $\mathcal{O}(D,N)_+^*$ has index 2 in $\mathcal{O}(D,N)^*$.*
(ii) *If H is definite, then there is no element in $\mathcal{O}(D, N)$ of reduced norm -1; thus, $\mathcal{O}(D,N)_+^* = \mathcal{O}(D,N)^*$. Moreover, $\mathcal{O}(D,1)^*$ is a cyclic group of order 2, 4 or 6, except for the following cases*:
  (a) *$H = \left(\frac{-1,-1}{\mathbb{Q}}\right)$, with $D_H = 2$. In this case, $\mathcal{O}(D,1)^*$ is isomorphic to $E_{24} := \{\pm 1, \pm i, \pm j, \pm ij, (\pm 1 \pm i \pm j \pm ij)/2\}$, the tetrahedral binary group, and hence it has order 24.*
  (b) *$H = \left(\frac{-1,-3}{\mathbb{Q}}\right)$, with $D_H = 3$. In this case, $\mathcal{O}(D,1)^* \simeq \langle s_6, j\rangle \simeq C_6 \times C_2$, a bicyclic group, with $s_6 = \cos 2\pi/6 + \iota \sin 2\pi/6$, and hence it has order 12.*

Since $\mathcal{O}(D,N)_+^*$ is finite when H is definite, we restrict ourselves to the case where H is an indefinite quaternion algebra.

Thus, assume D is the product of an even number of different primes and take an isomorphism $\Phi\colon \mathbb{R} \otimes H \to \mathrm{M}(2, \mathbb{R})$. Then, by Theorem 1.59, $\mathcal{O}(D,N)_+^*$ only depends on D and N, up to conjugation.

We will fix the monomorphism Φ given by Proposition 1.19. Thus, for an indefinite quaternion algebra $H = \left(\frac{a,b}{\mathbb{Q}}\right)$, assume $a > 0$ and consider $\Phi\colon H \hookrightarrow \mathrm{M}(2, \mathbb{R})$ given by

$$\Phi(x + y\sqrt{a} + z\sqrt{b} + t\sqrt{a}\sqrt{b}) = \begin{pmatrix} x + y\sqrt{a} & z + t\sqrt{a} \\ b(z - t\sqrt{a}) & x - y\sqrt{a} \end{pmatrix}.$$

Consider the discrete subgroup of $\mathrm{SL}(2, \mathbb{R})$ defined by

$$\Gamma(D,N) := \Phi(\mathcal{O}(D,N)_+^*).$$

The group $\Gamma(D, N)$ acts on the Poincaré half-plane. Its elements will be called quaternion transformations.

Therefore,

$$\Gamma(D,N) \subseteq \left\{ \begin{pmatrix} \alpha & \beta \\ b\beta' & \alpha' \end{pmatrix} : \alpha, \beta \in \mathbb{Q}(\sqrt{a}) \right\} \subseteq \mathrm{SL}(2, \mathbb{Q}(\sqrt{a})),$$

where $\alpha \mapsto \alpha'$ denotes the Galois conjugation in $\mathbb{Q}(\sqrt{a})$. The case $a = 1$ corresponds necessarily to a nonramified quaternion algebra; that is, $D = 1$. In this case, $\Gamma(1, N) \subseteq \mathrm{SL}(2, \mathbb{Q})$.

We make explicit the groups of quaternion transformations for the nonramified and small ramified quaternion algebras given in Chapter 1.

2.24. *Nonramified case.* Consider $H = \mathrm{M}(2, \mathbb{Q})$. In this case, Φ is the canonical embedding $\mathrm{M}(2, \mathbb{Q}) \hookrightarrow \mathrm{M}(2, \mathbb{R})$. For every N, consider the Eichler order of level N

$$\mathcal{O}_0(1, N) = \left\{ \begin{pmatrix} a & b \\ cN & d \end{pmatrix} : a, b, c, d \in \mathbb{Z} \right\}.$$

Then, $\Gamma(1, N) = \mathcal{O}(1, N)_+^*$ is the congruence group denoted usually by $\Gamma_0(N)$.

If we consider the nonramified quaternion algebra $H' = \left(\frac{1,-1}{\mathbb{Q}}\right)$ and the Eichler order $\mathcal{O}_M(1, N) = \mathbb{Z}[1, (j+ij)/2, N(-j+ij)/2, (1-i)/2]$, we also obtain $\Gamma(1, N) = \Phi(\mathcal{O}_M(1, N)_+^*) = \Gamma_0(N)$.

2.25. *Small ramified case of type* A. Consider the quaternion algebra $H_{\mathrm{A}}(p) = \left(\frac{p,-1}{\mathbb{Q}}\right)$ and the Eichler order $\mathcal{O}_{\mathrm{A}}(2p, N) = \mathbb{Z}[1, i, Nj, (1+i+j+ij)/2]$, $N \mid (p-1)/2$ square-free. Then we obtain the following equivalent descriptions for the group $\Gamma(2p, N)$:

(i) The matrix

$$\gamma = \frac{1}{2} \begin{pmatrix} (2x+t) + (2y+t)\sqrt{p} & (2Nz+t) + t\sqrt{p} \\ -(2Nz+t) + t\sqrt{p} & (2x+t) - (2y+t)\sqrt{p} \end{pmatrix}$$

lies in the group $\Gamma(2p, N)$ if and only if $x, y, z, t \in \mathbb{Z}$ and $\det(\gamma) = 1$.

(ii) The matrix

$$\gamma = \frac{1}{2} \begin{pmatrix} a + b\sqrt{p} & c + d\sqrt{p} \\ -c + d\sqrt{p} & a - b\sqrt{p} \end{pmatrix}$$

lies in the group $\Gamma(2p, N)$ if and only if a, b, c, d are integers, $a \equiv b \equiv c \equiv d \bmod 2$, $N \mid (c - d)$ and $\det(\gamma) = 1$.

(iii) The matrix

$$\gamma = \frac{1}{2} \begin{pmatrix} \alpha & \beta \\ -\beta' & \alpha' \end{pmatrix}$$

lies in the group $\Gamma(2p, N)$ if and only if $\alpha, \beta \in \mathbb{Z}[\sqrt{p}]$, $\alpha \equiv \beta \equiv \alpha\sqrt{p} \bmod 2$, $N \mid (\mathrm{tr}(\beta) - (\beta - \beta')/\sqrt{p})$, $\det(\gamma) = 1$.

2.26. *Small ramified case of type* B. Consider the quaternion algebra $H_{\mathrm{B}}(p, q) = \left(\frac{p,q}{\mathbb{Q}}\right)$ and the Eichler order $\mathcal{O}_{\mathrm{B}}(pq, N) = \mathbb{Z}[1, Ni, (1+j)/2, (i+ij)/2]$, N square-free, $N \mid (q-1)/4$, $\gcd(N, p) = 1$. Then, we obtain the following equivalent descriptions for the group $\Gamma(pq, N)$:

(i) The matrix

$$\gamma = \frac{1}{2} \begin{pmatrix} (2x+z) + (2Ny+t)\sqrt{p} & z + t\sqrt{p} \\ q(z + t\sqrt{p}) & (2x+z) - (2Ny+t)\sqrt{p} \end{pmatrix}$$

lies in the group $\Gamma(pq, N)$ if and only if $x, y, z, t \in \mathbb{Z}$ and $\det(\gamma) = 1$.

(ii) The matrix

$$\gamma = \frac{1}{2} \begin{pmatrix} a + b\sqrt{p} & c + d\sqrt{p} \\ q(c - d\sqrt{p}) & a - b\sqrt{p} \end{pmatrix}$$

lies in the group $\Gamma(pq, N)$ if and only if $a, b, c, d \in \mathbb{Z}$, $a \equiv c \bmod 2$, $b \equiv d \bmod 2$, $2N \mid (b - d)$ and $\det(\gamma) = 1$.

(iii) The matrix

$$\gamma = \frac{1}{2}\begin{pmatrix} \alpha & \beta \\ q\beta' & \alpha' \end{pmatrix}$$

lies in the group $\Gamma(pq, N)$ if and only if $\alpha, \beta \in \mathbb{Z}[\sqrt{p}]$, $\alpha \equiv \beta \bmod 2$, $2N \mid (\alpha - \alpha' - \beta + \beta')/(2\sqrt{p})$ and $\det(\gamma) = 1$.

REMARK 2.27. All the groups $\Gamma(D, N)$ of quaternion transformations satisfy the hypothesis of Lemma 2.20. Thus, we have symmetry conditions for the elliptic points and the $\Gamma(D, N)$-equivalent points in $\mathcal{H}$.

It is also true that all the groups $\Gamma(D, N)$ of quaternion transformations always contain $-\mathrm{Id}$, since the element -1 is always a unit in the order $\mathcal{O}(D, N)$ and $\Phi(-1) = -\mathrm{Id}$. Hence, by Lemma 2.10, the elliptic transformations defined by elements in $\Gamma(D, N)$ are of order 2 or 3.

2.4. The Shimura curves $X(D, N)$

Let D, N be natural numbers. Assume that D is a product of an even number of different primes and that $N \geq 1$ is a natural number such that $\gcd(D, N) = 1$.

Fix the following objects: an indefinite quaternion $\mathbb{Q}$-algebra H of discriminant $D_H = D$; an Eichler order $\mathcal{O}(D, N)$ of level N in H, and a monomorphism $\Phi\colon H \hookrightarrow \mathrm{M}(2, \mathbb{R})$. Consider the group of quaternion transformations $\Gamma(D, N)$ associated with the order $\mathcal{O}(D, N)$ and Φ.

The group $\Gamma(D, N) \subseteq \mathrm{SL}(2, \mathbb{R})$ is a Fuchsian group of the first kind acting on the Poincaré half-plane. The quotient $\Gamma(D, N) \setminus \mathcal{H}$ is a Riemann surface.

The theory of Shimura provides a canonical model $X(D, N)$ for $\Gamma(D, N) \setminus \mathcal{H}$ and a modular interpretation.

The canonical model $X(D, N)$ has the following properties:

(i) $X(D, N)$ is a projective curve defined over $\mathbb{Q}$.

(ii) There exists a mapping $j_{D,N}\colon \mathcal{H} \to X(D, N)(\mathbb{C})$ that factorizes in an isomorphism between the analytic space $\Gamma(D, N) \setminus \mathcal{H}$ and a Zariski open set in $X(D, N)(\mathbb{C})$.

(iii) Let $F = \mathbb{Q}(\sqrt{d})$ be an imaginary quadratic field splitting the algebra H. Let φ be an embedding of F into H, and $z \in \mathcal{H}$ the unique common fixed point of all the elements in $\Phi\big(\varphi(F^*)\big)$. Then the coordinates of the point $j_{D,N}(z)$ are algebraic, more specifically, $j_{D,N}(z) \in X(D, N)(F_{\mathrm{ab}})$, where $F_{\mathrm{ab}} \subseteq \mathbb{C}$ denotes the maximal abelian extension of F.

$X(D, N)$ is called the Shimura curve associated with the subgroup $\Gamma(D, N)$.

The case $D = 1$ corresponds to a nonramified quaternion algebra $H \simeq \mathrm{M}(2, \mathbb{Q})$. In this case, $\Gamma(1, N) \setminus \mathcal{H}$ is a noncompact Riemann surface, with finite volume. From the explicit expressions given for the groups $\Gamma(1, N)$, it is clear that the corresponding compact Shimura curve $X(1, N)$ is precisely the modular curve $X_0(N)$.

If $D > 1$, the quaternion algebra H is ramified. In this case, the Riemann surface $\Gamma(D, N) \setminus \mathcal{H}$ is already compact. In Chapter 5 we give a very simple proof of this fact.

The modular interpretation of $X(D, N)$ runs as follows. A point in $X(D, N)(\mathbb{C})$ corresponds to an isomorphism class of triples (A, i, G), where A is a principally polarized abelian surface, $i\colon H \hookrightarrow \mathbb{Q} \otimes_{\mathbb{Z}} \mathrm{End}(A)$ is a monomorphism such that $i\big(\mathcal{O}(D, 1)\big) \subseteq \mathrm{End}(A)$, and G is a subgroup of the group of N-torsion points $A[N]$ of A which is a cyclic $\mathcal{O}(D, N)$-module.

We now present some results about the computation of certain constants attached to Shimura curves. Some computations can be done in a general way for the ramified and nonramified cases.

NOTATION 2.28. Given a Shimura curve $X(D,N)$ and a fundamental domain for $\Gamma(D,N)$, we consider the following invariants: $V_h(D,N)$ the hyperbolic volume; $e_i(D,N)$ the number of elliptic cycles of order i, and $g(D,N)$ the genus. We also consider a normalization of the hyperbolic measure, given by $dx\,dy/(2\pi y^2)$, and we denote by $V(D,N)$ the volume computed with this normalized measure. It can be proved that $V(D,N) = V_h(R)/(2\pi)$ is a rational number.

The following propositions provide an arithmetical way to compute these constants. They can be found in [**Shi71, Vig80**].

PROPOSITION 2.29. *Consider a Shimura curve $X(D,N)$. The elliptic cycles are of order 2 or 3. The number of elliptic cycles, volume and genus are given by:*

$$e_2(D,N) = \begin{cases} \prod_{p|D}\left(1-\left(\frac{-4}{p}\right)\right)\prod_{p|N}\left(1+\left(\frac{-4}{p}\right)\right) & \text{if } 4\nmid N, \\ 0 & \text{if } 4 \mid N. \end{cases}$$

$$e_3(D,N) = \begin{cases} \prod_{p|D}\left(1-\left(\frac{-3}{p}\right)\right)\prod_{p|N}\left(1+\left(\frac{-3}{p}\right)\right) & \text{if } 9\nmid N, \\ 0 & \text{if } 9 \mid N, \end{cases}$$

$$V(D,N) = \frac{N}{6}\prod_{p|D}(p-1)\prod_{p|N}\left(1+\frac{1}{p}\right),$$

$$2-2g(D,N) = -V(D,N) + \tfrac{1}{2}e_2(D,N) + \tfrac{2}{3}e_3(D,N) + e_\infty(D,N).$$

For the Shimura curves corresponding to the small ramified quaternion algebras of type A and of type B, we deduce the following results.

COROLLARY 2.30. *Let $X(2p,1)$ be the Shimura curve associated with a maximal order of a small ramified quaternion algebra of type* A *of discriminant $2p$. Then,*

$$e_2(2p,1) = 2, \qquad e_3(2p,1) = \begin{cases} 4 & \text{if } p \equiv 11 \bmod 12, \\ 2 & \text{if } p = 3, \\ 0 & \text{if } p \equiv 7 \bmod 12. \end{cases}$$

COROLLARY 2.31. *Let $X(pq,1)$ be the Shimura curve associated with a maximal order of a small ramified quaternion algebra of type B of discriminant pq. Then,*

$$e_2(pq,1) = 0, \qquad e_3(pq,1) = \begin{cases} 4 & \text{if } p \equiv 2 \bmod 3 \text{ and } q \equiv 5 \bmod 12, \\ 2 & \text{if } p = 3, \\ 0 & \text{otherwise.} \end{cases}$$

From the relation between these constants, we can deduce some properties about the possible fundamental domains for the Shimura curve $X(D,N)$.

PROPOSITION 2.32. *Let $X(D,N)$ be a Shimura curve with $D > 1$. Assume $X(D,N)$ has a fundamental domain such that all the vertices are elliptic. Then, the number of vertices of this fundamental domain is*

$$n_e(D,N) = 2 + 2V(D,N) + e_2(D,N) + \tfrac{2}{3}e_3(D,N).$$

If $N = 1$, *then*

$$n_e(D,1) = 2 + \frac{1}{3}\prod_{p|D}(p-1) + \prod_{p|D}\left(1 - \left(\frac{-4}{p}\right)\right) + \frac{2}{3}\prod_{p|D}\left(1 - \left(\frac{-3}{p}\right)\right).$$

PROOF. Assume we have a fundamental domain for $X(D,N)$ such that all the vertices are elliptic and let $n_e(D,N)$ be the number of those vertices. Then, the hyperbolic volume $V_h(D,N)$ is $V_h(D,N) = (n_e(D,N)-2)\pi - (\theta_1 + \cdots + \theta_n)$, where $\theta_1, \ldots, \theta_n$ are the angles in the vertices.

In general, we do not know the angles in the vertices. Assuming all the vertices are elliptic points, we can determine the sum from the number of elliptic cycles of order 2 and 3, since the sum of the angles of a cycle of order q is exactly $2\pi/q$. Thus, we have $2\pi V(D,N) = (n_e(D,N)-2)\pi - \left(\pi e_2(D,N) + \frac{2}{3}\pi e_3(D,N)\right)$ and we deduce the expression for $n_e(D,N)$ given in the statement. □

In Appendix A we present several tables with numerical data about Shimura curves. Table A.8 lists the constants V, e_2, e_3 and g for the Shimura curves corresponding to small ramified case with $D < 200$ and $N = 1$; cf. [**Vig80**], for first cases. Tables A.9–A.12 contain constants attached to the Shimura curves $X(6,N)$, $X(10,N)$, $X(14,N)$, and $X(15,N)$, corresponding to the small ramified quaternion algebras of type A and small ramified quaternion algebras of type B mentioned in the previous chapter. The constants attached to Shimura curves of discriminant equal to the product of four primes can be found in Table A.13, for $D < 1000$ and $N = 1$.

Table A.14 lists known equations for Shimura curves obtained previously by Ihara, Kurihara [**Kur79**], Jordan–Livné [**Jor81**] and Michon [**Mic81a**]. Table A.15 lists all the values $D > 1$ and N such that $X(D,N)$ is a hyperelliptic Shimura curve, with the hyperelliptic involution w calculated by Ogg [**Ogg83**].

CHAPTER 3

Quaternion Algebras and Quadratic Forms

This chapter focuses on quaternion algebras and quadratic forms, with special emphasis on the binary, ternary and quaternary forms, with the aim of unifying and generalizing notations and definitions that will be used in the sequel. In Section 3.1, we fix definitions and notations. Section 3.2 recovers and generalizes Brandt's definitions and results relative to the reciprocal forms and the property of principal and K-forms. In Section 3.3, general properties of quadratic forms associated to algebras over a field K are presented. In Section 3.4, the results on the previous section are applied to quadratic fields with respect to normic quadratic forms. Thus, we deal with the normic binary quadratic forms associated to the quadratic orders. In Section 3.5, we deal with quaternary and ternary quadratic forms obtained by interpreting a quaternion $\mathbb{Q}$-algebra H as a quadratic space and by considering the subspace of pure quaternions. The interpretation may be made from the norm or the trace defined in H, and we consider the normic forms $n_{H,4}$ and $n_{H,3}$. We obtain relations between invariants of these forms and invariants of the algebra H. The main results for the normic forms are that $n_{H,4}$ and $n_{H,3}$ are K-forms, cf. Theorem 3.76. Section 3.6 is devoted to the study of quadratic forms associated to quaternion orders $\mathcal{O}$, in particular the quaternary and ternary normic forms, which we denote by $n_{\mathcal{O},4}$ and $n_{\mathcal{O},3}$. We give results about the relation between invariants associated to the order and the quadratic forms. We characterize those normic forms which are K-forms in Theorem 3.92. In the case of Eichler orders of indefinite quaternion algebras, more explicit criteria are supplied by Proposition 3.98.

3.1. Quadratic forms over rings

In this section, R denotes a ring of characteristic different from 2 and K its field of fractions.

3.1.1. General definitions. A quadratic form f in n variables over R is an homogeneous polynomial $f \in R[X_1, \dots, X_n]$ of degree 2, $f(X_1, \dots, X_n) = \sum_{i,j=1}^{n} a_{ij} X_i X_j$, where $a_{ij} = a_{ji}$, $a_{ii} \in R$, $a_{ij} \in R[\frac{1}{2}]$, $1 \leq i \leq n$. Over a vector K-space V of dimension n, a quadratic form f is said to define a structure of quadratic space. Basic references for the study of quadratic forms are [**Gau01, Sie44, Jon67, Ser73**].

DEFINITION 3.1. A symmetric matrix $A = (a_{ij}) \in \mathrm{M}(n, R)$ is called even if $2 \mid a_{ii}$ for $1 \leq i \leq n$.

DEFINITION 3.2. To any quadratic form f we associate two symmetric matrices: $A_1(f) := (a_{ij})$ and $A_2(f) := 2A_1(f)$. Thus, for $X = (X_1, \dots, X_n)^t$, we have $f(X) = X^t A_1(f) X = \frac{1}{2} X^t A_2(f) X$. If f is defined over R, then $A_1(f) \in \mathrm{M}(n, R[\frac{1}{2}])$ and $A_2(f) \in \mathrm{M}(n, R)$ is an even matrix.

The determinant $\det_i(f)$ of f is the determinant of the associated matrix $A_i(f)$, $i = 1, 2$. We have $\det_2(f) = 2^n \det_1(f)$. If f is defined over R, then $\det_2(f) \in R$; but $\det_1(f) \in R[\frac{1}{2}]$. If $\det_i(f) = 0$, the form f is called singular; otherwise, regular. For a regular form f, the K-discriminant $\operatorname{disc}_K(f)$ of f is the class of the determinant $\det_1(f)$ in K^*/K^{*2}. Observe that, if f is a regular quadratic form in an even number of variables, then $\det_1(f)$ and $\det_2(f)$ define the same class in K^*/K^{*2}.

From now on, we deal only with regular quadratic forms.

DEFINITION 3.3. The adjoint matrix of an $n \times n$-matrix $A = (a_{ij})$ is the matrix $\operatorname{ad}(A) = (\alpha_{ij})$ where

$$\alpha_{ij} := \sum_{\substack{\sigma \in S(n) \\ \sigma(i)=j}} \epsilon(\sigma) a_{1\sigma(1)} \cdots a_{(i-1)\sigma(i-1)} a_{(i+1)\sigma(i+1)} \cdots a_{n\sigma(n)}.$$

The adjoint form $\operatorname{ad}(f)$ of a quadratic form f is the quadratic form such that $A_1(\operatorname{ad}(f)) = \operatorname{ad}(A_1(f))$. Note that the form $\operatorname{ad}(f)$ has coefficients in $R[1/2^{n-1}]$, since $A_1(\operatorname{ad}(f)) \in \mathrm{M}(n, R[1/2^{n-1}])$.

The following lemma justifies the choice of $A_1(f)$, instead of $A_2(f)$, to define the adjoint form of f.

LEMMA 3.4. *Let f be a regular quadratic form in n variables over R. Then*:
(i) $\det_1(\operatorname{ad}(f)) = (\det_1(f))^{n-1}$.
(ii) $\operatorname{ad}(A_2(f)) = 2^{n-1} \operatorname{ad}(A_1 f) \in \mathrm{M}(n, R)$.
(iii) *If n is even,* $\operatorname{ad}(A_2(f))$ *is an even matrix and* $\operatorname{ad}(A_2(f)) = A_2(2^{n-2} \operatorname{ad}(f))$.

3.1.2. Representation of forms by forms. We make precise the definitions related to the representation of numbers and quadratic forms by quadratic forms which will be used in following chapters. There are numerous references containing classical results for these subjects; for example [**Sie35, BS66, Ser73, Are87**].

DEFINITION 3.5. Let f, g be quadratic forms over R in n and r variables, respectively. Assume that $r \leq n$. The form f represents g over R if there exists a matrix P in $\mathrm{M}(n \times r, R)$, of rank r, such that $P^t A_1(f) P = A_1(g)$. It will be denoted by $f \xrightarrow{R} g$.

In particular, a form f represents an element $\alpha \in R$, over R, if there exist elements $\alpha_1, \ldots, \alpha_n \in R$ not simultaneously zero such that $f(\alpha_1, \ldots, \alpha_n) = \alpha$.

DEFINITION 3.6. A quadratic form f over R is called R-isotropic if it represents 0 over R. Otherwise, it is called R-anisotropic.

If K is the field of fractions of R, a quadratic form over R is isotropic over R if and only if it is isotropic over K. The following result is well known.

PROPOSITION 3.7. *Let f be a regular quadratic form over K. If f is K-isotropic, then f represents any element $\alpha \in K$.*

DEFINITION 3.8. Let Γ be a subgroup of $\mathrm{GL}(n, R)$. Two quadratic forms f and g are Γ-equivalent if there exists a matrix $\gamma \in \Gamma$ such that $A_1(g) = \gamma^t A_1(f) \gamma$. It is denoted by $f \overset{\Gamma}{\sim} f'$. To shorten notation, the $\mathrm{GL}(n, R)$-equivalence will also be denoted by $f \overset{R}{\sim} f'$.

REMARK 3.9. For a binary quadratic form f over R, the forms f and $\mathrm{ad}(f)$ are always $\mathrm{SL}(2,R)$-equivalent. It is only necessary to consider $\gamma = \left(\begin{smallmatrix} 0 & -1 \\ 1 & 0 \end{smallmatrix}\right)$.

Let f and g be two quadratic forms over R in the same number of variables. Let $\Gamma \subseteq \mathrm{GL}(n,R)$ be a subgroup. It is clear that if $f \xrightarrow{\Gamma} g$, then $f \overset{K}{\sim} g$. Moreover, $f \overset{\Gamma}{\sim} g$ if and only if $f \xrightarrow{\Gamma} g$ and $g \xrightarrow{\Gamma} f$.

REMARK 3.10. Let f and f' be two regular quadratic forms in n variables over R such that $f \xrightarrow{R} f'$. Let $\gamma \in \Gamma$ be such that $A_1(f') = \gamma^t A_1(f)\gamma$. Then $\det_1(f') = \det_1(f)(\det\gamma)^2$ and $\mathrm{disc}_K(f') = \mathrm{disc}_K(f)$.

3.11. *Binary-ternary case.* Let f_2 and f_3 be a binary and a ternary quadratic forms over R, respectively. If $f_3 \xrightarrow{R} f_2$, then we also have $\mathrm{ad}(f_3) \xrightarrow{R} \det_1(f_2)$. This representation is obtained from the minors of the matrix P corresponding to the representation of f_2 by f_3. It is called the adjoint representation of $f_3 \xrightarrow{R} f_2$.

Conversely, all the representations of an element d by a ternary form f_3 come from a representation of a binary form f_2' with $\det_1(f_2') = d$ by a ternary form f_3' with $\mathrm{ad}(f_3') = f_3$. The construction of the adjoint representation justifies the definition of primitive representation.

DEFINITION 3.12. Assume R is a principal ideal domain. Let f and g be quadratic forms over R in n and r variables, respectively. Suppose $f \xrightarrow{R} g$ and let $P \in \mathrm{M}(n \times r, R)$ be such that $P^t A_1(f) P = A_1(g)$. We say that P is a primitive representation if the greater common divisor of the minors $r \times r$ of the matrix P is 1. We say that f represents primitively g over R if there exists at least a primitive representation P of g by f.

In particular, if f is a quadratic form over R and $\alpha \in R$, then $f(\alpha_1, \dots, \alpha_n) = \alpha$ is a primitive representation of α by the quadratic form f over R if and only if $\gcd(\alpha_1, \dots, \alpha_n) = 1$.

EXAMPLE 3.13. Consider the quadratic form $f(X,Y,Z) = -aX^2 - bY^2 + abZ^2$, $a, b \in \mathbb{Z}$. It is clear that the form f represents primitively the diagonal binary forms $g_1(X,Y) = -aX^2 - bY^2$, $g_2(X,Y) = -bX^2 + abY^2$, $g_3(X,Y) = -aX^2 + abY^2$. It also represents primitively nondiagonal binary forms, as for example $g_4(X,Y) = \big(a(b-1), 2ab, b(a-1)\big)$ and $g_5(X,Y) = (-a-b+9ab, -4b+6ab, -4b+ab)$ (cf. 3.23). The matrix $P = \left(\begin{smallmatrix} 1 & 0 \\ 0 & 2 \\ 3 & 1 \end{smallmatrix}\right)$ is a nonprimitive representation of $g_6(X,Y) = (-a+9ab, 6ab, -4b+ab)$ by f. These representations give explicit representations $\mathrm{ad}(f) \xrightarrow{R} d_i$, with $d_i = \det_1 g_i$. For example, if $a = p$ and $b = -1$, from the representation of g_5 by f we obtain a primitive representation of the number $p(p-29)$ by $\mathrm{ad}(f) = -pX^2 + p^2Y^2 - pZ^2$, given by $(5,1,2)$.

NOTATION 3.14. Let f and g be quadratic forms over R. Consider the following sets:

$$\mathcal{R}(f,g;R) = \{P : P \in \mathrm{M}_{n\times r}(R), \operatorname{rank} P = r, P^t A_1(f) P = A_1(g)\},$$
$$\mathcal{R}^*(f,g;R) = \{P : P \in \mathcal{R}(f,g;R), P \text{ primitive}\}, \quad \text{if } R \text{ is a principal ideal domain},$$
$$\mathcal{D}(f;R) = \{\alpha \in R : \mathcal{R}(f,\alpha;R) \neq \varnothing\}.$$

Let Γ be a subgroup of $\mathrm{GL}(n,R)$. The Γ-isotropy group of a quadratic form f is $O(f;\Gamma) := \{\gamma \in \Gamma : \gamma^t A_1(f)\gamma = A_1(f)\}$. If $R \subseteq \mathbb{R}$, put $O^+(f;\Gamma) = \{\gamma \in O(f;\Gamma) : \det\gamma > 0\}$.

Note that in the case $R = \mathbb{Z}$, we have $O^+(f;\Gamma) \subseteq \mathrm{SL}(n,\mathbb{Z})$, where n is the number of variables of f. Let $\alpha \in \mathcal{D}(f;R)$ and let $(\alpha_1,\dots,\alpha_n)$ be a representation of α by f over R. If $\gamma \in O^+(f;\Gamma)$, then $(\alpha'_1,\dots,\alpha'_n) = \gamma^{-1}(\alpha_1,\dots,\alpha_n)$ is also a representation of α by f over R. This leads to the following definition.

DEFINITION 3.15. Two representations $P, P' \in \mathcal{R}(f,g;R)$ are called Γ-equivalent if there exists $\gamma \in O(f,\Gamma)$ such that $P = \gamma P'$. In particular, two representations $(\alpha_1,\dots,\alpha_n)$, $(\alpha'_1,\dots,\alpha'_n)$ belonging to $\mathcal{R}(f,\alpha;R)$ are Γ-equivalent if and only if there exists $\gamma \in O(f,\Gamma)$ such that $(\alpha_1,\dots,\alpha_n) = \gamma(\alpha'_1,\dots,\alpha'_n)$.

REMARK 3.16. If f and f' are $\mathrm{SL}(n,R)$-equivalent, then $\mathcal{D}(f;R) = \mathcal{D}(f';R)$. Moreover, $\alpha \in \mathcal{D}(f;K)$ if and only if $\alpha\beta^2 \in \mathcal{D}(f;K)$. Hence, in the case of a field K, it is enough to study the representations of elements of K^*/K^{*2}.

After fixing two quadratic forms f, g over R in n variables, and a subgroup Γ in $\mathrm{GL}(n,\mathbb{R})$, we consider the set of Γ-classes of representations. Usually only the case $\Gamma = \mathrm{SL}(n,\mathbb{Z})$ is considered, but in Chapter 4 we will obtain results in a wider sense.

3.1.3. Quadratic forms over $\mathbb{Z}$. We now restrict ourselves to quadratic forms over $\mathbb{Z}$, although most of the following results can be easily extended to quadratic forms over a principal ideal domain R.

DEFINITION 3.17. The content $\mathrm{cont}(f)$ of a quadratic form f over $\mathbb{Z}$ is the greatest common divisor, with positive sign, of its coefficients. A quadratic form f over $\mathbb{Z}$ is called primitive if $\mathrm{cont}(f) = 1$. For $A_1(f) = (a_{ij})$ to be primitive is equivalent to the condition that $\gcd\{a_{ii}, 2a_{ij} : i,j,= 1,\dots,n, i \neq j\} = 1$.

DEFINITION 3.18. Let f be a quadratic form in n variables over $\mathbb{Z}$. The polar form $\mathrm{pol}(f)$ is the form $2^{n-1}\,\mathrm{ad}(f)$ divided by its content, that is,

$$\mathrm{pol}(f) = \frac{2^{n-1}\,\mathrm{ad}(f)}{\mathrm{cont}\big(2^{n-1}\,\mathrm{ad}(f)\big)}.$$

The polar form is determined from the adjoint form by imposing the condition of being defined over $\mathbb{Z}$ and being primitive. Thus, it makes no difference if we use the matrix $\mathrm{ad}\big(A_1(f)\big)$ or $\mathrm{ad}\big(A_2(f)\big)$ to compute its matrix.

LEMMA 3.19. *Let f be a quadratic form in n variables over $\mathbb{Z}$. Put $m = \mathrm{cont}(f)$ and consider $f' := (1/m)f$. Then, f' is a primitive form and the following equalities hold:*

(i) $\det_1(f) = m^n \det_1(f')$, $\det_2(f) = m^n \det_2(f')$.
(ii) $\mathrm{ad}(f) = m^{n-1}\,\mathrm{ad}(f')$ *and* $\mathrm{pol}(f) = \mathrm{pol}(f')$.

PROPOSITION 3.20. *Let f be a quadratic form in n variables over $\mathbb{Z}$. Let $N \in \mathbb{N}$. Then the following conditions are equivalent:*

(a) *N is the minimal positive integer such that $NA_2(f)^{-1}$ is the even matrix associated to some quadratic form with rational integer coefficients. Equivalently, $N\frac{1}{4}A_1(f)^{-1}$ is the matrix associated to some quadratic form with rational integer coefficients.*
(b) *Let $s = 0$ if $\det_1(f) > 0$ and $s = 1$ if $\det_1(f) < 0$. N satisfies $NA_2(f)^{-1} = (-1)^s A_2\big(\mathrm{pol}(f)\big)$. Equivalently $N\frac{1}{4}A_1(f)^{-1} = (-1)^s A_1\big(\mathrm{pol}(f)\big)$.*
(c) $N = (2^{n+1}|\det_1(f)|)/\mathrm{cont}\big(2^{n-1}\,\mathrm{ad}(f)\big)$.

The number N fulfilling the above conditions is called the level $\mathrm{N}(f)$ *of* f.

PROOF. The equivalence between (a) and (b), is obtained from the definition of the polar form.

For the equivalence between (b) and (c), we have

$$\frac{N}{4}A_1(f)^{-1} = \frac{NA_1\big(\mathrm{ad}(f)\big)}{4\det A_1(f)} = \frac{Nm}{2^{n+1}\det_1(f)}A_1\big(\mathrm{pol}(f)\big),$$

where $m = \mathrm{cont}\big(2^{n-1}\,\mathrm{ad}(f)\big)$. Thus, $(N/4)A_1(f)^{-1} = \pm A_1\big(\mathrm{pol}(f)\big)$ if and only if $N = (2^{n+1}|\det_1(f)|)/\mathrm{cont}\big(2^{n-1}\,\mathrm{ad}(f)\big)$. $\square$

REMARK 3.21. Condition (a) yields the definition of the level given by Ogg [**Ogg69**] for positive definite forms in an even number of variables. Condition (c) generalizes a computation of the level by Lehman [**Leh92**] for positive definite ternary forms.

PROPOSITION 3.22. *Let f be a quadratic form in n variables over $\mathbb{Z}$. Then,*

(i) $|\det_2\big(\mathrm{pol}(f)\big)| = \mathrm{N}(f)^n/|\det_2(f)|$.

(ii) $\mathrm{N}(f)$ *and* $\det_2(f)$ *have the same odd prime factors. For $n = 2$,* $\mathrm{N}(f) = |\det_2(f)|$ *if and only if the form* $\mathrm{ad}(f)$ *is primitive.*

(iii) $\mathrm{N}\big(\mathrm{pol}(f)\big)$ *divides* $\mathrm{N}(f)$. *The equality holds if and only if the form f is primitive.*

PROOF. The statement in (i) follows from Proposition 3.20(b).

Let us prove (ii). On the one hand, since $\det_2\big(\mathrm{pol}(f)\big)$ is an integer, we deduce that $\det_2(f) \mid \mathrm{N}(f)^n$ from (i). Hence for each p prime $p \mid \det_2(f)$, we have $p \mid \mathrm{N}(f)$. On the other hand, again by Proposition 3.20(b), we have $(-1)^s A_2\big(\mathrm{pol}(f)\big) = \mathrm{N}(f)A_2(f)^{-1} = \big(\mathrm{N}(f)/\det_2(f)\big)\,\mathrm{ad}\big(A_2(f)\big)$. Hence

$$\mathrm{ad}\big(A_2(f)\big) = \frac{(-1)^s\det_2(f)}{\mathrm{N}(f)}A_2\big(\mathrm{pol}(f)\big).$$

We know the matrix $\mathrm{ad}\big(A_2(f)\big)$ has rational integer entries, $\det_2(f)$ is an integer and $\mathrm{pol}(f)$ is a primitive form. Therefore, for any odd prime p, $p \mid \mathrm{N}(f)$, we have $p \mid \det_2(f)$.

For a binary form f, $\mathrm{ad}\big(A_2(f)\big) = A_2\big(\mathrm{ad}(f)\big)$. Thus,

$$A_2\big(\mathrm{ad}(f)\big) = \frac{(-1)^s\det_2(f)}{\mathrm{N}(f)}A_2\big(\mathrm{pol}(f)\big).$$

Hence, we have that $\mathrm{N}(f) = |\det_2(f)|$ if and only if $A_2\big(\mathrm{ad}(f)\big) = A_2\big(\mathrm{pol}(f)\big)$, which is equivalent to $\mathrm{ad}(f)$ being a primitive form.

Finally, let us see (iii). We assume $\det_1(f) > 0$ to simplify notation, since the divisibility conditions do not depend on the sign. Applying Proposition 3.20,

$$\mathrm{N}\big(\mathrm{pol}(f)\big)\tfrac{1}{4}A_1\big(\mathrm{pol}(f)\big)^{-1} = \mathrm{N}\big(\mathrm{pol}(f)\big)\tfrac{1}{4}(\mathrm{N}(f)\tfrac{1}{4}A_1(f)^{-1})^{-1} = \frac{\mathrm{N}\big(\mathrm{pol}(f)\big)}{\mathrm{N}(f)}A_1(f)$$

is the matrix of a primitive form with rational integer coefficients. Since the form f has its coefficients in $\mathbb{Z}$, we deduce the divisibility relation.

The above computations show that the equality $\mathrm{N}\big(\mathrm{pol}(f)\big) = \mathrm{N}(f)$ is equivalent to $\mathrm{N}\big(\mathrm{pol}(f)\big)\frac{1}{4}A_1\big(\mathrm{pol}(f)\big)^{-1} = A_1(f)$. By Proposition 3.20 this happens if and only if f is a primitive form. $\square$

In particular, we are interested in quadratic forms in $n = 2$, 3, or 4 variables. We now present some notations and results.

3.23. *Binary case.* Consider a binary quadratic form over $\mathbb{Z}$, $f(X,Y) = aX^2 + bXY + cY^2$, also denoted by $f = (a,b,c)$. According to the above definitions,

$$A_1(f) = \begin{pmatrix} a & b/2 \\ b/2 & c \end{pmatrix}, \qquad A_2(f) = \begin{pmatrix} 2a & b \\ b & 2c \end{pmatrix},$$

$$A_1\big(\mathrm{ad}(f)\big) = \begin{pmatrix} c & -b/2 \\ -b/2 & a \end{pmatrix}, \qquad \mathrm{ad}\big(A_2(f)\big) = \begin{pmatrix} 2c & -b \\ -b & 2a \end{pmatrix}.$$

LEMMA 3.24. *Given a binary quadratic form $f = (a,b,c)$, the following properties are satisfied:*

(i) $\det_1(f) = \det_1\big(\mathrm{ad}(f)\big) = ac - b^2/4$; $\det_2(f) = \det_2\big(\mathrm{ad}(f)\big) = 4ac - b^2$, $-\det_2(f) \equiv 0, 1 \bmod 4$. *Thus* $\mathrm{disc}_{\mathbb{Q}}(f) = \mathrm{disc}_{\mathbb{Q}}\big(\mathrm{ad}(f)\big) = 4ac - b^2$.
(ii) $\mathrm{ad}\big(A_2(f)\big) = A_2\big(\mathrm{ad}(f)\big)$; $\mathrm{cont}(f) = \mathrm{cont}\big(\mathrm{ad}(f)\big)$.
(iii) $\mathrm{pol}(f) = \big(1/\mathrm{cont}(f)\big)\,\mathrm{ad}(f)$; *hence f is primitive if and only if* $\mathrm{pol}(f) = \mathrm{ad}(f)$.
(iv) $\mathrm{N}(f) = 4|\det_1(f)|/\mathrm{cont}(f) = |\det_2(f)|/\mathrm{cont}(f)$.

3.25. *Ternary case.* Consider a ternary quadratic form

$$f(X,Y,Z) = aX^2 + bY^2 + cZ^2 + a'YZ + b'XZ + c'XY, \quad a,b,c,a',b',c' \in \mathbb{Z}.$$

The following notation has also been used (cf. [**Gau01**]):

$$f = \begin{pmatrix} a & b & c \\ a'/2 & b'/2 & c'/2 \end{pmatrix},$$

$$A_1(f) = \begin{pmatrix} a & c'/2 & b'/2 \\ c'/2 & b & a'/2 \\ b'/2 & a'/2 & c \end{pmatrix}, \qquad A_2(f) = \begin{pmatrix} 2a & c' & b' \\ c' & 2b & a' \\ b' & a' & 2c \end{pmatrix},$$

$$\det_1(f) = abc + \tfrac{1}{4}(a'b'c' + aa'^2 + bb'^2 + cc'^2).$$

EXAMPLE 3.26. Consider the ternary form: $f = X^2 + Y^2 - XZ$. Then $\mathrm{ad}(f) = -\frac{1}{4}Y^2 + Z^2 + XZ$,

$$f = \begin{pmatrix} 1 & 1 & 0 \\ 0 & -\frac{1}{2} & 0 \end{pmatrix}, \qquad \mathrm{ad}(f) = \begin{pmatrix} 0 & -\frac{1}{4} & 1 \\ 0 & \frac{1}{2} & 0 \end{pmatrix},$$

$$A_1(f) = \begin{pmatrix} 1 & 0 & -\frac{1}{2} \\ 0 & 1 & 0 \\ -\frac{1}{2} & 0 & 0 \end{pmatrix}, \qquad A_1\big(\mathrm{ad}(f)\big) = \begin{pmatrix} 0 & 0 & \frac{1}{2} \\ 0 & -\frac{1}{4} & 0 \\ \frac{1}{2} & 0 & 1 \end{pmatrix}.$$

It makes no sense to think about the even matrix associated to $\mathrm{ad}(f)$, because it does not have rational integer coefficients. Note that, considering the even matrix of f, its adjoint matrix is not an even matrix:

$$A_2(f) = \begin{pmatrix} 2 & 0 & -1 \\ 0 & 2 & 0 \\ -1 & 0 & 0 \end{pmatrix}, \qquad \mathrm{ad}\big(A_2(f)\big) = \begin{pmatrix} 0 & 0 & 2 \\ 0 & -1 & 0 \\ 2 & 0 & 4 \end{pmatrix}.$$

In this case we have:

(i) $\det_1(f) = -\frac{1}{4}$, $\det_2(f) = -2$, $\mathrm{disc}_{\mathbb{Q}}(f) = -1$.

(ii) $\det_1\bigl(\mathrm{ad}(f)\bigr) = 1/16$, $\mathrm{disc}_{\mathbb{Q}}\bigl(\mathrm{ad}(f)\bigr) = 1$, $\mathrm{ad}\bigl(A_2(f)\bigr) = 2^2 A_1\bigl(\mathrm{ad}(f)\bigr)$.
(iii) $\mathrm{pol}(f) = -Y^2 + 4Z^2 + 4XZ$.
(iv) $\mathrm{N}(f) = 2^2$.

3.27. *Quaternary case.* Consider a quaternary form over $\mathbb{Z}$, $f(X,Y,Z,T)$ equal to

$$aX^2 + bY^2 + cZ^2 + dT^2 + b'XY + c'XZ + d'XT + d''YZ + c''YT + b''ZT.$$

The following notation has also been used (cf. [**Bra24**]):

$$\boxed{\begin{matrix} & b' & c' & d' \\ a & b & c & d \\ & b'' & c'' & d'' \end{matrix}}$$

$$A_1(f) = \begin{pmatrix} a & b'/2 & c'/2 & d'/2 \\ b'/2 & b & d''/2 & c''/2 \\ c'/2 & d''/2 & c & b''/2 \\ d'/2 & c''/2 & b''/2 & d \end{pmatrix}, \qquad A_2(f) = \begin{pmatrix} 2a & b' & c' & d \\ b' & 2b & d' & c'' \\ c' & d' & 2c & b'' \\ d & c'' & b'' & 2d \end{pmatrix}.$$

EXAMPLE 3.28. Consider the form $f(X,Y,Z,T) = X^2 - 2Y^2 + T^2 + XZ - 2YT$. Then $\mathrm{ad}(f)(X,Y,Z,T) = -\frac{1}{4}Y^2 - 3Z^2 + \frac{1}{2}T^2 + 3XZ - \frac{1}{2}YT$.

$$f = \boxed{\begin{matrix} & 0 & \frac{1}{2} & 0 \\ 1 & -2 & 0 & 1 \\ & 0 & -1 & 0 \end{matrix}}, \qquad \mathrm{ad}(f) = \boxed{\begin{matrix} & 0 & \frac{3}{2} & 0 \\ 0 & -\frac{1}{4} & -3 & \frac{1}{2} \\ & 0 & -\frac{1}{4} & 0 \end{matrix}},$$

$$A_1(f) = \begin{pmatrix} 1 & 0 & \frac{1}{2} & 0 \\ 0 & -2 & 0 & -1 \\ \frac{1}{2} & 0 & 0 & 0 \\ 0 & -1 & 0 & 1 \end{pmatrix}, \qquad A_2(f) = \begin{pmatrix} 2 & 0 & 1 & 0 \\ 0 & -4 & 0 & -2 \\ 1 & 0 & 0 & 0 \\ 0 & -2 & 0 & 2 \end{pmatrix},$$

$$A_1\bigl(\mathrm{ad}(f)\bigr) = \begin{pmatrix} 0 & 0 & \frac{3}{2} & 0 \\ 0 & -\frac{1}{4} & 0 & -\frac{1}{4} \\ \frac{3}{2} & 0 & -3 & 0 \\ 0 & -\frac{1}{4} & 0 & \frac{1}{2} \end{pmatrix}, \qquad \mathrm{ad}\bigl(A_2(f)\bigr) = \begin{pmatrix} 0 & 0 & 12 & 0 \\ 0 & -2 & 0 & -2 \\ 12 & 0 & -24 & 0 \\ 0 & -2 & 0 & 4 \end{pmatrix}.$$

In this case:

(i) $\det_1(f) = \frac{3}{4}$, $\det_2(f) = 12$, $\mathrm{disc}_{\mathbb{Q}}(f) = 3$.
(ii) $\det_1\bigl(\mathrm{ad}(f)\bigr) = 3^3/4^3$, $\mathrm{disc}_{\mathbb{Q}}\bigl(\mathrm{ad}(f)\bigr) = 3$. The matrix $\mathrm{ad}\bigl(A_2(f)\bigr) = 2^3 \times A_1\bigl(\mathrm{ad}(f)\bigr)$ is even; in fact, $\mathrm{ad}\bigl(A_2(f)\bigr) = A_2\bigl(2^2\,\mathrm{ad}(f)\bigr)$.
(iii) $\mathrm{pol}(f) = -Y^2 - 12Z^2 + 2T^2 + 12XZ - 2YT$.
(iv) $\mathrm{N}(f) = 12$.

3.2. Principal forms

In this section we deal with quadratic forms from another point of view. We generalize the concepts of reciprocal form, K-form and principal form considered by Brandt [**Bra24**] in the case of quaternary quadratic forms. We also give precise descriptions for them.

DEFINITION 3.29. Let f be a quadratic form over $\mathbb{Q}$ given by $f = (1/\sigma) \times \sum_{i,j} c_{ij} X_i X_j$, where $\sigma = 1$ or 2 and $c_{ij} \in \mathbb{Q}$, $c_{ij} = c_{ji}$. We say that the pair $\bigl(f, (c_{ij})\bigr)$ is of the first kind if $\sigma = 1$; we say that the pair $\bigl(f, (c_{ij})\bigr)$ is of the second kind if $\sigma = 2$. The central coefficients of $\bigl(f, (c_{ij})\bigr)$ are $a_{ii} = (1/\sigma)c_{ii}$ and its lateral coefficients are $2a_{ij} = (2/\sigma)c_{ij}$, $i \neq j$.

Given a pair $\bigl(f, (c_{ij})\bigr)$ of kind σ, consider the determinants:

$$\det(c_{ij}) \quad \text{and} \quad \Delta(f) := \det\left(\frac{\partial^2 f}{\partial X_i \partial X_j}\right).$$

REMARK 3.30. Let $\bigl(f, (c_{ij})\bigr)$ be a pair of kind σ. We have the following relations between quadratic forms, their attached matrices and kinds:

(a) With the form f fixed, we have $(c_{ij}) = A_1(f)$ if $\sigma = 1$ and $(c_{ij}) = A_2(f)$ if $\sigma = 2$.
(b) With the matrix $(c_{ij}) = A_1(g)$ fixed, we have $g = f$ if $\sigma = 1$ and $g = 2f$ if $\sigma = 2$.

LEMMA 3.31. *For a pair $\bigl(f, (c_{ij})\bigr)$ of kind σ, the following properties are satisfied:*

(i) $A_1(2f) = A_2(f)$, $\operatorname{ad}(2f) = 2^{n-1}\operatorname{ad}(f)$, $\operatorname{pol}(2f) = \operatorname{pol}(f)$.
(ii) $\det(c_{ij}) = \det\bigl(\sigma A_1(f)\bigr) = \det_\sigma(f) = \det_1(\sigma f)$.
(iii) $(\partial^2 f/\partial X_i \partial X_j) = A_2(f)$.
(iv) $\Delta(f) = \det_2(f) = \det_1(2f)$.

PROOF. Properties (i), (ii) are easy to prove. The equalities in (iv) follow from (iii).

To prove (iii), we compute

$$\frac{\partial^2 f}{\partial X_i \partial X_j} = \frac{1}{\sigma}\frac{\partial}{\partial X_i}\left(2c_{jj}X_j + \sum_{k \neq j} c_{kj}X_k\right) = \begin{cases} (1/\sigma)c_{ij} = a_{ij} & \text{if } i \neq j, \\ (2/\sigma)c_{ii} = 2a_{ii} & \text{if } i = j. \end{cases}$$

Thus, we obtain the equality of matrices $(\partial^2 f/\partial X_i \partial X_j) = A_2(f)$ for both kinds. □

From the remark and the lemma, it is clear that to deal with the two kinds of a quadratic form f is equivalent to dealing with the two matrices attached to f, or with the two forms f and $2f$.

Next we define the concept of σ-reciprocal form, $\sigma = 1, 2$, for quadratic forms f in n variables such that $\det_\sigma(f)$ is a perfect square.

DEFINITION 3.32. Let $f = (1/\sigma)\sum_{i,j} c_{ij}X_iX_j$ be a quadratic form over $\mathbb{Q}$ of kind σ such that $\det_\sigma(f)$ is a perfect square. Put $\mathrm{d}_\sigma(f) = +\sqrt{\det_\sigma(f)}$ if the quadratic form is positive definite and $\mathrm{d}_\sigma(f) = -\sqrt{\det_\sigma(f)}$ if the quadratic form is indefinite; we will not consider negative definite quadratic forms.

The σ-reciprocal form of f is the quadratic form

$$\operatorname{rec}_\sigma(f) = \frac{1}{\sigma}\sum_{i,j} t_{ij}X_iX_j, \quad \text{where } t_{ij} = \frac{1}{\mathrm{d}_\sigma(f)}\frac{\partial \det_\sigma(f)}{\partial c_{ij}}.$$

The following proposition shows technical details and relates the σ-reciprocal form to the adjoint form.

PROPOSITION 3.33. *Let f be a quadratic form over $\mathbb{Q}$ in n variables such that $\det_\sigma(f)$ is a perfect square. Then:*

(i) *The matrix $(\partial \det_\sigma(f)/\partial c_{ij})_{i,j=1}^n$ is $A_1\bigl(\operatorname{ad}(\sigma f)\bigr)$, equal to $\operatorname{ad}\bigl(A_\sigma(f)\bigr)$.*

(ii) *The σ-reciprocal form* $\operatorname{rec}_\sigma(f)$ *is equal to* $\big(1/\operatorname{d}_\sigma(f)\big)\operatorname{ad}(\sigma f)$. *Thus its associated matrices are* $A_1\big(\operatorname{rec}_1(f)\big) = \big(1/\operatorname{d}_1(f)\big)\operatorname{ad}\big(A_1(f)\big)$ *and* $A_1\big(\operatorname{rec}_2(f)\big) = \big(1/\operatorname{d}_2(f)\big)\operatorname{ad}\big(A_1(2f)\big) = \big(1/\operatorname{d}_2(f)\big)\operatorname{ad}\big(A_2(f)\big)$.

(iii) $\det_1\big(\operatorname{rec}_\sigma(f)\big) = \big(\operatorname{d}_\sigma(f)\big)^{n-2}$. *In particular, for* $n = 4$, $\det_1\big(\operatorname{rec}_\sigma(f)\big) = \det_\sigma(f)$.

PROOF. Put $B = (b_{ij}) = (\partial \det_\sigma(f)/\partial c_{ij})$. Expanding the determinant $\det_\sigma(f) = \det(c_{ij})$ by the row i, we have

$$\det(c_{ij}) = (-1)^{i+1}c_{i1}C_{i1} + \cdots + (-1)^{i+j}c_{ij}C_{ij} + \cdots + (-1)^{i+n}c_{in}C_{in},$$

where C_{ij} denotes the minor obtained removing the row i and the column j. We obtain b_{ij} deriving the above expression with respect to c_{ij}, $b_{ij} = (-1)^{i+j}C_{ij}$. Thus b_{ij} coincides with the adjoint of c_{ij} and we have the matrix equality $(b_{ij}) = \operatorname{ad}(c_{ij}) = \operatorname{ad}\big(A_1(\sigma f)\big)$, which is equal to $A_1\big(\operatorname{ad}(\sigma f)\big)$. This proves (i).

The properties in (ii) follow from (i), adding the factor $1/\operatorname{d}_\sigma(f)$ to obtain the σ-reciprocal form. The first statement expresses the result for the forms, and the second one for the matrices attached to the forms.

To see (iii), expand

$$\det_1\big(\operatorname{rec}_\sigma(f)\big) = \frac{1}{\operatorname{d}_\sigma^n(f)}\det_1\operatorname{ad}(\sigma f) = \frac{1}{\operatorname{d}_\sigma^n(f)}\det_1(\sigma f)^{n-1} = \operatorname{d}_\sigma^{n-2}(f). \qquad \square$$

The following corollary makes explicit the relation between both σ-reciprocal forms, whenever this makes sense, that is for n even.

COROLLARY 3.34. *Let f be a quadratic form over $\mathbb{Q}$ in an even number n of variables such that* $\det_1(f)$ *is a perfect square. Then,* $\operatorname{rec}_2(f) = 2^{n/2-1}\operatorname{rec}_1(f)$.

More specifically, for the binary quadratic forms, the associated reciprocal forms coincide. For the quaternary quadratic forms, we have $\operatorname{rec}_2(f) = 2\operatorname{rec}_1(f)$.

DEFINITION 3.35. Let f be a quadratic form over $\mathbb{Z}$ such that $\det_\sigma(f)$ is a perfect square. The form f is called a K_σ-form if $\operatorname{rec}_\sigma(f)$ also has coefficients in $\mathbb{Z}$.

The following result relates the reciprocal form with the polar form defined in the first section and gives necessary and/or sufficient conditions for a quadratic form to be a K_σ-form.

LEMMA 3.36. *Let f be a quadratic form in n variables over $\mathbb{Z}$, such that* $\det_\sigma(f)$ *is a perfect square. Then, f is a K_σ-form if and only if there exists $\lambda \in \mathbb{Z}$ such that* $\operatorname{rec}_\sigma(f) = \lambda\operatorname{pol}(f)$.

PROOF. On the one hand, the polar form is defined from the adjoint form, multiplying in such a way that it becomes a primitive form with rational integer coefficients. On the other hand, by Proposition 3.33(ii), the reciprocal form is also a multiple of the adjoint form, $\operatorname{rec}_\sigma(f) = (1/d_\sigma)\operatorname{ad}(\sigma f)$. Thus both the σ-reciprocal form and the polar form come from the adjoint form. Moreover, both forms have rational integer coefficients and the polar form is primitive. Therefore we obtain the relation $\operatorname{rec}_\sigma(f) = \lambda\operatorname{pol}(f)$ for some $\lambda \in \mathbb{Z}$. Conversely, if we have this relation, necessarily f is a K_σ-form, since the σ-reciprocal form is a rational integer multiple of a form having rational integer coefficients. $\square$

PROPOSITION 3.37. *Let f be a quadratic form in n variables with rational integer coefficients such that* $\det_\sigma(f)$ *is a perfect square. Let* $\mathrm{N}(f)$ *be its level. Then:*

(i) *f is a* K_1*-form if and only if* $4\,\mathrm{d}_1(f)$ *is an integer and* $\mathrm{N}(f) \mid 4\,\mathrm{d}_1(f)$.
(ii) *f is a* K_2*-form if and only if* $\mathrm{N}(f) \mid 2\,\mathrm{d}_2(f)$.
(iii) $\mathrm{N}(f) = 4|\,\mathrm{d}_1(f)|$ *if and only if* $\mathrm{pol}(f) = \pm\,\mathrm{rec}_1(f)$.
(iv) $\mathrm{N}(f) = 2|\,\mathrm{d}_2(f)|$ *if and only if* $\mathrm{pol}(f) = \pm\,\mathrm{rec}_2(f)$.
(v) *Assume n is even. If f is a* K_1*-form, then f is a* K_2*-form.*
(vi) *If $n = 2$, then f is a* K_1*-form if and only if f is a* K_2*-form.*
(vii) *If $n = 4$, then f is a* K_1*-form if and only if* $\mathrm{N}(f) \mid \mathrm{d}_2(f)$.

PROOF. Use Proposition 3.33 for the relation between the matrices attached to the reciprocal and the adjoint forms. Next we make explicit the relation, given in Proposition 3.20, between the matrices attached to the polar and the adjoint forms in terms of a function involving the level of f:

$$A_1\big(\mathrm{pol}(f)\big) = \frac{\mathrm{N}(f)}{4\det_1(f)} A_1\big(\mathrm{ad}(f)\big), \qquad A_2\big(\mathrm{pol}(f)\big) = \frac{\mathrm{N}(f)}{\det_2(f)}\,\mathrm{ad}\big(A_2(f)\big).$$

If $\sigma = 1$, by the above lemma we have that f is a K_1-form if and only if there exists $\lambda \in \mathbb{Z}$ such that

$$\frac{1}{\mathrm{d}_1(f)}\,\mathrm{ad}(f) = \lambda \frac{\mathrm{N}(f)}{4\,\mathrm{d}_1^2(f)}\,\mathrm{ad}(f).$$

This is clearly equivalent to the fact that $4\,\mathrm{d}_1(f) \in \mathbb{Z}$ and $\mathrm{N}(f) \mid 4\,\mathrm{d}_1(f)$; this proves (i).

In the case $\sigma = 2$, the matrix of the polar form in terms of $\mathrm{d}_2(f)$ is:

$$A_1\big(\mathrm{pol}(f)\big) = \frac{1}{2} A_2\big(\mathrm{pol}(f)\big) = \frac{1}{2}\frac{\mathrm{N}(f)}{\det_2(f)}\,\mathrm{ad}\big(A_2(f)\big) = \frac{\mathrm{N}(f)}{2\,\mathrm{d}_2^2(f)}\,\mathrm{ad}\big(A_2(f)\big).$$

Applying again the above lemma, we obtain that f is a K_2-form if and only if $\mathrm{N}(f) \mid 2\,\mathrm{d}_2(f)$, which proves (ii).

From the above formulas we deduce that the equalities $N(f) = 4|\,\mathrm{d}_1(f)|$ and $N(f) = 2|\,\mathrm{d}_2(f)|$ are equivalent to the equalities $\mathrm{pol}(f) = \pm\,\mathrm{rec}_1(f)$ and $\mathrm{pol}(f) = \pm\,\mathrm{rec}_2(f)$, respectively. This proves statements (iii) and (iv).

Assume n is even. We have $2\,\mathrm{d}_2(f) = 4\,\mathrm{d}_1(f)2^{(n-2)/2}$. Hence, if $4\,\mathrm{d}_1(f)$ is an integer and $\mathrm{N}(f) \mid 4\,\mathrm{d}_1(f)$, then $\mathrm{N}(f) \mid 2\,\mathrm{d}_2(f)$. Thus, by using (i) and (iii) above, the condition K_1-form implies K_2-form.

In the binary case, we have $\mathrm{rec}_1(f) = \mathrm{rec}_2(f)$, by Corollary 3.34. Therefore, Lemma 3.36 implies (vi).

In the case of $n = 4$, we have that $4\,\mathrm{d}_1(f) = \mathrm{d}_2(f)$. Hence, applying (i), we obtain directly that the condition K_1-form is equivalent to $\mathrm{N}(f) \mid \mathrm{d}_2(f)$. □

COROLLARY 3.38. *Let f be a quadratic form in n variables, over $\mathbb{Z}$. Suppose that* $\det_2(f)$ *is a perfect square.*

(i) *If the level* $\mathrm{N}(f)$ *is square-free, then f is a* K_2*-form.*
(ii) *If* $\mathrm{d}_2(f)$ *is square-free, then* $\mathrm{d}_2(f) \mid N(f)$. *In particular, for $n = 4$, f is a* K_1*-form if and only if* $\mathrm{N}(f) = \mathrm{d}_2(f)$.

PROOF. Since $\mathrm{N}(f)$ is square-free, by applying Proposition 3.22(ii), we have $\mathrm{N}(f) \mid 2\det_2(f)$ and we deduce $\mathrm{N}(f) \mid 2\,\mathrm{d}_2(f)$. This is equivalent to being K_2-form by Proposition 3.37(ii) above.

By Proposition 3.22(i) we know $\det_2(f) \mid N(f)^n$; hence, directly, $\mathrm{d}_2(f) \mid N(f)^n$. If $\mathrm{d}_2(f)$ is square-free, we deduce $\mathrm{d}_2(f) \mid N(f)$. For $n = 4$, together with the statement (vii) in the previous proposition, we obtain the equivalence stated in (ii). □

PROPOSITION 3.39. *Let f be a quadratic form over $\mathbb{Q}$ in an even number n of variables. Let $\lambda \in \mathbb{Z}$ and consider the form λf. If f is a K_σ-form, then λf is also a K_σ-form.*

PROOF. First, note that n must be even because each $\det_\sigma(\lambda f)$ and $\det_\sigma(\lambda f)$ must be simultaneously a perfect square.

On the one hand, since $\det_\sigma(\lambda f) = \lambda^n \det_\sigma(\lambda f)$, we have $\mathrm{d}_\sigma(\lambda f) = \lambda^{n/2}\,\mathrm{d}_\sigma(\lambda f)$. On the other hand, $\mathrm{ad}(\lambda f) = \lambda^{n-1}\,\mathrm{ad}(f)$. Thus, we have the following equalities:

$$\mathrm{rec}_\sigma(\lambda f) = \frac{1}{\mathrm{d}_\sigma(\lambda f)}\,\mathrm{ad}(\lambda f) = \lambda^{n-1-n/2}\frac{1}{\mathrm{d}_\sigma(f)}\,\mathrm{ad}(f) = \lambda^{n/2-1}\,\mathrm{rec}_\sigma(f).$$

The result now follows by the definition of K_σ-form. □

REMARK 3.40. For binary forms, we have $\mathrm{rec}_\sigma(\lambda f) = \mathrm{rec}_\sigma(f)$; for quaternary forms, $\mathrm{rec}_\sigma(\lambda f) = \lambda\,\mathrm{rec}_\sigma(f)$. Hence, it is clear that the converse of the previous proposition is not true.

3.41. *Binary case.* Consider a binary quadratic form over $\mathbb{Z}$, $f = aX^2 + bXY + cY^2$, such that its determinant $\det_1(f) = ac - \frac{1}{4}b^2$ is a perfect square; that is, $\mathrm{disc}_{\mathbb{Q}}(f) = 1$, which is equivalent to the fact that the form f is a product of two linear factors. In this case, we have $2\,\mathrm{d}_1(f) = \mathrm{d}_2(f) = \sqrt{4ac - b^2}$. Therefore,

$$\mathrm{rec}_1(f)(X, Y) = \mathrm{rec}_2(f)(X, Y) = \frac{2}{\sqrt{4ac - b^2}}(cX^2 - bXY + aY^2).$$

For example, consider $f = X^2 + 4XY + 8Y^2$. We obtain $\mathrm{d}_1(f) = 2$, $\mathrm{d}_2(f) = 4$, $\mathrm{ad}(f) = \mathrm{pol}(f) = 8X^2 - 4XY + Y^2$, $\mathrm{rec}_\sigma(f) = 4X^2 - 2XY + \frac{1}{2}Y^2$, and $\mathrm{N}(f) = \det_2(f) = 16$. It is clear that f is not a K_σ-form, for $\sigma = 1$ nor $\sigma = 2$.

3.42. *Ternary case.* Consider for example the form $f = -245X^2 - 35Y^2 - 422Z^2 + 182YX + 644ZX - 238YZ$. Then we have:

(i) $\det_1(f) = 441$, $\mathrm{d}_1(f) = -21$ and $\mathrm{N}(f) = 84$.
(ii) $\mathrm{ad}(f) = 609X^2 + 168YX + 882ZX - 294Y^2 + 294YZ + 294Z^2$.
(iii) $\mathrm{pol}(f) = 29X^2 - 14Y^2 + 14Z^2 + 8XY + 42XZ + 14YZ$.
(iv) $\mathrm{rec}_1(f) = -29X^2 + 14Y^2 - 14Z^2 - 8XY - 42XZ - 14YZ$.

Note that f is a K_1-form.

3.43. *Quaternary case.* Consider the quaternary form $f = X^2 - 44Y^2 - 5Z^2 - 131T^2 - 2XZ - XT + 36YZ - 150YT + 61ZT$. We obtain:

(i) $\det_1(f) = 225$, $\det_2(f) = 3600$.
(ii) $\mathrm{d}_1(f) = -15$, $\mathrm{d}_2(f) = -60$.
(iii) $\mathrm{N}(f) = 120$.
(iv) $\mathrm{ad}(f) = 330X^2 + 315XY + 270XZ - 120XT - (225/2)Y^2 + 225YZ + 180YT + 150Z^2 - 60ZT$.
(v) $\mathrm{pol}(f) = 44X^2 - 15Y^2 + 20Z^2 - 8T^2 + 42XY + 36XZ - 16XT + 30YZ + 24YT - 8ZT$.
(vi) $\mathrm{rec}_1(f) = -22X^2 - 21XY - 18XZ + 8XT + (15/2)Y^2 - 15YZ - 12YT - 10Z^2 + 4ZT + 4T^2$.
(vii) $\mathrm{rec}_2(f) = -44X^2 - 42XY - 36XZ + 16XT + 15Y^2 - 30YZ - 24YT - 20Z^2 + 8ZT + 8T^2$.

Note that f is a K_2-form, but is not a K_1-form.

The following definition recovers the concept of *Hauptform* used by Brandt [**Bra24**]. We restrict ourselves to the case $\sigma = 1$.

DEFINITION 3.44. A K_1-form f is called principal if it represents 1 over $\mathbb{Z}$.

PROPOSITION 3.45. *Let f and f' be two quadratic forms over $\mathbb{Z}$, with square determinants. If f and f' are $\mathbb{Z}$-equivalent, then:*

(i) $\mathrm{d}_\sigma(f) = \mathrm{d}_\sigma(f')$; $\mathrm{N}(f) = \mathrm{N}(f')$.
(ii) $\mathrm{ad}(f) \overset{\mathbb{Z}}{\sim} \mathrm{ad}(f')$; $\mathrm{pol}(f) \overset{\mathbb{Z}}{\sim} \mathrm{pol}(f')$; $\mathrm{rec}(f) \overset{\mathbb{Z}}{\sim} \mathrm{rec}(f')$.
(iii) *f is a K_σ-form if and only if f' is a K_σ-form.*
(iv) *f is principal if and only if f' is principal.*

PROOF. The equalities in (i) are clear, because the $\mathbb{Z}$-equivalence of forms gives the equality for their determinants: $\det_\sigma(f) = \det_1(\sigma f) = \det_1(\sigma f') = \det_\sigma(f')$.

Assume $A_1(f') = P^t A_1(f) P$, with $P \in \mathrm{GL}(n, \mathbb{Z})$. Then, we have $\mathrm{ad}\big(A_1(f')\big) = Q^t \,\mathrm{ad}\big(A_1(f)\big) Q$, with $Q = (P^t)^{-1}$, hence $Q \in \mathrm{GL}(n, \mathbb{Z})$. Thus, Q gives the $\mathbb{Z}$-equivalence between the adjoint forms.

To see the equality of levels, apply Proposition 3.20(c), the equality for the determinants and the equivalence of adjoint forms, and note that the contents of $\mathbb{Z}$-equivalent forms coincide.

The equivalence of the polar and reciprocal forms follows from the definitions and the previous equalities and equivalences.

The conditions that the determinant is a perfect square and that the form and its reciprocal form have rational integer coefficients are preserved by $\mathbb{Z}$-equivalence; hence, the condition of being K_σ-form is also preserved. Finally, it is only necessary to remark that $\mathbb{Z}$-equivalent forms represent the same integers over $\mathbb{Z}$. □

LEMMA 3.46. *If f is a principal quaternary form over $\mathbb{Z}$, then $f(X, Y, Z, T) \overset{\mathbb{Z}}{\sim} X^2 + g(X, Y, Z, T)$, where g is a quaternary quadratic form not containing the term in X^2.*

PROOF. Since f is principal over $\mathbb{Z}$, there exist integers $\alpha_1, \ldots, \alpha_n$ such that $f(\alpha_1, \ldots, \alpha_n) = 1$. Note the representations of 1 are always primitive; that is, $\gcd(\alpha_1, \ldots, \alpha_n) = 1$. This allows us to construct $S \in \mathrm{GL}(n, \mathbb{Z})$ by taking $\alpha_1, \ldots, \alpha_n$ as its first column. Then, the matrix $S^t A_1(f) S$ determines a quadratic form f', $\mathbb{Z}$-equivalent to f, which contains X^2. □

LEMMA 3.47. *Let $f(X, Y, Z, T) = X^2 + g(Y, Z, T)$ a principal quaternary form over $\mathbb{Z}$, with g a ternary quadratic form. Then,*

$$\mathrm{rec}_\sigma(f)(X, Y, Z, T) = \mathrm{d}_\sigma(f) X^2 \oplus \mathrm{rec}_\sigma(g)(Y, Z, T).$$

PROOF. It is clear that $\mathrm{d}_\sigma(f) = \mathrm{d}_\sigma(g)$. Moreover it is easy to check that

$$\mathrm{ad}\big(A_1(\sigma f)\big) = \left(\begin{array}{c|c} \det_\sigma(f) & 0 \\ \hline 0 & \mathrm{ad}\big(A_1(\sigma g)\big) \end{array}\right).$$

Then the result follows from Proposition 3.33. □

3.3. Quadratic forms associated to K-algebras

In this section we consider K-algebras with an additional structure of quadratic space. The results will be applied to quadratic fields over K and to quaternion K-algebras. As a general reference for K-algebras and orders see [**Rei75**].

Let R be a ring and K its field of fractions. Let A be a finitely generated K-algebra, $n = [A{:}K]$, and let B be a bilinear symmetric form on A:

$$\begin{aligned} B\colon A \times A &\to K \\ (\alpha, \beta) &\mapsto B(\alpha, \beta). \end{aligned}$$

Consider the associated mapping denoted by Q_A:

$$\begin{aligned} Q_A\colon A &\to K \\ \alpha &\mapsto Q_A(\alpha) = B(\alpha, \alpha). \end{aligned}$$

The expression of Q_A with respect to a fixed K-basis $\{e_1, \dots, e_n\}$ of A leads to a quadratic form $f_A = f_A(X_1, \dots, X_n)$, such that for a generic element $\alpha = \sum_{i=1}^n x_i e_i$, $x_i \in K$, $f_A(\alpha) = f_A(x_1, \dots, x_n) = Q_A(\sum_{i=1}^n x_i e_i)$. If we change the K-basis of A, we obtain a K-equivalent quadratic form.

DEFINITION 3.48. A quadratic form f_A is unitary if $f_A(1_A) = 1_A$. A quadratic form f_A is multiplicative if $f_A(\alpha\beta) = f_A(\alpha) f_A(\beta)$, for α, $\beta \in A$. These concepts are independent of the choice of the basis in A used to express the quadratic form f_A.

For any quadratic form f_A, $\big(1/f_A(1_A)\big) f_A$ is a unitary form. If f_A is a unitary quadratic form, then, $f_A(x) = x^2$, for every $x \in K$. Fix a K-basis of an order in A. If f_A is unitary, then it is clear that f_A represents 1_A over R.

As for examples of multiplicative quadratic forms, we have the normic forms, which we will analyze in following sections.

From the quadratic form f_A we also obtain other quadratic forms. On the one hand, using the structure of A as a vector space over K, we can restrict ourselves to a subspace A', $[A'{:}K] = n'$ and consider the associated quadratic form in n' variables. In particular, we can consider the case in which A' is a K-subalgebra of A.

On the other hand, we can consider the R-orders of A; in this case, we obtain quadratic forms in n variables. Let $\mathcal{O}$ be an R-order in A. Fixing an R-basis $\{v_1, \dots, v_n\}$ for the order $\mathcal{O}$, a generic element $\omega \in \mathcal{O}$ is written as $\omega = X_1 v_1 + \cdots + X_n v_n$. Restricting the mapping Q_A to the elements in the order, we obtain a quadratic form in the variables $X_1, \dots, X_n$. We denote this quadratic form by $f_{\mathcal{O},n}$. The quadratic form depends on the fixed basis of $\mathcal{O}$, consequently this notation could be ambiguous. But, by the following lemma, we can avoid the ambiguity by considering the R-equivalence class of quadratic forms.

LEMMA 3.49. *The quadratic forms $f_{\mathcal{O},n}$ attached to an R-order $\mathcal{O}$ in different bases are R-equivalent.*

PROPOSITION 3.50. *Let $\mathcal{O}$ be an order in a K-algebra A. Fix bases $\mathcal{B}_{\mathcal{O}}$ and $\mathcal{B}_A$ of $\mathcal{O}$ and A, respectively. Consider the associated quadratic forms $f_{\mathcal{O},n}$ and f_A. Let P be the base change matrix from the $\mathcal{B}_{\mathcal{O}}$-coordinates to the $\mathcal{B}_A$-coordinates. Then $f_{\mathcal{O},n} \overset{K}{\sim} f_A$. In particular,* $\operatorname{disc}_K(f_{\mathcal{O},n}) = \operatorname{disc}_K(f_A)$; *more concretely* $\det_1(f_{\mathcal{O},n}) = (\det P)^2 \det_1(f_A)$. *Moreover, if f_A is a unitary quadratic form, then $f_{\mathcal{O},n}$ represents 1 over R.*

PROOF. The basis $\mathcal{B}_{\mathcal{O}}$ is also a K-basis of A. Thus the form $f_{\mathcal{O}}$ is also the quadratic form of A, but with respect to another basis. Hence, $f_{\mathcal{O}}$ and f_A are K-equivalent.

The unit of the algebra A belongs to $\mathcal{O}$, because every order is a subring. Since f_A is a unitary form, we have that $f_{\mathcal{O},n}(1_A) = 1_A$. The representation is over R because 1_A is expressed in terms of the basis of $\mathcal{O}$ with coefficients in R. □

COROLLARY 3.51. *Let K be a number field with a real place, and we fix an embedding $\phi\colon K \hookrightarrow \mathbb{R}$. This allows us to define the signature and the definite or indefinite character of the quadratic forms with coefficients in K. Then, for any R-order $\mathcal{O}$ of A, the forms f_A and $f_{\mathcal{O},n}$ have the same signature, hence the same definite or indefinite character.*

PROPOSITION 3.52. *Assume f_A is a multiplicative and unitary form. Let $\mathcal{O}$ and $\mathcal{O}'$ be two conjugate R-orders in A. Then, $f_{\mathcal{O},n} \overset{R}{\sim} f_{\mathcal{O}',n}$.*

PROOF. Let $\{v_i\}_{1\leq i\leq n}$ be an R-basis of $\mathcal{O}$. Assume $\mathcal{O}' = u^{-1}\mathcal{O}u$. Then, $\{u^{-1}v_iu\}_{1\leq i\leq n}$ is an R-basis of $\mathcal{O}'$. For any $\omega \in A$, we have $f_A(\omega) = f_A(u^{-1}\omega u)$, since it is a multiplicative and unitary quadratic form. We are going to deduce that the quadratic forms associated to the orders $\mathcal{O}$ and $\mathcal{O}'$, in the bases $\{v_i\}_{1\leq i\leq n}$ and $\{u^{-1}v_iu\}_{1\leq i\leq n}$, respectively, are exactly the same. We have

$$\begin{aligned} B(v_i + v_j, v_i + v_j) &= f_A(v_i + v_j) = f_A(u^{-1}(v_i + v_j)u) \\ &= B(u^{-1}(v_i + v_j)u, u^{-1}(v_i + v_j)u). \end{aligned}$$

Now we expand and apply bilinear and multiplicative properties, and we obtain $B(v_i, v_j) = B(u^{-1}v_iu, u^{-1}v_ju)$. This proves the equality of the forms. But, inside an order we already have R-equivalence of forms, by Lemma 3.49, hence, effectively, the forms are R-equivalent. □

PROPOSITION 3.53. *Let $\mathcal{O}' \subseteq \mathcal{O}$ be two orders of A, with fixed bases $\mathcal{B}'$ and $\mathcal{B}$, respectively. Put $r = \det P$, where P is the base change matrix from the $\mathcal{B}'$-coordinates to the $\mathcal{B}$-coordinates. Then:*

(i) $f_{\mathcal{O},n} \overset{R}{\longrightarrow} f_{\mathcal{O}',n}$.
(ii) $\det_1(f_{\mathcal{O}',n}) = r^2 \det_1(f_{\mathcal{O},n})$.
(iii) $\det_1(f_{\mathcal{O}',n}) = \det_1(f_{\mathcal{O},n})$ *if and only if* $f_{\mathcal{O},n} \overset{R}{\sim} f_{\mathcal{O}',n}$, *if and only if* $\mathcal{O} = \mathcal{O}'$.

3.4. Quadratic forms associated to quadratic fields and orders

In this section we fix some notations and recall the relation between the quadratic fields and the binary quadratic forms. This will serve as the prelude to the parallel results relating quaternion algebras and ternary and quaternary quadratic forms. These results will be used to relate the embeddings of quadratic fields in quaternion algebras with representations of quadratic forms by quadratic forms in the next chapter.

First, we give a brief outline of the notation about quadratic fields.

Consider a quadratic field $F = \mathbb{Q}(\sqrt{d})$, $d \in \mathbb{Z}$ square-free. We denote by Λ_F the ring of integers of F. A $\mathbb{Z}$-basis of Λ_F is $\{1, w\}$, where $w = \sqrt{d}$ if $d \equiv 2, 3 \bmod 4$ and $w = (1+\sqrt{d})/2$ if $d \equiv 1 \bmod 4$. For an element $\lambda \in F$, λ' denotes its conjugate by the action of the Galois group, and $\mathrm{n}(\lambda)$ and $\mathrm{tr}(\lambda)$ denote its norm and its trace, respectively. We denote by F_0 the elements of F of trace equal to 0.

A quadratic order Λ in a quadratic field is a free $\mathbb{Z}$-module of rank 2, with integral elements. The discriminant of an order Λ is denoted by D_Λ. The ring of integers Λ_F is the unique maximal order in F. The fundamental discriminant D_F of the field F is the discriminant of the maximal order, that is $D_F := D_{\Lambda_F}$. We have $D_F = 4d$ if $d \equiv 2,3 \bmod 4$, and $D_F = d$ if $d \equiv 1 \bmod 4$; thus, $F = \mathbb{Q}(\sqrt{D_F})$ and $D_F \equiv 0,1 \bmod 4$. Any order Λ is contained in the maximal order and the conductor of Λ is the index $[\Lambda_F{:}\Lambda]$, as $\mathbb{Z}$-modules. In fact, for every $m \in \mathbb{N}$, the quadratic order $\Lambda(d,m) := \mathbb{Z}[1, mw]$ is the unique order of conductor m in the quadratic field $\mathbb{Q}(\sqrt{d})$, and its discriminant is $D_{\Lambda(d,m)} = m^2 D_F$. Thus, the conductor determines the order. The quadratic order $\mathbb{Z}[1, \sqrt{d}]$ is the order of conductor 1 if $d \equiv 2,3 \bmod 4$, and the order of conductor 2 if $d \equiv 1 \bmod 4$.

NOTATION 3.54. Let $F = \mathbb{Q}(\sqrt{d})$ be an imaginary quadratic field, where d is square-free. We denote by $h(d)$ the ideal class number of F. Let $\Lambda(d,m) \subseteq F$ be the quadratic order with conductor m. We denote by $h(d,m)$ the ideal class number of $\Lambda(d,m)$.

PROPOSITION 3.55 (cf. [**Zag81**]). *Let $\Lambda(d,m)$ be the quadratic order of conductor $m > 1$ in $F = \mathbb{Q}(\sqrt{d})$, with d square-free, $d < 0$. Let D_F be the fundamental discriminant of F. Then,*

$$h(d,m) = \frac{mh(d)}{\nu} \prod_{p \mid m} \left(1 - \left(\frac{D_F}{p}\right) p^{-1}\right), \quad \text{where } \nu = \begin{cases} 2 & \text{if } d = -1, m > 1 \\ 3 & \text{if } d = -3, m > 1 \\ 1 & \text{otherwise.} \end{cases}$$

Using the trace map defined over F, we define the bilinear symmetric form

$$\begin{aligned} B\colon\ & F \times F \to \mathbb{Q} \\ & (\lambda,\ \mu) \mapsto \tfrac{1}{2}\operatorname{tr}(\lambda\mu'). \end{aligned}$$

The quadratic space (F, B) is regular. The mapping associated to this bilinear form is precisely the norm,

$$\begin{aligned} & F \to \mathbb{Q} \\ & \lambda \mapsto B(\lambda,\lambda) = \tfrac{1}{2}\operatorname{tr}(\lambda\lambda') = \mathrm{n}(\lambda). \end{aligned}$$

With the basis $\{1, \sqrt{d}\}$ fixed, which is a $\mathbb{Q}$-orthogonal basis in (F, B), the associated binary quadratic form is called the normic form of F and is denoted by $n_{F,2}$. We have $n_{F,2}(X,Y) = X^2 - dY^2$. If we choose another basis for F, we will obtain a $\mathbb{Q}$-equivalent form. Some of its well-known properties are collected in the following lemma.

LEMMA 3.56. *Let $F = \mathbb{Q}(\sqrt{d})$ be a quadratic field with a fixed basis and $n_{F,2}$ the associated normic form. Then:*

(i) $\det_1(n_{F,2}) = -d$, $\det_2(n_{F,2}) = -4d$, *and* $\operatorname{disc}_{\mathbb{Q}}(n_{F,2}) = -d = -D_F$ *in* $\mathbb{Q}^*/\mathbb{Q}^{*2}$.
(ii) $n_{F,2}$ *is positive definite if F is an imaginary quadratic field.* $n_{F,2}$ *is indefinite if F is a real quadratic field.*
(iii) $n_{F,2}$ *is a multiplicative and unitary form.*
(iv) $n_{F,2}$ *is primitive and it has rational integer coefficients.*
(v) $\operatorname{ad}(n_{F,2})(X,Y) = \operatorname{pol}(n_{F,2})(X,Y) = -dX^2 + Y^2$ *and this form is* $\mathrm{SL}(2,\mathbb{Z})$-*equivalent to* $n_{F,2}$.
(vi) $\mathrm{N}(n_{F,2}) = 4d$.

Given a quadratic order Λ with a fixed basis, we consider the binary normic form associated to Λ, denoted by $n_{\Lambda,2}$. Note that, by Lemma 3.49, normic forms obtained using different bases inside an order are $\mathrm{SL}(2,\mathbb{Z})$-equivalent. On the other hand, recall that the normic forms of orders are related with the normic form associated to the algebra via $\mathbb{Q}$-equivalence, by Proposition 3.50.

Next we provide the normic forms attached to the order of conductor m in $\mathbb{Q}(\sqrt{d})$ with respect to some standard bases.

REMARK 3.57. Let Λ be the order of $\mathbb{Q}(\sqrt{d})$ of conductor m. We obtain the following normic forms of Λ, with respect to the given basis:

$$\begin{cases} X^2-(D_F/4)m^2Y^2 & \text{if } D_F\equiv 0 \bmod 4,\\ X^2+mXY+m^2((1-D_F)/4)Y^2 & \text{if } D_F\equiv 1 \bmod 4, \end{cases} \qquad \text{basis } \{1, mw\}$$

$$X^2+mD_FXY+m^2D_F\left(\frac{D_F-1}{4}\right)Y^2, \qquad \left\{1, m\frac{D_F+\sqrt{D_F}}{2}\right\}$$

$$X^2+XYD_\Lambda+Y^2\left(\frac{D_\Lambda^2-D_\Lambda}{4}\right), \qquad \left\{1, \frac{D_\Lambda+\sqrt{D_\Lambda}}{2}\right\}.$$

PROPOSITION 3.58. *The* $\mathrm{SL}(2,\mathbb{Z})$*-reduced normic form (cf.* **[Coh95]**) *of the order* $\Lambda=\Lambda(d,m)$ *is*

$$\begin{cases} X^2-(D_\Lambda/4)Y^2 & \text{if } D_\Lambda\equiv 0 \bmod 4,\\ X^2+XY+((1-D_\Lambda)/4)Y^2 & \text{if } D_\Lambda\equiv 1 \bmod 4. \end{cases}$$

It is obtained using the basis

$$\begin{cases} \{1, mw\} & \text{if } d\equiv 2,3 \bmod 4,\\ \{1, -[m/2]+mw\} & \text{if } d\equiv 1 \bmod 4. \end{cases}$$

The normic forms of the orders contain information about the quadratic field. We collect in the following lemma some properties and relations between invariants attached to the orders and to the quadratic forms.

LEMMA 3.59. *Let* Λ *be the order of conductor* m *in the quadratic field* $F=\mathbb{Q}(\sqrt{d})$. *Let* $n_{\Lambda,2}$ *and* $n_{F,2}$ *be the normic forms attached to the order and to the quadratic field respectively, with respect to fixed bases. Then*:

(i) $\mathrm{disc}_{\mathbb{Q}}(n_{\Lambda,2})=\mathrm{disc}_{\mathbb{Q}}(n_{F,2})$; $\det_1(n_{\Lambda,2})=-D_\Lambda/4$; $\det_2(n_{\Lambda,2})=-D_\Lambda$.
(ii) $n_{\Lambda,2}$ *has rational integer coefficients and it is unitary.*
(iii) $n_{\Lambda,2}$ *is primitive and the* $\mathrm{SL}(2,\mathbb{Z})$*-equivalence* $n_{\Lambda,2}\sim\mathrm{ad}(n_{\Lambda,2})$ *holds.*
(iv) $\mathrm{N}(n_{\Lambda,2})=|\det_2(n_{\Lambda,2})|=D_\Lambda$.

We are interested in the fundamental units of the quadratic fields $F=\mathbb{Q}(\sqrt{d})$, $d\in\mathbb{Z}$ square-free, and in those of the quadratic orders $\Lambda(d,m)$. In relation to quadratic forms, this is equivalent to the study of the representations of ±1 by the normic form.

If F is an imaginary quadratic field, the results about the units are easier to describe. For $\mathbb{Q}(\iota)$, $\iota^2=-1$, the group of units is formed by the fourth roots of unity. For $\mathbb{Q}(\sqrt{-3})$, the group of units is formed by the sixth roots of unity. For the other imaginary quadratic fields, the group of the units is $\{1,-1\}$.

Assume F is a real quadratic field; that is, $d>0$. By the Dirichlet unit theorem, given an order Λ in $\mathbb{Q}(\sqrt{d})$, there exists a unit $\varepsilon\in\Lambda$ such that each unit of Λ has a unique expression $\pm\varepsilon^n$. For $\varepsilon = a + b\sqrt{d}$, the set of units $\{\varepsilon, -\varepsilon, 1/\varepsilon, -1/\varepsilon\}$ corresponds to the four elements $\pm a \pm b\sqrt{d}$. The unique element $\varepsilon = a + b\sqrt{d}$ with $a, b > 0$ satisfying the condition $\varepsilon > 1$ is called the fundamental unit of Λ. The fundamental unit of the maximal order is called the fundamental unit of the field. In particular, the fundamental units of the quadratic orders are powers of the fundamental unit of the field. The fundamental unit of the field is the solution, with the minimal positive integers x and y, of $\mathrm{n}(x+yw) = \pm 1$; that is, of $x^2 - dy^2 = \pm 1$ if $d \equiv 2, 3 \bmod 4$, or of $x^2 + xy + \big((1-d)/4\big)y^2 = \pm 1$ if $d \equiv 1 \bmod 4$. Talking about normic forms, the fundamental unit of the quadratic order $\Lambda(d, m)$ is the representation $(x, y) \in \mathcal{R}(n_{\Lambda(d,m),2}, 1; \mathbb{Z})$ with x and y minimal positive integers.

3.5. Normic forms of quaternion algebras

In this section we deal with quaternion $\mathbb{Q}$-algebras as quadratic spaces and we will apply the results found for quadratic forms attached to K-algebras stated in Section 3.3. The restriction of the quadratic form associated to the quaternion algebra to the subspace of pure quaternions H_0 allows us to define a ternary quadratic form associated to the quaternion algebra which plays an important role in the forthcoming results.

Let $H = \left(\frac{a,b}{\mathbb{Q}}\right)$ be a quaternion $\mathbb{Q}$-algebra, $a, b \in \mathbb{Z}$, square-free. Fix a canonical basis $\mathcal{C} = \{1, i, j, ij\}$ in H.

Using the trace mapping defined over H, we define the symmetric bilinear form

$$\begin{aligned} B\colon H \times H &\to \mathbb{Q} \\ (\alpha, \beta) &\mapsto \tfrac{1}{2}\operatorname{tr}(\alpha\overline{\beta}) \end{aligned}$$

that gives to H and H_0 a natural structure of quadratic spaces: (H, B) and $(H_0, B_{|H_0})$, respectively.

REMARK 3.60. For $u, v \in H_0$ we have $B_{|H_0}(u, v) = \frac{1}{2}(u\overline{v} + v\overline{u}) = -\frac{1}{2}(uv + vu)$; hence, u and v are orthogonal in $(H_0, B_{|H_0})$ if and only if u and v anticommute in H_0. Thus, $\{i, j, k\}$ is an orthogonal basis of H_0. Moreover, for $u \in H_0$ we have $B(1, u) = 0$; hence, the subspace $\mathbb{Q}$ is orthogonal to H_0. Therefore, the basis $\mathcal{C}$ is an orthogonal basis of the regular quadratic space (H, B).

The following lemma, easy to prove, gives an explicit relation between the bilinear form associated to the quaternion $\mathbb{Q}$-algebra and the noncommutative quaternion product. In particular, it could be used to implement the quaternion product for pure quaternions.

LEMMA 3.61. *Let $H = \left(\frac{a,b}{\mathbb{Q}}\right)$ be a quaternion algebra with canonical basis $\{1, i, j, ij\}$ and B the former bilinear form associated to H. Let $u, v \in H_0$ and consider the vector product $P(u, v)$ given by the determinant*

$$P\big((u_1, u_2, u_3), (v_1, v_2, v_3)\big) = \begin{vmatrix} -bi & -aj & ij \\ u_1 & u_2 & u_3 \\ v_1 & v_2 & v_3 \end{vmatrix}.$$

Then, $u \cdot v = P(u, v) - B(u, v)$.

3.5.1. Definitions and elementary properties. The quadratic mapping attached to the bilinear form B is precisely the norm,

$$\begin{aligned} H &\to \mathbb{Q} \\ \alpha &\mapsto B(\alpha,\alpha) = \tfrac{1}{2}\operatorname{tr}(\alpha\overline{\alpha}) = \operatorname{n}(\alpha). \end{aligned}$$

DEFINITION 3.62. We call normic forms attached to the quaternion $\mathbb{Q}$-algebra H the normic forms over H, and over H_0, respectively, which are calculated in the bases $\mathcal{C} = \{1, i, j, ij\}$ and $\mathcal{C}_0 = \{i, j, ij\}$, respectively. They are denoted by $n_{H,4}$ and $n_{H,3}$, respectively.

Consider $H = \left(\frac{a,b}{\mathbb{Q}}\right)$ and fix the canonical basis. The normic forms are:

- the quaternary form $n_{H,4}(X,Y,Z,T) = X^2 - aY^2 - bZ^2 + abT^2$,
- the ternary form $n_{H,3}(Y,Z,T) = -aY^2 - bZ^2 + abT^2$.

3.63. *Nonramified case.* For $H = \left(\frac{1,-1}{\mathbb{Q}}\right)$, we obtain the normic forms:

$$n_{H,4}(X,Y,Z,T) = X^2 - Y^2 + Z^2 - T^2, \qquad n_{H,3}(Y,Z,T) = -Y^2 + Z^2 - T^2.$$

We can also consider the isomorphic algebra $H' = M(2,\mathbb{Q})$. Then, dealing directly with the norm, we obtain the following quadratic forms, $\mathbb{Q}$-equivalent to the previous ones,

$$n_{H',4}(X,Y,Z,T) = XT - YZ, \qquad n_{H',3}(Y,Z,T) = -YZ - T^2.$$

LEMMA 3.64. *Let H be a quaternion $\mathbb{Q}$-algebra. We have the following properties.*

(i) *For every $\lambda \in \mathbb{Q}$, $n_{H,4}(\lambda) = \lambda^2$.*
(ii) *For every $u \in H_0$, $n_{H,4}(u) = n_{H,3}(u) = -u^2$.*
(iii) *$n_{H,4}$ is a multiplicative unitary quaternary form.*
(iv) *$n_{H,4} \sim \langle 1 \rangle \oplus n_{H,3}$.*
(v) *The normic forms $n_{H,4}$, $n_{H,3}$ have rational integer coefficients.*

3.5.2. Relations between the invariants. Since isomorphisms between algebras preserve the norm, we have the following well-known result about the equivalence between isomorphisms of algebras and isometries of quadratic spaces.

PROPOSITION 3.65. *Let H and H' be two quaternion $\mathbb{Q}$-algebras. Then, the following conditions are equivalent:*

(i) *H and H' are isomorphic $\mathbb{Q}$-algebras.*
(ii) *$(H, n_{H,4})$ and $(H', n_{H',4})$ are isometric quadratic $\mathbb{Q}$-spaces.*

These normic forms attached to the quaternion algebra contain a great amount of information about the algebra. We will interpret some invariants and definitions of the quaternion algebra in terms of the associated normic forms.

First, we list some properties of these normic quadratic forms, determined basically by the numbers a and b.

LEMMA 3.66. *Let $H = \left(\frac{a,b}{\mathbb{Q}}\right)$ be a quaternion algebra. Consider the normic forms $n_{H,4}$ and $n_{H,3}$ attached to H. Then:*

(i) $\operatorname{disc}_{\mathbb{Q}}(n_{H,4}) = \operatorname{disc}_{\mathbb{Q}}(n_{H,3}) = 1$ *and* $\operatorname{disc}_{\mathbb{Q}}(n_{H,2}) = ab$.
(ii) *The quadratic forms $n_{H,4}$ and $n_{H,3}$ have the same character, definite or indefinite, and they are never negative definite.*

(iii) *The form* $n_{H,4}$ *is of signature* $(4,0)$ *or* $(2,2)$; *the form* $n_{H,3}$ *is of signature* $(3,0)$ *or* $(1,2)$.
(iv) $\varepsilon_v(n_{H,4}) = \varepsilon_v(n_{H,3}) = (-1,-1)_v(a,b)_v$, *for any place* v *of* $\mathbb{Q}$.
(v) $n_{H,4} \xrightarrow{\mathbb{Z}} n_{H,3}$.
(vi) $n_{H,3}$ *and* $n_{H,4}$ *are* $\mathrm{SL}(n,\mathbb{Z})$*-reduced forms.*
(vii) $n_{H,4}$ *is a primitive form; if* $\gcd(a,b)=1$, *then* $n_{H,3}$ *is also primitive.*

COROLLARY 3.67. *Let* H *and* H' *be quaternion* $\mathbb{Q}$*-algebras. Then,* $H \overset{\mathbb{Q}}{\simeq} H'$ *if and only if* $D_H = D_{H'}$.

PROOF. Put $H = \left(\frac{a,b}{\mathbb{Q}}\right)$ and $H' = \left(\frac{a',b'}{\mathbb{Q}}\right)$. By the above proposition, the algebras H and H' are $\mathbb{Q}$-isomorphic if the corresponding ternary forms are $\mathbb{Q}$-equivalent. Now, the diagonal quadratic forms $n_{H,3} = \langle -a,-b,ab\rangle$ and $n_{H',3} = \langle -a',-b',a'b'\rangle$ are $\mathbb{Q}$-equivalent if and only if they have the same discriminant over $\mathbb{Q}$, the same Hasse–Witt invariant and the same signature (cf. [**Ser73**]).

Directly we have the equality of discriminants, since $\mathrm{disc}_{\mathbb{Q}}(n_{H,3}) = a^2b^2 = 1 = (a')^2(b')^2 = \mathrm{disc}_{\mathbb{Q}}(n_{H',3})$. By Lemma 3.66(iii), a ternary normic form such as the previous one can have only two possible signatures, $(3,0)$ and $(1,2)$, according to the local Hilbert symbol at the infinity takes the value 1 or -1. Thus, the equality of signatures is in correspondence with the equality $(a,b)_\infty = (a',b')_\infty$. By Lemma 3.66(iv), the equality of the Hasse–Witt invariants, for every finite place, is equivalent to the equality of the local Hilbert symbols $(a,b)_v = (a',b')_v$.

Thus, the forms $n_{H,3}$ and $n_{H',3}$ are $\mathbb{Q}$-equivalent if and only if the symbols $(a,b)_v = (a',b')_v$ coincide for every place v. By definition of the discriminant of the quaternion algebra, this is exactly equivalent to the equality of discriminants $D_H = D_{H'}$. □

In the following corollary we review the well-known characterization of the nonramified quaternion algebras in terms of the properties of the attached normic forms.

COROLLARY 3.68. *Let* $H = \left(\frac{a,b}{\mathbb{Q}}\right)$ *be a quaternion algebra and let* $n_{H,4}$ *and* $n_{H,3}$ *be the associated normic forms. The following are equivalent:*

(i) $H \simeq \mathrm{M}(2,\mathbb{Q})$.
(ii) $n_{H,4}$ *is* $\mathbb{Q}$*-isotropic.*
(iii) $n_{H,3}$ *is* $\mathbb{Q}$*-isotropic.*
(iv) $a \in N_{L|\mathbb{Q}}(L)$, *where* $L = \mathbb{Q}(\sqrt{b})$.

The following lemma states the relation between the Hasse invariant of the quaternion algebra (cf. Chapter 1) and the Hasse–Witt invariant of normic forms (cf. [**Ser73**]).

LEMMA 3.69. *Let* $H = \left(\frac{a,b}{\mathbb{Q}}\right)$ *be a quaternion algebra and consider the ternary normic form* $n_{H,3}$ *attached to* H. *Then,* $\varepsilon_v(n_{H,3}) = \varepsilon\left(\frac{a,b}{\mathbb{Q}}\right)_v$ *if* $v \neq 2,\infty$, *and* $\varepsilon_v(n_{H,3}) = -\varepsilon\left(\frac{a,b}{\mathbb{Q}}\right)_v$ *if* $v = 2,\infty$.

Using this lemma, we can extend Corollary 3.68 to ramified algebras, in such a way as to characterize the discriminant of a quaternion algebra from the properties of the associated normic forms.

NOTATION 3.70. Let f be a quadratic form over $\mathbb{Z}$. Put

$$S(f) := \{p \text{ prime} : f_p \text{ is } \mathbb{Q}_p\text{-anisotropic}\}.$$

If f and f' are two $\mathbb{Q}$-equivalent quadratic forms over $\mathbb{Z}$, then $S(f) = S(f')$.

PROPOSITION 3.71. *Let H be a quaternion $\mathbb{Q}$-algebra and p a prime number. Then, $p \mid D_H$ if and only if the form $n_{H,4}$ is anisotropic in $\mathbb{Q}_p$, equivalently if and only if $n_{H,3}$ is anisotropic in $\mathbb{Q}_p$. Thus,*

$$D_H = \prod_{p \in S(n_{H,4})} p = \prod_{p \in S(n_{H,3})} p.$$

PROOF. On the one hand, for the primes $p \mid D_H$, we have $(a,b)_p = -1$. On the other hand, the form $n_{H,4}$ is $\mathbb{Q}_p$-isotropic if and only if $\varepsilon_p(n_{H,4}) = (-1,-1)_p$, since $\operatorname{disc}_{\mathbb{Q}_p}(n_{H,4}) = 1$. By Lemma 3.66(iv) we have that $\varepsilon_p(n_{H,4}) = (-1,-1)_p(a,b)_p$. Hence being anisotropic over $\mathbb{Q}_p$ is equivalent to $(a,b)_p = -1$.

For the ternary form $n_{H,3}$, the isotropy condition is also expressed in terms of the Hasse–Witt invariant: $\varepsilon_p(n_{H,3}) = \left(-1, -\operatorname{disc}_{\mathbb{Q}_p}(n_{H,3})\right)_p$. We also have $\operatorname{disc}_{\mathbb{Q}_p}(n_{H,3}) = 1$. Thus, we obtain a condition analogous to the previous one. □

The following lemma relates the definite or indefinite character of the quaternion algebra with the definite or indefinite character of the associated normic quadratic forms.

LEMMA 3.72. *Let H be a quaternion $\mathbb{Q}$-algebra and $n_{H,i}$, for $i = 3, 4$, the attached normic forms. Then:*

(i) *H is definite if and only if $n_{H,4}$ is positive definite; equivalently, if and only if $n_{H,3}$ is positive definite.*
(ii) *H is indefinite if and only if $n_{H,4}$ is indefinite; equivalently, if and only if $n_{H,3}$ is indefinite.*

3.5.3. The normic forms and the $\mathbf{K}_\sigma$-forms. We collect in the following two lemmas the computation of the determinants, the adjoint, reciprocal and polar forms, and the levels of the normic forms $n_{H,4}$ and $n_{H,3}$.

LEMMA 3.73. *Consider the ternary normic form $n_{H,3}$ associated to the quaternion algebra $H = \left(\frac{a,b}{\mathbb{Q}}\right)$ with respect to a canonical basis $C = \{1, i, j, ij\}$. Assume H is indefinite and $a > 0$. Then:*

(i) $\det_1(n_{H,3}) = a^2b^2$ *and* $\mathrm{d}_1(n_{H,3}) = -a|b|$.
(ii) $\operatorname{ad}(n_{H,3})(Y,Z,T) = -ab^2Y^2 - a^2bZ^2 + abT^2$; *in particular,* $\operatorname{ad}(n_{H,3}) \overset{\mathbb{Q}}{\sim} n_{H,3}$.
(iii) $\operatorname{rec}_1(n_{H,3})(Y,Z,T) = \begin{cases} bY^2 + aZ^2 - T^2 & \text{if } b > 0, \\ -bY^2 - aZ^2 + T^2 & \text{if } b \le 0. \end{cases}$
(iv) $\mathrm{N}(n_{H,3}) = 4a|b|$.
(v) $\operatorname{pol}(n_{H,3})(Y,Z,T) = -bY^2 - aZ^2 + T^2$.

LEMMA 3.74. *Let $H = \left(\frac{a,b}{\mathbb{Q}}\right)$ be a quaternion algebra. Consider the quaternary normic form $n_{H,4}$ associated to H with respect to a canonical basis $C = \{1, i, j, ij\}$. Assume H is indefinite and $a > 0$. Then:*

(i) $\det_1(n_{H,4}) = a^2b^2$, $\mathrm{d}_1(n_{H,4}) = -a|b|$, $\mathrm{d}_2(n_{H,4}) = -4a|b|$.

(ii) $\mathrm{ad}(n_{H,4})(X,Y,Z,T) = a^2b^2X^2 - ab^2Y^2 - a^2bZ^2 + abT^2$; *thus*, $\mathrm{ad}(n_{H,4}) \overset{\mathbb{Q}}{\sim} n_{H,4}$.

(iii) $\mathrm{rec}_1(n_{H,4})(X,Y,Z,T) = \begin{cases} -abX^2 + bY^2 + aZ^2 - T^2 & \text{if } b > 0, \\ abX^2 - bY^2 - aZ^2 + T^2 & \text{if } b < 0. \end{cases}$

$\mathrm{rec}_2(n_{H,4}) = 2\,\mathrm{rec}_1(n_{H,4})$.

Moreover, $\mathrm{rec}_1(n_{H,4})(X,Y,Z,T) = \mathrm{d}_1(n_{H,4})X^2 \oplus \mathrm{rec}_1(n_{H,3})(Y,Z,T)$.

(iv) $\mathrm{N}(n_{H,4}) = 4a|b|$.

(v) $\mathrm{pol}(n_{H,4})(X,Y,Z,T) = abX^2 - bY^2 - aZ^2 + T^2$.

In the case that H is a definite quaternion algebra, we obtain analogous results.

Next we make explicit the relation between the signature of the normic forms and the signature of their adjoint and reciprocal forms. Note that the condition definite or indefinite is preserved, but there is some change in the signs.

PROPOSITION 3.75. *Let H be a quaternion $\mathbb{Q}$-algebra. We have the following relations.*

(i) $\mathrm{sign}(n_{H,i}) = \mathrm{sign}(\mathrm{ad}(n_{H,i})) = \mathrm{sign}(\mathrm{pol}(n_{H,i}))$, *for* $i = 3, 4$.

(ii) $\mathrm{sign}(n_{H,4}) = \mathrm{sign}(\mathrm{rec}_\sigma(n_{H,4}))$.

(iii) *The form* $\mathrm{rec}_1(n_{H,3})$ *has the same character, definite or indefinite, as the form* $n_{H,3}$. *Moreover, if* $n_{H,3}$ *is an indefinite form of signature* $(1,2)$, *then* $\mathrm{sign}(\mathrm{rec}_1(n_{H,3})) = (2,1)$.

PROOF. From the explicit computations of the adjoint forms of these normic forms, it is clear that the signature of the adjoint form coincides with the signature of the form. However, in general, the signatures are preserved only for the binary forms, since the form and its adjoint form are equivalent. For any quadratic form, the polar form is obtained by multiplying a positive rational number to the adjoint form. This proves (i).

To study the signature of the reciprocal form it is only necessary to take into account the sign of $\mathrm{d}_1(n_{H,3})$ and $\mathrm{d}_\sigma(n_{H,4})$. This sign is positive if the form is positive definite and is negative if the form is indefinite. Note that, in the quaternary case, if the signature is $(2,2)$, the sign does not change when we divide by $\mathrm{d}_\sigma(n_{H,4})$. Instead, if $n_{H,3}$ is indefinite of signature $(1,2)$, we obtain $\mathrm{sign}(\mathrm{rec}_1(n_{H,3})) = (2,1)$. □

Using Lemmas 3.73 and 3.74, Proposition 3.75, and the characterization of the K_σ-forms, cf. Definition 3.35 and Proposition 3.37, we obtain our main result in this section.

THEOREM 3.76. *Let* $H = \left(\frac{a,b}{\mathbb{Q}}\right)$ *be an indefinite quaternion algebra, and let* $n_{H,4}$ *and* $n_{H,3}$ *be the attached normic forms. Then:*

(i) $n_{H,4}$ *is a* K_σ*-form, for* $\sigma = 1, 2$; *moreover, it is a principal form.*

(ii) $n_{H,3}$ *is a* K_1*-form.*

Table A.16 in Appendix A lists the ternary normic forms attached to quaternion algebras, definite or indefinite, of discriminant $D \leq 100$, and some of their invariants. From these invariants it its easy to check that they are K_1-forms.

3.5.4. Correspondence between quaternion algebras and quadratic forms. We have constructed mappings:

$$\{\text{Quaternion } \mathbb{Q}\text{-algebras}\}/\text{isom} \to \{\text{Quadratic forms over } \mathbb{Q} \text{ in } n \text{ variables}\}$$

for $n = 3, 4$, by means of normic forms. The next goal is to show that these mappings are bijective.

NOTATION 3.77. Consider the following sets of regular quadratic forms over $\mathbb{Z}$:

$$Q_4^+ := \{f : f \text{ quaternary form, nonnegative definite}, \det_1(f) = \lambda^2 \text{ in } (\mathbb{Q}^*)^2\},$$
$$Q_3^+ := \{f : f \text{ ternary form}, \det_1(f) = \lambda^2 \text{ in } (\mathbb{Q}^*)^2\}.$$

We denote by $\mathcal{C}(Q_i^+)$, $i = 3, 4$, the sets of $\mathbb{Q}$-equivalent class of quadratic forms in Q_i^+.

PROPOSITION 3.78. *There is a bijective mapping*

$$\mathcal{C}(Q_3^+) \to \mathcal{C}(Q_4^+),$$
$$f(Y, Z, T) \mapsto X^2 + f(Y, Z, T).$$

PROOF. The mapping is well defined. The injectivity of the mapping is also clear, using Witt's cancellation theorem. Thus, we only need to prove that it is surjective. This will be done by using classical results about representation and equivalence of forms (cf. [**Ser73, BS66**]).

Consider a quadratic form $g \in \mathcal{C}(Q_4^+)$. We will prove that $g \xrightarrow{\mathbb{Q}} 1$. First, we have that $g \xrightarrow{\mathbb{Q}_p} 1$ for every finite place p, since it is a form in 4 variables. If g is an indefinite form, it is clear that $g \xrightarrow{\mathbb{R}} 0$ and, by Proposition 3.7, we have that $g \xrightarrow{\mathbb{R}} 1$. If g is a positive definite form, then $g \xrightarrow{\mathbb{R}} 1$ if and only if $g(X_1, \ldots, X_4) - X_0^2$ is a form in 5 variables that represents 0 over $\mathbb{R}$. Now, this form is indefinite, since g is positive definite; hence it represents 0. In any case, we have that the quadratic form g in 4 variables is regular, it is not negative definite, and it represents 0 over $\mathbb{R}$ and over $\mathbb{Q}_p$, for all p. Thus, we deduce $g \xrightarrow{\mathbb{Q}} 1$. Since g represents 1 over $\mathbb{Q}$, we have that g is $\mathbb{Q}$-equivalent to a form $X^2 + f(Y, Z, T)$. It is clear that $\det_1(f) = \det_1(g)$; hence, $f \in \mathcal{C}(Q_3^+)$. Since $\det_1(f) > 0$, f cannot be negative definite and it has the same character as g. This proves that the mapping from $\mathcal{C}(Q_3^+)$ to $\mathcal{C}(Q_4^+)$ is surjective. $\square$

The following theorem shows a bijective mapping between the Brauer group $\mathrm{Br}_2(\mathbb{Q})$ and the sets of quadratic forms $\mathcal{C}(Q_3^+)$ and $\mathcal{C}(Q_4^+)$. Given a ternary or quaternary quadratic form satisfying predetermined conditions, we construct a quaternion algebra whose normic form is equivalent to the prescribed ternary or quaternary form. In particular, we deduce a method of constructing quaternion algebras with prescribed discriminant from invariants of the initial quadratic form.

THEOREM 3.79. *Notations as above. We have the bijective mappings*

$$\nu_3\colon \mathrm{Br}_2(\mathbb{Q}) \to \mathcal{C}(Q_3^+), \qquad \nu_4\colon \mathrm{Br}_2(\mathbb{Q}) \to \mathcal{C}(Q_4^+)$$
$$H \mapsto n_{H,3}, \qquad H \mapsto n_{H,4}.$$

PROOF. The mappings ν_i, $i = 3, 4$, are well defined and injective. We must prove they are also surjective.

To see that ν_3 is surjective, consider a quadratic form $f \in Q_3^+$. Up to $\mathbb{Q}$-equivalence, we may assume that f is diagonal; thus, $f(X, Y, Z) = a_1X^2 + a_2Y^2 + a_3Z^2$ with $a_1a_2a_3 = \lambda^2$, $a_i \in \mathbb{Z}$ square-free. Put $a = -a_1$ and $b = -a_2$. Then we obtain $ab = a_1a_2 = (a_1a_2/a_3)a_3$ where $a_1a_2/a_3 = \det_1(f)/a_3^2 = (\lambda/a_3)^2 \in \mathbb{Q}^2$. Then we have $f(X, Y, Z) \overset{\mathbb{Q}}{\sim} -aX^2 - bY^2 + abZ^2 = n_{H,3}(X, Y, Z)$, for $H = \left(\frac{a,b}{\mathbb{Q}}\right)$.

Consider now the mapping ν_4. It is clear that ν_4 is obtained composing ν_3 with the bijective mapping from $\mathcal{C}(Q_3^+)$ to $\mathcal{C}(Q_4^+)$, defined in Proposition 3.78. Thus, ν_4 is automatically bijective. □

This construction of a quaternion algebra with a prescribed discriminant can be made explicit. It is implemented in the `Poincare` package.

3.6. Normic forms of quaternion orders

To each order in a quaternion algebra, we may associate, by means of normic forms, a quaternary and a ternary quadratic form, which we call normic forms of the order.

In this section we study these normic forms and their properties, especially in the case of Eichler orders.

3.6.1. Construction and properties. Let $\mathcal{O}$ be an order in a quaternion $\mathbb{Q}$-algebra H, and consider a normalized basis $\mathcal{B}$ in $\mathcal{O}$ (cf. Section 1.2).

DEFINITION 3.80. The quaternary normic form attached to the order $\mathcal{O}$, with respect to $\mathcal{B}$, is the quadratic form obtained from the normic form of the algebra expressed in the basis $\mathcal{B}$. Let us denote it by $n_{\mathcal{O},4}$.

To define the ternary normic form, we restrict ourselves to the subspace of pure quaternions. The norm applied to $\mathcal{O}\cap H_0$ is obtained from $n_{\mathcal{O},4}$ with the restriction $\operatorname{tr}(\alpha) = 0$. But the quadratic form obtained in this way could have nonintegral coefficients, depending on the parity of the order. We therefore construct our ternary quadratic form using a nonintegral change of basis, to obtain rational integer coefficients.

DEFINITION 3.81. Consider the quaternary normic form $n_{\mathcal{O},4}$ attached to $\mathcal{O}$ with respect to $\mathcal{B}$. The ternary normic form associated to $\mathcal{O}$, with respect to $\mathcal{B}$ is the ternary quadratic form obtained by substituting in $n_{\mathcal{O},4}$ the expression $T = -2X$ if $\mathcal{O}$ is an odd order, and $X = 0$ if $\mathcal{O}$ is an even order. We denote it by $n_{\mathcal{O},3}$.

Next, we make explicit the relations between the matrices of the normic forms attached to the quaternion algebra H and the matrices of the normic forms attached to a quaternion order in H.

NOTATION 3.82. Put $k = 0$ if $\mathcal{O}$ is an even order, and $k = 1$ if $\mathcal{O}$ is an odd order. We are going to define the matrices P_k, R_k and $R_{k,i}$ for $k = 0, 1$ and $i = 3, 4$.

We denote by P_k the base change matrix from the basis of the algebra to the basis of the order.

Assume $\mathcal{O}$ is an even order. We define the matrices P_0, R_0, $R_{0,4}$, $R_{0,3}$ as follows:

$$P_0 = \left(\begin{array}{c|ccc} 1 & 0 & 0 & 0 \\ \hline 0 & & & \\ 0 & & P_{0,3} & \\ 0 & & & \end{array}\right), \qquad R_0 = \begin{pmatrix} 0 & 0 & 0 \\ 1 & 0 & 0 \\ 0 & 1 & 0 \\ 0 & 0 & 1 \end{pmatrix},$$

$$R_{0,4} = \begin{pmatrix} 0 & 0 & 0 & 1 \\ 1 & 0 & 0 & 0 \\ 0 & 1 & 0 & 0 \\ 0 & 0 & 1 & 0 \end{pmatrix}, \qquad R_{0,3} = \begin{pmatrix} 1 & 0 & 0 \\ 0 & 1 & 0 \\ 0 & 0 & 1 \end{pmatrix} = \mathrm{Id}.$$

Assume $\mathcal{O}$ is an odd order. We define the matrices $P_{1,3}$, R_1, $R_{1,4}$, $R_{1,3}$ as follows.

$$P_1 = \left(\begin{array}{c|ccc} 1 & 0 & 0 & \frac{1}{2} \\ \hline 0 & & & \\ 0 & & P_{1,3} & \\ 0 & & & \end{array}\right), \qquad R_1 = \begin{pmatrix} 1 & 0 & 0 \\ 0 & 1 & 0 \\ 0 & 0 & 1 \\ -2 & 0 & 0 \end{pmatrix},$$

$$R_{1,4} = \begin{pmatrix} 1 & 0 & 0 & 0 \\ 0 & 1 & 0 & 0 \\ 0 & 0 & 1 & 0 \\ -2 & 0 & 0 & 1 \end{pmatrix}, \qquad R_{1,3} = \begin{pmatrix} 0 & 1 & 0 \\ 0 & 0 & 1 \\ -2 & 0 & 0 \end{pmatrix}.$$

LEMMA 3.83. *We have that:*

(i) $A_1(n_{\mathcal{O},4}) = P_k^t A_1(n_{H,4}) P_k$.
(ii) $A_1(n_{\mathcal{O},3}) = R_k^t A_1(n_{\mathcal{O},4}) R_k$.
(iii) $A_1(n_{\mathcal{O},3}) = (R_{k,3})^t (P_{k,3})^t A_1(n_{H,3}) P_{k,3} R_{k,3}$.

Using the above results, we deduce the following properties.

PROPOSITION 3.84. *Let $\mathcal{O}$ be a quaternion order in a $\mathbb{Q}$-algebra H. Then:*

(i) *The algebra H is definite if and only if the form $n_{\mathcal{O},4}$ is positive definite; if and only if the form $n_{\mathcal{O},3}$ is positive definite.*
(ii) *The algebra H is indefinite if and only if the form $n_{\mathcal{O},4}$ is indefinite, of signature $(2,2)$; if and only if the form $n_{\mathcal{O},3}$ is indefinite, of signature $(1,2)$.*
(iii) $\det_1(n_{\mathcal{O},4})$ *and* $\det_1(n_{\mathcal{O},3})$ *are perfect square numbers. Explicitly, if P is the base change matrix from the normalized basis of the algebra to the basis of order $\mathcal{O}$, we have that*

$$\det_1(n_{\mathcal{O},4}) = (ab \det P)^2,$$

$$\det_1(n_{\mathcal{O},3}) = \begin{cases} (ab \det P)^2 & \text{if } \mathcal{O} \text{ is an even order,} \\ (2ab \det P)^2 & \text{if } \mathcal{O} \text{ is an odd order.} \end{cases}$$

(iv) $|\,\mathrm{d}_1(n_{\mathcal{O},4})| = \frac{1}{4} D_{\mathcal{O}}$, $|\,\mathrm{d}_2(n_{\mathcal{O},4})| = D_{\mathcal{O}}$ *and*

$$|\,\mathrm{d}_1(n_{\mathcal{O},3})| = \begin{cases} \frac{1}{4} D_{\mathcal{O}} & \text{if } \mathcal{O} \text{ is an even order,} \\ \frac{1}{2} D_{\mathcal{O}} & \text{if } \mathcal{O} \text{ is an odd order.} \end{cases}$$

LEMMA 3.85. *Let $\mathcal{O}$ be an order of a quaternion $\mathbb{Q}$-algebra H and fix a normalized basis of $\mathcal{O}$. Then the normic forms $n_{\mathcal{O},4}$ and $n_{\mathcal{O},3}$ have coefficients in $\mathbb{Z}$.*

PROPOSITION 3.86. (i) *The normic forms attached to a $\mathbb{Z}$-order in different normalized bases are $\mathbb{Z}$-equivalent.*
(ii) *The normic forms attached to conjugate $\mathbb{Z}$-orders are $\mathbb{Z}$-equivalent.*

PROOF. For the quaternary normic form $n_{\mathcal{O},4}$ the result in (i) follows from Lemma 3.49.

We need to study the ternary case. Consider two normalized bases of a $\mathbb{Z}$-order $\mathcal{O}$, $\{v_i\}$ and $\{v_i'\}$, and let Q be the base change matrix from $\{v_i'\}$ to $\{v_i\}$; thus, $Q \in \mathrm{GL}(4,\mathbb{Z})$ and satisfies the conditions given in Lemma 1.40. Put $n_{\mathcal{O},3}$, $n'_{\mathcal{O},3}$, $n_{\mathcal{O},4}$ and $n'_{\mathcal{O},4}$ for the ternary and quaternary forms attached to $\mathcal{O}$ in the two bases, respectively. We use the notation of 3.82 to write the relations between the matrices. We want to see that there exists $S \in \mathrm{GL}(3,\mathbb{Z})$ such that $A_1(n'_{\mathcal{O},3}) = S^t A_1(n_{\mathcal{O},3}) S$. We have $A_1(n'_{\mathcal{O},4}) = Q^t A_1(n_{\mathcal{O},4}) Q$. Hence, on the one hand, $A_1(n'_{\mathcal{O},3}) = R_k^t A_1(n'_{\mathcal{O},4}) R_k = R_k^t Q^t A_1(n_{\mathcal{O},4}) Q R_k$. On the other hand, $A_1(n_{\mathcal{O},3}) = R_k^t A_1(n_{\mathcal{O},4}) R_k$. Hence, it is enough to find $S \in \mathrm{GL}(3,\mathbb{Z})$ such that $QR_k = R_k S$. Consider the following matrices in $\mathrm{M}(3,\mathbb{Z})$ defined from $Q = (q_{ij})$:

$$S_0 = \begin{pmatrix} q_{22} & q_{23} & q_{24} \\ q_{32} & q_{33} & q_{34} \\ q_{42} & q_{43} & q_{44} \end{pmatrix}, \qquad S_1 = \begin{pmatrix} 1-2q_{14} & q_{12} & q_{13} \\ -2q_{24} & q_{22} & q_{23} \\ -2q_{34} & q_{32} & q_{33} \end{pmatrix}.$$

Using the conditions satisfied by Q it is easy to check that $QR_k = R_k S_k$. Moreover, it follows that $\det S_k = \det Q$. Hence, $S_k \in \mathrm{GL}(3,\mathbb{Z})$. Now take $S = S_k$, $k = 0, 1$ according to whether the order $\mathcal{O}$ is even or odd respectively.

Statement (ii) follows from Proposition 3.52 for the quaternary normic form. For the ternary form, it is only necessary to reduce to quaternary case and use (i). □

REMARK 3.87. Note that it also makes sense to define the quaternary normic form of the order with respect to a not necessarily normalized basis. In this case, by using Lemma 3.49, it is possible to obtain the $\mathbb{Z}$-equivalence for quaternary forms from different bases. Moreover, since the quaternary normic form $n_{\mathcal{O},4}$ is multiplicative, we can apply the general result for K-algebras Proposition 3.52, and we would have that the quaternary normic forms attached to conjugate orders are $\mathbb{Z}$-equivalent. In fact, the proof of Proposition 3.52 consists in the reduction to the case of different bases of the same order, seeing that the quadratic forms corresponding to conjugate bases are the same. Actually, the same argument could be applied to the quaternary normic form defined from normalized bases, since the conjugate of a normalized normic form is also normalized.

EXAMPLE 3.88. Let $H = \left(\frac{a,b}{\mathbb{Q}}\right)$ be any quaternion algebra and consider the order $\mathcal{O} = \mathbb{Z}[1, i, j, ij]$, which is an even order. Then:

(i) $n_{\mathcal{O},4} = X^2 - aY^2 - bZ^2 + abT^2$.
(ii) $n_{\mathcal{O},3} = -aX^2 - bY^2 + abZ^2$.
(iii) $\det_1(n_{\mathcal{O},4}) = a^2b^2$, $D_{\mathcal{O}} = 4ab$.

3.89. *Nonramified case.* Consider the nonramified quaternion algebra $H = \left(\frac{1,-1}{\mathbb{Q}}\right)$ and the order $\mathcal{O}_M(1,N) = \mathbb{Z}[1, (j+ij)/2, N(-j+ij)/2, (1-i)/2]$, which is an odd order, with respect to a normalized basis. Then we have:

(i) $n_{\mathcal{O}_M(1,N),4} = X^2 + XT - NYZ$.
(ii) $n_{\mathcal{O}_M(1,N),3} = -X^2 - NYZ$.
(iii) $\det_1(n_{\mathcal{O}_M(1,N),4}) = N^2/16$, $D_{\mathcal{O}_M(1,N)} = 1$.

Let us consider now the nonramified quaternion algebra $H = \mathrm{M}(2,\mathbb{Q})$ and the Eichler order $\mathcal{O}_0(1,N)$. When we fix in $\mathcal{O}_0(1,N)$ the following normalized basis:

$$\left\{ \begin{pmatrix} 1 & 0 \\ 0 & 1 \end{pmatrix}, \begin{pmatrix} 0 & 1 \\ 0 & 0 \end{pmatrix}, \begin{pmatrix} 0 & 0 \\ N & 0 \end{pmatrix}, \begin{pmatrix} 0 & 0 \\ 0 & 1 \end{pmatrix} \right\},$$

the normic forms that we obtain are exactly the same as the ones above.

3.90. *Small ramified case of type* A. Consider the quaternion order $\mathcal{O}_{\mathrm{A}}(2p,N) = \mathbb{Z}[1, i, Nj, (1+i+j+ij)/2]$ in $H_{\mathrm{A}}(p)$ given in Proposition 1.60. Then $\mathcal{O}_{\mathrm{A}}(2p,N)$ is an odd order and the given basis is normalized. The normic forms attached to the order $\mathcal{O}_{\mathrm{A}}(2p,N)$ in this basis are:

(i) $n_{\mathcal{O}_{\mathrm{A}}(2p,N),4} = X^2 - pY^2 + N^2Z^2 + (1-p)T^2/2 + XT - pYT + NZT$.
(ii) $n_{\mathcal{O}_{\mathrm{A}}(2p,N),3} = (1-2p)X^2 - pY^2 + N^2Z^2 + 2pXY - 2NXZ$.

(iii) $\det_1(n_{\mathcal{O}_{\mathrm{A}}(2p,N),4}) = p^2N^2/4$, $D_{\mathcal{O}_{\mathrm{A}}(2p,N)} = 2pN$.

3.91. *Small ramified case of type* B. Consider the quaternion order $\mathcal{O}_{\mathrm{B}}(pq,N)=\mathbb{Z}[1, Ni, (1+j)/2, (i+ij)/2]$ in $H_{\mathrm{B}}(p,q)$ given in Proposition 1.60. Then $\mathcal{O}_{\mathrm{B}}(pq,N)$ is an odd order and the given basis is not normalized; we transform it into a normalized basis by transposing the third and fourth element. Then we have:

(i) $n_{\mathcal{O}_{\mathrm{B}}(pq,N),4} = X^2 - pN^2Y^2 - p(1-q)Z^2/4 + (1-q)/4T^2/4 + XT - pNYZ$.
(ii) $n_{\mathcal{O}_{\mathrm{B}}(pq,N),3} = -qX^2 - pN^2Y^2 - p(1-q)Z^2/4 - pNYZ$.
(iii) $\det_1(n_{\mathcal{O}_{\mathrm{B}}(pq,N),4}) = p^2q^2N^2/16$, $D_{\mathcal{O}_{\mathrm{B}}(pq,N)} = pqN$.

Next we will characterize the conditions under which normic forms are K_σ-forms, by using Propositions 3.37 and 3.84.

THEOREM 3.92. *Let $\mathcal{O}$ be a quaternion order.*

(i) *$n_{\mathcal{O},4}$ is a K_1-form if and only if $\mathrm{N}(n_{\mathcal{O},4}) \mid D_{\mathcal{O}}$.*
(ii) *$n_{\mathcal{O},4}$ is a K_2-form if and only if $\mathrm{N}(n_{\mathcal{O},4}) \mid 2D_{\mathcal{O}}$.*
(iii) *If $\mathcal{O}$ is an even order, then $n_{\mathcal{O},3}$ is a K_1-form if and only if $\mathrm{N}(n_{\mathcal{O},3}) \mid D_{\mathcal{O}}$.*
(iv) *If $\mathcal{O}$ is an odd order, then $n_{\mathcal{O},3}$ is a K_1-form if and only if $\mathrm{N}(n_{\mathcal{O},3}) \mid 2D_{\mathcal{O}}$.*

PROOF. The quaternary normic form $n_{\mathcal{O},4}$ is a K_1-form if the divisibility relation $\mathrm{N}(n_{\mathcal{O},4}) \mid \mathrm{d}_2(n_{\mathcal{O},4})$ is satisfied. Since $\mathrm{d}_2(n_{\mathcal{O},4}) = D_{\mathcal{O}}$, we obtain that $n_{\mathcal{O},4}$ is a K_1-form if and only if $\mathrm{N}(n_{\mathcal{O},4}) \mid D_{\mathcal{O}}$.

The $n_{\mathcal{O},4}$ is a K_2-form if $\mathrm{N}(n_{\mathcal{O},4}) \mid 2\,\mathrm{d}_2(n_{\mathcal{O},4})$ and this condition can be written as stated in (ii).

The results in (iii) and (iv), for the ternary form $n_{\mathcal{O},3}$, are also deduced from Propositions 3.37 and 3.84. □

REMARK 3.93. Let H be a quaternion $\mathbb{Q}$-algebra, $\mathcal{O} \subseteq H$ an order, $\mathcal{B}$ a fixed normalized basis of $\mathcal{O}$, $n_{\mathcal{O},4}$ and $n_{\mathcal{O},3}$ the normic forms attached to the order $\mathcal{O}$ and δ an integer. Consider the basis $\mathcal{B}$ as a $\mathbb{Q}$-basis of H and put $\omega = (x,y,z,t) \in H$ for the coordinates of a quaternion with respect to the basis $\mathcal{B}$. It is clear that, by construction, we have the following facts.

(i) $\omega \in \mathcal{O}$ if and only if $x,y,z,t \in \mathbb{Z}$; in this case, $n(\omega) = n_{\mathcal{O},4}(x,y,z,t)$. Thus, finding an element $\omega \in \mathcal{O}$ of norm equal to δ is equivalent to finding a representation of δ over $\mathbb{Z}$ by the quaternary normic form $n_{\mathcal{O},4}$.

(ii) If $\mathcal{O}$ is an even order, then $\omega \in \mathcal{O} \cap H_0$ if and only if $x = 0$ and $y,z,t \in \mathbb{Z}$; in this case, $n(\omega) = n_{\mathcal{O},3}(y,z,t)$. If $\mathcal{O}$ is an odd order, then $\omega \in \mathcal{O} \cap H_0$ if and only if $x,y,z \in \mathbb{Z}$ and $t = -2x$; in this case, $n(\omega) = n_{\mathcal{O},3}(x,y,z)$. Thus, independently of the character of the order $\mathcal{O}$, finding an element $\omega \in \mathcal{O} \cap H_0$ of norm equal δ is equivalent to finding a representation of δ over $\mathbb{Z}$ by the ternary normic form $n_{\mathcal{O},3}$.

Since the results about the representations of quadratic forms of rational integer coefficients do not depend on the class of $\mathbb{Z}$-equivalence, these comments only depend on the conjugacy class of the order.

3.6.2. Normic forms of Eichler orders. For the normic forms attached to Eichler orders we have more explicit descriptions.

Recall that, for the indefinite rational quaternion algebras of discriminant D we denoted by $\mathcal{O}(D,N)$ a representative of the conjugacy class of Eichler orders of level N. Thus, for the indefinite rational quaternion algebras of discriminant D, let us denote by $n_{\mathcal{O}(D,N),4}$ and $n_{\mathcal{O}(D,N),3}$ the normic forms for the Eichler orders of level N, up to $\mathbb{Z}$-equivalence.

For the small ramified quaternion algebras of type A and of type B, the tables in Section A.3 yield the normic forms attached to Eichler orders.

From Eichler's result, cf. Theorem 2.23, we have the following property of normic forms of Eichler orders.

PROPOSITION 3.94. *Let $\mathcal{O}$ be an Eichler order in an indefinite quaternion $\mathbb{Q}$-algebra. Then,* $n_{\mathcal{O},4} \overset{\mathbb{Z}}{\to} -1$.

PROPOSITION 3.95. *Let $\mathcal{O}(D,N)$ be an Eichler order in a quaternion algebra H of discriminant D.*

(i) *If $\mathcal{O}(D,N)$ is included in another Eichler order $\mathcal{O}(D,M)$, then their levels are related by* $N = M[\det_1(n_{\mathcal{O}(D,N),4})/\det_1(n_{\mathcal{O}(D,M),4})]^{1/2}$.
(ii) *The quotient* $\det_2(n_{\mathcal{O}(D,N),4})/N^2 = D^2$ *does not depend on $\mathcal{O}(D,N)$ and is an invariant of H.*

PROOF. Fix a basis $\mathcal{B}$ in $\mathcal{O}(D,N)$ and a basis $\mathcal{B}'$ in $\mathcal{O}(D,M)$. Let Q be the base change matrix from the $\mathcal{B}'$-coordinates to the $\mathcal{B}$-coordinates. On the one hand, by Proposition 1.55 we have that $N/M = [\mathcal{O}(D,M) : \mathcal{O}(D,N)] = \det Q$. On the other hand, applying Proposition 3.53, we have that $\det_1(n_{\mathcal{O}(D,N),4}) = (\det Q)^2 \det_1(n_{\mathcal{O}(D,M),4})$. Thus, solving $\det Q$ in the two expressions, we deduce the equality in (i).

To see (ii), take $M = 1$ in the equality in (i). Thus we have

$$\det_1(n_{\mathcal{O}(D,1),4}) = \frac{\det_1(n_{\mathcal{O}(D,N),4})}{N^2}.$$

The value of $\det_1(n_{\mathcal{O}(D,1),4})$ can be computed using the properties of the normic form stated in Proposition 3.84. Thus, we have that $\det_1(n_{\mathcal{O}(D,1),4}) = (\det P)^2 a^2 b^2$, where P is the matrix describing the basis of the order $\mathcal{O}(D,1)$ in terms of the basis of the quaternion algebra H. Now, if we apply Corollary 1.33, $\det_1(n_{\mathcal{O}(D,1),4}) = (D_{\mathcal{O}(D,1)}/4)^2$.

Since $\mathcal{O}(D,1)$ is a maximal order, we have that $D_{\mathcal{O}(D,1)} = D$. Hence, we have $(D_H/4)^2 = \det_1(n_{\mathcal{O}(D,N),4})/N^2$.

Note that $\det_2(n_{\mathcal{O}(D,N),4}) = 2^4 \det_1(n_{\mathcal{O}(D,N),4})$. Thus,

$$\frac{\det_2(n_{\mathcal{O},4})}{N_{\mathcal{O}}^2} = D_H^2. \qquad \square$$

THEOREM 3.96. *Let $\mathcal{O}$ and $\mathcal{O}'$ be two Eichler orders in a quaternion $\mathbb{Q}$-algebra H.*

(i) *If $N_{\mathcal{O}} = N_{\mathcal{O}'}$, then $n_{\mathcal{O},4} \overset{\mathbb{Z}_p}{\sim} n_{\mathcal{O}',4}$, for any prime p.*
(ii) *If $n_{\mathcal{O},4} \overset{\mathbb{Z}}{\sim} n_{\mathcal{O}',4}$, then, $N_{\mathcal{O}} = N_{\mathcal{O}'}$.*
(iii) *If H is indefinite, then $n_{\mathcal{O},4} \overset{\mathbb{Z}}{\sim} n_{\mathcal{O}',4}$ if and only if $N_{\mathcal{O}} = N_{\mathcal{O}'}$.*

PROOF. The statement in (i) follows from the general properties of Eichler orders and Proposition 3.52. Namely, Eichler orders of the same level in any quaternion $\mathbb{Q}$-algebra are conjugate over $\mathbb{Z}_p$.

To prove (ii), assume we have the equivalence of forms $n_{\mathcal{O},4} \overset{\mathbb{Z}}{\sim} n_{\mathcal{O}',4}$. We denote by P and P' the base change matrices from the basis of the algebra to the bases of the orders $\mathcal{O}$ and $\mathcal{O}'$ respectively. Then, the two forms have the same determinant and $\det_1(n_{\mathcal{O},4})(ab \det P)^2 = (ab \det P')^2 = \det_1(n_{\mathcal{O}',4})$, by Proposition 3.84.

Hence, $\det P = \det P'$. Now, by Corollary 1.33, we also have the equality between discriminants $D_{\mathcal{O}} = D_{\mathcal{O}'}$ and, by Proposition 1.54, the equality of levels.

Note that to see that the orders $\mathcal{O}$ and $\mathcal{O}'$ have the same level, we could also use the fact that the attached quaternary normic forms are equivalent and Proposition 3.95.

Assume that H is indefinite. We need to prove that the equality of levels implies the equivalence of the normic forms. By Theorem 1.59, we know that all the Eichler $\mathbb{Z}$-orders of the same level are conjugate. Hence it follows that their normic forms are equivalent. □

As a corollary we obtain the converse of Proposition 3.52 in the case of Eichler orders in indefinite $\mathbb{Q}$-algebras.

COROLLARY 3.97. *Let $\mathcal{O}$ and $\mathcal{O}'$ be two Eichler orders in an indefinite quaternion $\mathbb{Q}$-algebra H. Then, $\mathcal{O}$ and $\mathcal{O}'$ are conjugate if and only if $n_{\mathcal{O},4} \overset{\mathbb{Z}}{\sim} n_{\mathcal{O}',4}$.*

PROPOSITION 3.98. *Let $\mathcal{O}$ be an Eichler order of level $N_{\mathcal{O}}$ square-free in a quaternion $\mathbb{Q}$-algebra H. Then $n_{\mathcal{O},4}$ is a K_1-form if and only if $\mathrm{N}(n_{\mathcal{O},4}) = D_{\mathcal{O}}$.*

PROOF. On the one hand, since $\mathcal{O}$ is an Eichler order of level $N_{\mathcal{O}}$ square-free, we have that $D_{\mathcal{O}} = D_H N_{\mathcal{O}}$ is square-free. On the other hand, by Proposition 3.84 we have $D_{\mathcal{O}} = \mathrm{d}_2(n_{\mathcal{O},4})$. Thus, we also have that $\mathrm{d}_2(n_{\mathcal{O},4})$ is square-free. This allows us to apply Corollary 3.38 (ii) to the quaternary form $n_{\mathcal{O},4}$. Thus, putting together the above equalities, we obtain $n_{\mathcal{O},4}$ is a K_1-form if and only if $\mathrm{N}(n_{\mathcal{O},4}) = D_{\mathcal{O}}$. □

It is plausible that for Eichler orders the normic form $n_{\mathcal{O},4}$ is a K_σ-form, for $\sigma = 1, 2$, and $n_{\mathcal{O},3}$ is a K_1-form. In fact, by using the above results and the `Poincare` package, we obtain the following proposition, which provides numerical evidence to this claim.

PROPOSITION 3.99. *Let $\mathcal{O}$ be an Eichler order of level $N < 1000$ of a small ramified quaternion algebra H equal to one of the following: $H_{\mathrm{A}}(3)$, $H_{\mathrm{B}}(2, 5)$, $H_{\mathrm{A}}(7)$ and $H_{\mathrm{B}}(3, 5)$. Then, $n_{\mathcal{O},4}$ is a K_σ-form, for $\sigma = 1, 2$, and $n_{\mathcal{O},3}$ is a K_1-form.*

CHAPTER 4

Embeddings and Quadratic Forms

In this chapter, we deal with the embeddings of quadratic fields into quaternion algebras, taking into account the arithmetic of quaternion algebras and of quadratic fields. We study the embeddings of quadratic orders into quaternion orders and we relate them with the quadratic forms associated to the orders in the previous chapter.

Throughout this chapter, $F = \mathbb{Q}(\sqrt{d})$ will denote a quadratic field, d a square-free integer, and D_F the fundamental discriminant of F. We shall denote quaternion orders in a quaternion $\mathbb{Q}$-algebra H by $\mathcal{O}$ and $\mathcal{O}'$, and quadratic orders in F by Λ and Λ'.

In the first section, we characterize the existence of embeddings of F into H by means of representations of quadratic forms. In the second section, we recall the definitions of embeddings relative to quadratic and quaternion orders, and we state Eichler's results on their classification. In the third section, we state bijective mappings between these sets of embeddings and the sets of representations by certain ternary normic forms, and we obtain a classification of these representations. In the fourth section, we introduce binary forms associated to the elements in a quaternion order. In the fifth section we prove results for bijective mappings between sets of embeddings and sets of binary quadratic forms. From this, we obtain a classification result of these binary forms. In particular, for the nonramified and the small ramified algebras of type A or B, we make explicit the corresponding sets of binary forms.

4.1. Embeddings of quadratic fields into quaternion algebras

The existence of embeddings of a quadratic field F into a quaternion algebra H was characterized in Proposition 1.14. In this section, we shall use quadratic forms to construct those embeddings.

NOTATION 4.1. We fix the following notations for the corresponding sets of embeddings and representations. For the quadratic forms, we use the notation in Chapter 3. The number of variables of the quadratic form is indicated by a subscript, e.g., f_4, g_2, etc.

$$\begin{aligned} \mathcal{E}(H,F) &:= \{\varphi : \varphi \text{ embedding of } F \text{ into } H\}, \\ \mathcal{R}(f_3,\delta_1;\mathbb{Q}) &:= \{\alpha \in \mathbb{Q}^3 : \alpha^t A_1(f_3)\alpha = \delta_1\}, \\ \mathcal{R}(f_4,g_2;\mathbb{Q}) &:= \{P \in M(4\times 2,\mathbb{Q}) : P^t A_1(f_4) P = A_1(g_2)\}. \end{aligned}$$

PROPOSITION 4.2. *Let $H = \left(\frac{a,b}{\mathbb{Q}}\right)$ be a quaternion algebra and $F = \mathbb{Q}(\sqrt{d})$ a quadratic field. Let $n_{H,4}$ and $n_{H,3}$ be the quaternary and ternary normic forms associated to H, and $n_{F,2}$ the binary normic form associated to F. Then the following assertions hold:*

(i) *There is a bijective mapping between* $\mathcal{E}(H,F)$ *and* $\mathcal{R}(n_{H,3},-d;\mathbb{Q})$.
(ii) *There is an injective mapping between* $\mathcal{E}(H,F)$ *and* $\mathcal{R}(n_{H,4},n_{F,2};\mathbb{Q})$.
(iii) *If* $\mathcal{R}(n_{H,4},n_{F,2};\mathbb{Q}) \neq \varnothing$, *then* $\mathcal{E}(H,F) \neq \varnothing$.

PROOF. Consider $\omega = \varphi(\sqrt{d})$. The set of embeddings $\varphi\colon F \hookrightarrow H$ is in one to one correspondence with the set of elements $\omega \in H \setminus \mathbb{Q}$ such that $n(\omega) = n(\varphi(\sqrt{d})) = n(\sqrt{d}) = -d$ and $\operatorname{tr}(\omega) = \operatorname{tr}(\varphi(\sqrt{d})) = \operatorname{tr}(\sqrt{d}) = 0$; that is, $\omega \in H_0$ and $n(\omega) = n_{H,3}(\omega)$. If $\omega = yi + zj + tij$, then the above set is in one to one correspondence with the set of $\mathbb{Q}$-representations (y,z,t) of $-d$ by the ternary quadratic form $n_{H,3}$. It is clear that different embeddings lead to different representations. This proves (i).

Given an embedding $\varphi \in \mathcal{E}(H,F)$, consider the element $\omega = \varphi(\sqrt{d}) \in H \setminus \mathbb{Q}$, with $n(\omega) = -d$ and $\operatorname{tr}(\omega) = 0$. We have that $\{1,\omega\}$ is a basis of $\varphi(F)$ and $n_{H,4}(\varphi(\alpha)) = n_{F,2}(\alpha)$, for all $\alpha \in F$. Consider the matrix P, formed with the coordinates of 1 and ω with respect to the quaternion algebra basis. Then, P has the coefficients in $\mathbb{Q}$ and satisfies $P^t A_1(n_{H,4}) P = A_1(n_{F,2})$. Hence, $P \in \mathcal{R}(n_{H,4},n_{F,2};\mathbb{Q})$. It is clear that different embeddings give different representation matrices. Thus, we have proved (ii).

Let us prove (iii). Assume the quaternary form $n_{H,4}$ represents the binary form $n_{F,2}$. For each $p \mid D_H$, by Proposition 3.71, we have that $n_{H,4}$ is not $\mathbb{Q}_p$-isotropic. We deduce that, for each $p \mid D_H$, the form $n_{F,2}$ is not $\mathbb{Q}_p$-isotropic; that is, $x^2 - dy^2$ does not represent 0 in $\mathbb{Q}_p^*$; this is equivalent to saying that d is not a square in $\mathbb{Q}_p^*$. Thus, for each $p \mid D_H$ we have that p does not split in $\mathbb{Q}(\sqrt{d})$, and this is equivalent to the existence of an embedding of F into H, cf. Proposition 1.14. □

Next we find the effective conditions for the existence of embeddings of F into H, in terms of the discriminant of H and of properties of the ternary normic form $n_{H,3}$.

LEMMA 4.3. *The following statements are equivalent.*
(a) *There exists an embedding of* F *into* H.
(b) *For every prime* p *such that* $n_{H,3}$ *is* $\mathbb{Q}_p$*-anisotropic,* p *does not split in* F.
(c) *For every prime* p *such that* $p \mid D_H$, *we have* $\left(\frac{D_F}{p}\right) \neq 1$.

PROOF. Recall that $p \mid D_H$ if and only if $n_{H,3}$ is $\mathbb{Q}_p$-anisotropic, by Proposition 3.71. Hence, we deduce the equivalence between (a) and (b) directly from Proposition 1.14.

The equivalence between (b) and (c) comes from the explicit conditions for the representation of numbers by ternary forms, which coincide with the conditions for the splitting of primes. □

COROLLARY 4.4. *There are no embeddings of real quadratic fields into definite quaternion algebras.*

PROOF. If F is a real quadratic field, then $n_{F,2}$ is an indefinite quadratic form. If H is definite, then $n_{H,4}$ is a positive definite form, by Lemma 3.72. It is clear that definite quadratic forms cannot represent indefinite forms; hence, $\mathcal{R}(n_{H,4},n_{F,2};\mathbb{Q}) = \varnothing$. By Proposition 4.2 (ii), we obtain $\mathcal{E}(H,F) = \varnothing$. □

COROLLARY 4.5. *Let* $H = \left(\frac{a,b}{\mathbb{Q}}\right)$ *be a quaternion algebra. Then:*
(i) *There exist embeddings of* $\mathbb{Q}(\sqrt{a})$, $\mathbb{Q}(\sqrt{b})$ *and* $\mathbb{Q}(\sqrt{-D_H})$ *into* H.

(ii) *There exist embeddings of* $\mathbb{Q}(\sqrt{D_H})$ *into* H *if and only if* H *is an indefinite quaternion algebra.*

Let us consider now the orders and the attached quadratic forms. From Proposition 4.2, we immediately obtain the following result about $\mathbb{Q}$-representations.

COROLLARY 4.6. *Let* $\mathcal{O}$ *and* Λ *be orders in* H *and* F *respectively. If* $\mathcal{E}(H,F) \neq \varnothing$, *then* $n_{\mathcal{O},4} \xrightarrow{\mathbb{Q}} n_{\Lambda,2}$.

4.2. Embeddings of quadratic orders into quaternion orders

DEFINITION 4.7. Let $\mathcal{O}$ and Λ be orders in H and F, respectively. An embedding $\varphi\colon F \to H$ is an embedding of the quadratic order Λ into the quaternion order $\mathcal{O}$ if $\varphi(\Lambda) \subseteq \mathcal{O}$.

DEFINITION 4.8. Let $\mathcal{O}$ and Λ be orders of H and F respectively. An embedding $\varphi\colon \Lambda \to \mathcal{O}$ is called optimal if $\varphi(F) \cap \mathcal{O} = \varphi(\Lambda)$; equivalently, if $\Lambda = \varphi^{-1}(\varphi(F) \cap \mathcal{O})$.

NOTATION 4.9. As in the previous section, consider the following sets of embeddings:

$$\mathcal{E}(\mathcal{O},\Lambda) := \{\varphi : \varphi \text{ embedding of } \Lambda \text{ into } \mathcal{O}\},$$
$$\mathcal{E}^*(\mathcal{O},\Lambda) := \{\varphi : \varphi \text{ optimal embedding of } \Lambda \text{ into } \mathcal{O}\}.$$

REMARK 4.10. Let Λ be the quadratic order of conductor m in $F = \mathbb{Q}(\sqrt{d})$; that is, $\Lambda = \mathbb{Z}[1, mw]$, where $w = \sqrt{d}$ if $d \equiv 2,3 \bmod 4$, and $w = (1+\sqrt{d})/2$ if $d \equiv 1 \bmod 4$. Let $\mathcal{O}$ be a quaternion order in H. To define an embedding φ of Λ into $\mathcal{O}$ is equivalent to fixing $\omega = \varphi(mw) \in \mathcal{O}$ such that $n\big(\varphi(mw)\big) = m^2 n(w)$ and $\operatorname{tr}\big(\varphi(mw)\big) = m \operatorname{tr}(w)$. In an equivalent way, we could also fix $\omega \in \mathcal{O}$ such that $(\omega - \overline{\omega})^2 = D_\Lambda$. Note that, given an order Λ in F, an embedding $\varphi\colon F \to H$ is determined by $\varphi_{|\Lambda}$.

The following lemmas state certain technical conditions about orders and embeddings.

LEMMA 4.11. *Let* $\varphi\colon F \to H$ *be an embedding and* $\mathcal{O}$ *a quaternion order into* H.
(i) *Then* $\varphi \in \mathcal{E}^*\big(\mathcal{O}, \varphi^{-1}(\varphi(F) \cap \mathcal{O})\big)$.
(ii) *If* $\varphi \in \mathcal{E}(\mathcal{O},\Lambda)$, *then* $\Lambda \subseteq \varphi^{-1}(\varphi(F) \cap \mathcal{O})$.

COROLLARY 4.12. *Let* $\mathcal{O} \subseteq H$ *be a quaternion order and* Λ_F *be the ring of integers of* F. *Then,* $\mathcal{E}(\mathcal{O},\Lambda_F) = \mathcal{E}^*(\mathcal{O},\Lambda_F)$.

LEMMA 4.13. *Let* $\mathcal{O}' \subsetneq \mathcal{O}$ *be quaternion orders in* H *and* $\Lambda' \subsetneq \Lambda$, *quadratic orders in* F.
(i) *If* $\varphi \in \mathcal{E}(\mathcal{O},\Lambda)$, *then* $\varphi \in \mathcal{E}(\mathcal{O},\Lambda') \setminus \mathcal{E}^*(\mathcal{O},\Lambda')$.
(ii) *If* $\varphi \in \mathcal{E}(\mathcal{O}',\Lambda)$, *then* $\varphi \in \mathcal{E}(\mathcal{O},\Lambda)$.
(iii) *If* $\varphi \in \mathcal{E}^*(\mathcal{O},\Lambda')$, *then* $\varphi \notin \mathcal{E}(\mathcal{O},\Lambda)$.

The following results give conditions for the existence of optimal embeddings between prescribed orders.

LEMMA 4.14. *Consider the maximal order* $\mathcal{O}_p$ *in a local division quaternion* $\mathbb{Q}_p$*-algebra* H_p. *Let* F_p *be such that* $\mathcal{E}(H_p,F_p) \neq \varnothing$ *and consider a quadratic order* Λ_p *in* F_p. *Then,* $\mathcal{E}^*(\mathcal{O}_p,\Lambda_p) \neq \varnothing$ *if and only if* Λ_p *is maximal.*

PROPOSITION 4.15. *Let $\mathcal{O} \subseteq H$ be an Eichler order of level N in a quaternion $\mathbb{Q}$-algebra H of discriminant D. Let $F = \mathbb{Q}(\sqrt{d})$ be a quadratic field such that $\mathcal{E}(H, F) \neq \varnothing$ and let Λ be the quadratic order of conductor m in F.*

(i) *If $\mathcal{E}^*(\mathcal{O}, \Lambda) \neq \varnothing$, then* $\gcd(D, m) = 1$.
(ii) *If $N = 1$ and H is an indefinite quaternion algebra, then $\mathcal{E}^*(\mathcal{O}, \Lambda) \neq \varnothing$ if and only if* $\gcd(D, m) = 1$.

PROOF. Assume $\mathcal{E}^*(\mathcal{O}, \Lambda) \neq \varnothing$ and $p \mid D$. Then, $\mathcal{O}_p \subseteq H_p$ is the only maximal order in the division quaternion algebra H_p. Note that $\mathcal{E}^*(\mathcal{O}, \Lambda) \neq \varnothing$ implies $\mathcal{E}^*(\mathcal{O}_p, \Lambda_p) \neq \varnothing$. By using Lemma 4.14, we have Λ_p is the maximal order of F_p; thus $p \nmid m$. This proves (i).

If $N = 1$ and H is indefinite, we also have the converse. Indeed, assume $\gcd(D, m) = 1$ and let $\varphi \in \mathcal{E}(H, F)$. Up to conjugation, we can assume $\varphi \in \mathcal{E}(\mathcal{O}, \Lambda)$, since $\varphi(\Lambda)$ is contained in a maximal quaternion order, and maximal orders in an indefinite quaternion $\mathbb{Q}$-algebra are all conjugate. It is clear that φ is optimal if and only if φ_p is optimal for all the places p. For the places $p \mid D$, we have $p \nmid m$, and by applying Lemma 4.14, we deduce that φ_p is an optimal embedding. For the places $p \nmid Dm$, we have that $\mathcal{O}$ is a maximal order in $H \simeq \mathrm{M}(2, \mathbb{Q})$ and the embedding is optimal automatically, by Corollary 4.12. If $p \nmid D$ and $p \mid m$, we have that $\mathcal{O}$ is a maximal order in $\mathrm{M}(2, \mathbb{Q})$ and Λ_p is a nonmaximal order; in this case there are also optimal embeddings, cf. [**Vig80**]. □

Consider the group H^* of the units of H and the set of embeddings $\mathcal{E}(H, F)$. We define an action of H^* on $\mathcal{E}(H, F)$ in the following way:

$$\mathcal{E}(H, F) \times H^* \to \mathcal{E}(H, F)$$
$$(\varphi, \sigma) \mapsto \varphi^\sigma,$$

where $\varphi^\sigma(\alpha) := \sigma^{-1}\varphi(\alpha)\sigma$, for all $\alpha \in F$.

LEMMA 4.16. *Let $\varphi \in \mathcal{E}(H, F)$. Then, for all $\sigma \in H^*$:*

(i) *$\varphi \in \mathcal{E}(\mathcal{O}, \Lambda)$ if and only if $\varphi^\sigma \in \mathcal{E}(\sigma^{-1}\mathcal{O}\sigma, \Lambda)$.*
(ii) *$\varphi \in \mathcal{E}^*(\mathcal{O}, \Lambda)$ if and only if $\varphi^\sigma \in \mathcal{E}^*(\sigma^{-1}\mathcal{O}\sigma, \Lambda)$.*

If H is an indefinite quaternion $\mathbb{Q}$-algebra, then all the Eichler orders of the same level are conjugate. Thus, by the above lemma, we can reduce ourselves to studying the local Eichler order for each level.

COROLLARY 4.17. *Let G be a subgroup of* $\mathrm{Nor}(\mathcal{O})$. *Then, the action of H^* on $\mathcal{E}(H, F)$ induces an action of G on $\mathcal{E}(\mathcal{O}, \Lambda)$ and on $\mathcal{E}^*(\mathcal{O}, \Lambda)$, which gives rise to an equivalence relation. Namely $\varphi, \varphi' \in \mathcal{E}^*(\mathcal{O}, \Lambda)$ are G-equivalent if there exists $\sigma \in G$ such that $\varphi' = \varphi^\sigma$.*

Given a basis $\{1, \lambda\}$ in the quadratic order Λ, any embedding is determined by the image of λ. Thus, two embeddings $\varphi, \varphi' \in \mathcal{E}^*(\mathcal{O}, \Lambda)$ are G-equivalent if and only if there exists $\sigma \in G$ such that $\varphi'(\lambda) = \sigma^{-1}\varphi(\lambda)\sigma$.

NOTATION 4.18. Consider the quotient set $\mathcal{E}^*(\mathcal{O}, \Lambda)/G$. The class number of G-equivalent optimal embeddings of Λ into $\mathcal{O}$ is defined by the cardinality of $\mathcal{E}^*(\mathcal{O}, \Lambda)/G$. If it is finite, it will be denoted by $\nu(\mathcal{O}, \Lambda; G)$; that is,

$$\nu(\mathcal{O}, \Lambda; G) := \sharp\, \mathcal{E}^*(\mathcal{O}, \Lambda)/G.$$

For $G = \mathcal{O}^*$, we consider also the local case. Let us denote by $\nu_p(\mathcal{O}, \Lambda; \mathcal{O}^*)$ the corresponding class number for the local orders; that is,

$$\nu_p(\mathcal{O}, \Lambda; \mathcal{O}^*) := \sharp\, \mathcal{E}^*(\mathcal{O}_p, \Lambda_p)/\mathcal{O}_p^*.$$

Consider an Eichler order $\mathcal{O} = \mathcal{O}(D, N)$ of level N in an indefinite quaternion $\mathbb{Q}$-algebra of discriminant D, and a quadratic order $\Lambda = \Lambda(d, m)$ in $\mathbb{Q}(\sqrt{d})$ with conductor m. Then the class number of G-equivalent optimal embeddings of $\Lambda(d, m)$ in $\mathcal{O}(D, N)$, if finite, will be denoted by $\nu(D, N, d, m; G)$. Analogously, in the local case, for the corresponding class number we write

$$\nu_p(D, N, d, m; \mathcal{O}^*) := \nu_p(\mathcal{O}_p, \Lambda_p; \mathcal{O}_p^*).$$

Let us denote by $h(\Lambda)$ or $h(d, m)$ the ideal class number of a quadratic order $\Lambda(d, m)$ (cf. Section 3.4).

Next, we list known results about the class number $\nu(D, N, d, m; G)$ in the case $G = \mathcal{O}^*$. Theorem 4.19 (cf. [**Eic55b**]) provides the relation between the class numbers of local and global embeddings, and the formulas for the class number of local embeddings given in [**Ogg83, Vig80**]. Consider ψ_p the multiplicative function given by $\psi_p(p^k) = p^k(1 + 1/p)$, $\psi_p(a) = 1$ if $p \nmid a$.

THEOREM 4.19. *Let $\mathcal{O} = \mathcal{O}(D, N)$ be an Eichler order of level N in an indefinite quaternion $\mathbb{Q}$-algebra H of discriminant D. Let $\Lambda(d, m)$ be the quadratic order of conductor m in $\mathbb{Q}(\sqrt{d})$. Assume that $\mathcal{E}(H, \mathbb{Q}(\sqrt{d})) \neq \varnothing$ and $\gcd(m, D) = 1$. Then,*

$$\nu(D, N, d, m; \mathcal{O}^*) = h(d, m) \prod_{p \mid DN} \nu_p(D, N, d, m; \mathcal{O}^*).$$

The local class numbers of embeddings $\nu_p(D, N, d, m; \mathcal{O}^*)$, for the primes $p \mid DN$, are given by

(i) *If $p \mid D$, then $\nu_p(D, N, d, m; \mathcal{O}^*) = 1 - \left(\frac{D_F}{p}\right)$.*
(ii) *If $p \parallel N$, then $\nu_p(D, N, d, m; \mathcal{O}^*)$ is equal to $1 + \left(\frac{D_F}{p}\right)$ if $p \nmid m$, and equal to 2 if $p \mid m$.*
(iii) *Assume $N = p^r u_1$, with $p \nmid u_1$, $r \geq 2$. Put $m = p^k u_2$, $p \nmid u_2$.*
   (a) *If $r \geq 2k+2$, then $\nu_p(D, N, d, m; \mathcal{O}^*)$ is equal to $2\psi_p(m)$ if $\left(\frac{D_F}{p}\right) = 1$, and equal to 0 otherwise.*
   (b) *If $r = 2k+1$, then $\nu_p(D, N, d, m; \mathcal{O}^*)$ is equal to $2\psi_p(m)$ if $\left(\frac{D_F}{p}\right) = 1$, equal to p^k if $\left(\frac{D_F}{p}\right) = 0$, and equal to 0 if $\left(\frac{D_F}{p}\right) = -1$.*
   (c) *If $r = 2k$, then $\nu_p(D, N, d, m; \mathcal{O}^*) = p^{k-1}\left(p + 1 + \left(\frac{D_F}{p}\right)\right)$.*
   (d) *If $r \leq 2k - 1$, then $\nu_p(D, N, d, m; \mathcal{O}^*)$ is equal to $p^{k/2} + p^{k/2-1}$ if k is even, and equal to $2p^{k-1/2}$ if k is odd.*

In the following corollaries, we elaborate on the computation of this class number for Eichler orders of prime level N.

COROLLARY 4.20. *Let $\mathcal{O}$ be an Eichler order of level N square-free in an indefinite quaternion $\mathbb{Q}$-algebra H of discriminant D. Let $F = \mathbb{Q}(\sqrt{d})$ be a quadratic field such that $\mathcal{E}(H, F) \neq \varnothing$, and let Λ be the order of conductor m in F, with $\gcd(m, D) = 1$. Then,*

$$\nu(D, N, d, m; \mathcal{O}^*) = \begin{cases} 0 & \text{if there exists } p \mid N, p \nmid m, p \text{ inert a } F, \\ h(d, m) \cdot 2^{s+t} & \text{otherwise,} \end{cases}$$

where $s = \sharp\{p \mid D : p \text{ inert a } F\}$ *and* $t = \sharp\{p \mid N : p \text{ splits in } F \text{ or } p \mid m\}$.

COROLLARY 4.21. *Consider* $\mathcal{O} = \mathcal{O}_0(1, N) \subseteq M(2, \mathbb{Q})$, *an Eichler order of prime level* N. *Let* $\Lambda(d, m)$ *be the order of* $F = \mathbb{Q}(\sqrt{d})$ *with conductor* m. *Then,*

$$\nu(1, N, d, m; \mathcal{O}^*) = \begin{cases} 0 & \text{if } N \text{ is inert in } F \text{ and } N \nmid m, \\ h(d, m) & \text{if } N \text{ ramifies in } F \text{ and } N \nmid m, \\ 2h(d, m) & \text{if } N \text{ splits in } F \text{ or if } N = m. \end{cases}$$

COROLLARY 4.22. *Assume* $F = \mathbb{Q}(\sqrt{d})$ *splits* H. *Consider* $\mathcal{O} = \mathcal{O}(pq, N)$, *an Eichler order of level* N *in* H, *where* N *is a prime number different from* p *and* q. *Let* $\Lambda(d, m)$ *be the order of conductor* m *in* F.

(i) *If* $\nu(pq, N, d, m; \mathcal{O}^*) = 0$, *then* $\gcd(pq, m) \neq 1$.
(ii) *Assume* $\gcd(pq, m) = 1$; *then,* $\nu(pq, N, d, m; \mathcal{O}^*)$ *is equal to*

$$\begin{cases} 0 & \text{if } N \text{ is inert in } F \text{ and } N \nmid m, \\ h(d, m) & \text{if } N, p \text{ and } q \text{ ramify in } F \text{ and } N \nmid m, \\ 2h(d, m) & \text{if } N, p \text{ ramify, and } q \text{ is inert in } F, \\ & \text{either } N \text{ splits and } p, q \text{ ramify in } F, \\ & \text{either } N = m \text{ and } p, q \text{ ramify in } F, \\ 4h(d, m) & \text{if } N \text{ ramifies, } N \nmid m \text{ and } p, q \text{ are inert in } F, \\ & \text{either } N \text{ splits, } N \nmid m, \ p \text{ ramifies and } q \text{ is inert in } F, \\ & \text{either } N \mid m, \ p \text{ ramifies and } q \text{ is inert in } F, \\ 8h(d, m) & \text{if } N \text{ splits or } N \nmid m \text{ and } p, q \text{ are inert in } F. \end{cases}$$

Next, we consider the group $\mathcal{O}_+^*$, used in the definition of the groups of quaternion transformations, cf. Chapter 2. More generally, given a group G, $\mathcal{O}^* \subseteq G \subseteq \operatorname{Nor}(\mathcal{O})$, we will define the group $G_+ := \{\sigma \in G : \mathrm{n}(\sigma) > 0\}$. The group G_+ acts on $\mathcal{E}^*(\mathcal{O}, \Lambda)$ and we can consider the quotient set $\mathcal{E}^*(\mathcal{O}, \Lambda)/G_+$.

In the case that $\mathcal{O}$ is an Eichler order in an indefinite quaternion algebra, by the results in Section 2.3, we have that $\mathcal{O}_+^*$ is a subgroup of index 2 in $\mathcal{O}^*$ and, therefore, $[G{:}G_+] = 2$. This leads to the following proposition.

PROPOSITION 4.23. *Let* $\mathcal{O}$ *be an Eichler order of level* N *in an indefinite quaternion* $\mathbb{Q}$*-algebra* H *of discriminant* D. *Let* G *be a group,* $\mathcal{O}^* \subseteq G \subseteq \operatorname{Nor}(\mathcal{O})$. *Then,*

$$\nu(D, N, d, m; G_+) = 2\nu(D, N, d, m; G).$$

Finally, in the case of an indefinite quaternion $\mathbb{Q}$-algebra H, we note the role played by the embeddings of quadratic orders into quaternion orders of H in relation with the embedding of H into $M(2, \mathbb{R})$. Fix $\Phi\colon H \to M(2, \mathbb{R})$. Then, given an embedding φ of F into H, the composition $\phi := \Phi \circ \varphi$ is an embedding of F into $M(2, \mathbb{R})$ called quaternion embedding of F into $M(2, \mathbb{R})$. Thus, we can translate the results on embeddings of quadratic orders Λ into quaternion orders $\mathcal{O}$ into embeddings of quadratic orders Λ into $\Phi(\mathcal{O}) \subseteq M(2, \mathbb{R})$. In particular, we can consider Γ-equivalent quaternion embeddings for $\Gamma = \Phi(G_+) \subseteq \mathrm{GL}(2, \mathbb{R})$, $\mathcal{O}^* \subseteq G \subseteq \operatorname{Nor}(\mathcal{O})$.

4.3. Classification of representations by normic ternary forms

We now characterize the existence of embeddings of quadratic orders into quaternion orders. We will also construct such embeddings using representations related to quadratic forms.

NOTATION 4.24. Let f, g be quadratic forms with rational integer coefficients in n and r variables, respectively. Recall the notation for the following sets of representations over $\mathbb{Z}$, defined in Section 3.1.2.

$$\mathcal{R}(f, g; \mathbb{Z}) := \{P \in M(n \times r, \mathbb{Z}) : P^t A_1(f) P = A_1(g)\},$$
$$\mathcal{R}^*(f, g; \mathbb{Z}) := \{P \in M(n \times r, \mathbb{Z}) : P^t A_1(f) P = A_1(g), P \text{ primitive}\}.$$

The following theorem gives the bijective mapping between the set of embeddings and the set of representations. Its proof contains the explicit construction of the bijection. First we need a lemma stating a technical condition for defining the normic forms.

LEMMA 4.25. *Let $\{1, v_2, v_3, v_4\}$ be a normalized basis of a quaternion order $\mathcal{O}$ in a quaternion $\mathbb{Q}$-algebra. Then, the order $\mathbb{Z} + 2\mathcal{O}$ is even and a normalized basis of this order is $\{1, 2v_2, 2v_3, 2v_4 - \operatorname{tr}(v_4)\}$.*

THEOREM 4.26. *Let H be a quaternion $\mathbb{Q}$-algebra and $F = \mathbb{Q}(\sqrt{d})$. Let $\mathcal{O}$ and Λ be orders in H and F, respectively. Then:*

(i) *There is a bijective mapping $\rho\colon \mathcal{E}(\mathcal{O}, \Lambda) \to \mathcal{R}(n_{\mathbb{Z}+2\mathcal{O},3}, -D_\Lambda; \mathbb{Z})$.*
(ii) *There is an injective mapping $\varrho\colon \mathcal{E}(\mathcal{O}, \Lambda) \to \mathcal{R}(n_{\mathcal{O},4}, n_{\Lambda,2}; \mathbb{Z})$.*

PROOF. Let m be the conductor of the quadratic order Λ. We know that $\{1, mw\}$ is a basis of Λ, where $\{1, w\}$ is a $\mathbb{Z}$-basis of the ring of integers Λ_F, and that $D_\Lambda = D_F m^2 = -\det_1(n_{\Lambda,2})$. Fix a normalized basis $\mathcal{B} = \{1, v_2, v_3, v_4\}$ in the quaternion order $\mathcal{O}$. Then, $\mathcal{B}' = \{1, 2v_2, 2v_3, 2v_4 - \operatorname{tr}(v_4)\}$ is a normalized basis of the order $\mathbb{Z} + 2\mathcal{O}$, which is an even order.

First, we will define the mapping $\rho\colon \mathcal{E}(\mathcal{O}, \Lambda) \to \mathcal{R}(n_{\mathbb{Z}+2\mathcal{O},3}, -D_\Lambda; \mathbb{Z})$, depending on d in $\mathbb{Z}/4\mathbb{Z}$.

Assume $d \equiv 2, 3 \bmod 4$. Let $\varphi \in \mathcal{E}(\mathcal{O}, \Lambda)$ be given by $\varphi(m\sqrt{d}) = \omega$; that is, $\omega \in \mathcal{O}$ such that $n(\omega) = -dm^2$ and $\operatorname{tr}(\omega) = 0$. Consider $e = \varphi(2m\sqrt{d}) = 2\omega$. Then, $e \in \mathbb{Z} + 2\mathcal{O}$ satisfies $n(e) = -4dm^2$ and $\operatorname{tr}(e) = 0$. Hence, with respect to the normalized basis of $\mathbb{Z} + 2\mathcal{O}$, we have $e = (0, x, y, z)_{\mathcal{B}'}$, where (x, y, z) is a representation of $-4dm^2 = -D_F m^2 = -D_\Lambda$ by $n_{\mathbb{Z}+2\mathcal{O},3}$. Then put $\rho(\varphi) := (x, y, z) \in \mathcal{R}(n_{\mathbb{Z}+2\mathcal{O},3}, -D_\Lambda; \mathbb{Z})$. It is clear that, if $\rho(\varphi) = \rho(\varphi')$, then $\varphi(2m\sqrt{d}) = \varphi'(2m\sqrt{d})$; thus, necessarily it will be $\varphi = \varphi'$.

Let us construct the inverse mapping. Let $(x, y, z) \in \mathcal{R}(n_{\mathbb{Z}+2\mathcal{O},3}, -D_\Lambda; \mathbb{Z})$. For $e = x(2v_2) + y(2v_3) + z\big(2v_4 - \operatorname{tr}(v_4)\big)$, we have $e \in \mathbb{Z} + 2\mathcal{O}$, with $n(e) = -4dm^2$ and $\operatorname{tr}(e) = 0$. Since $\mathbb{Z} + 2\mathcal{O} \subseteq \mathcal{O}$, with respect to the normalized basis of $\mathcal{O}$, we obtain $e = (-z\operatorname{tr}(v_4), 2x, 2y, 2z)_{\mathcal{B}}$. Note that $2 \mid z\operatorname{tr}(v_4)$, since $n_{\mathcal{O},4}(e) = z^2\operatorname{tr}(v_4) + g(x, y, z) = -4dm^2$ for some quadratic expression $g(x, y, z)$ satisfying $2 \mid g(x, y, z)$. Hence, $\omega = e/2 = (-z\operatorname{tr}(v_4)/2, x, y, z)_{\mathcal{B}}$ belongs to $\mathcal{O}$ and satisfies $n(\omega) = -dm^2$ and $\operatorname{tr}(\omega) = 0$. Thus, putting $\varphi(m\sqrt{d}) := \omega$, we obtain an embedding $\varphi \in \mathcal{E}(\mathcal{O}, \Lambda)$. It is clear that, if we consider a different representation, it will give rise to a different embedding.

Assume $d \equiv 1 \bmod 4$. Consider an embedding $\varphi \in \mathcal{E}(\mathcal{O}, \Lambda)$. It is given by $\varphi(m(1 + \sqrt{d})/2) = \omega$; that is, given by $\omega \in \mathcal{O}$ such that $n(\omega) = m^2(1 - d)/4$

and $\mathrm{tr}(\omega) = m$. Consider $e = \varphi(m\sqrt{d}) = 2\omega - m$. Then, $e \in \mathbb{Z} + 2\mathcal{O}$ satisfies $n(e) = -dm^2$ and $\mathrm{tr}(e) = 0$. Hence, $e = (0, x, y, z)_{\mathcal{B}'}$ and $n_{\mathbb{Z}+2\mathcal{O},3}(x, y, z) = n(e)$. Thus, (x, y, z) is a representation of $-dm^2$ by $n_{\mathbb{Z}+2\mathcal{O},3}$; therefore, $\rho(\varphi) := (x, y, z) \in \mathcal{R}(n_{\mathbb{Z}+2\mathcal{O},3}, -D_\Lambda; \mathbb{Z})$. By construction ρ is injective.

In the reverse case, consider $(x, y, z) \in \mathcal{R}(n_{\mathbb{Z}+2\mathcal{O},3}, -D_\Lambda; \mathbb{Z})$. Put $e = x(2v_2) + y(2v_3) + z\big(2v_4 - \mathrm{tr}(v_4)\big)$. Then, we have that $e \in \mathbb{Z} + 2\mathcal{O}$ satisfies $n(e) = -dm^2$ and $\mathrm{tr}(e) = 0$. Note that $e = (-z\,\mathrm{tr}(v_4), 2x, 2y, 2z)_{\mathcal{B}}$, with respect to the normalized basis of $\mathcal{O}$. Consider now the element $\omega = (m+e)/2 = \big((m - z\,\mathrm{tr}(v_4))/2, x, y, z\big)_{\mathcal{B}}$, which satisfies $n(\omega) = (1-d)m^2/4$ and $\mathrm{tr}(\omega) = m$. We have that $\omega \in \mathcal{O}$ if and only if $2 \mid \big(m - z\,\mathrm{tr}(v_4)\big)$. If $\mathcal{O}$ is an odd order, we have $\mathrm{tr}(v_4) = 1$ and the condition is equivalent to the fact that m and z are of the same parity. Indeed, if $2 \mid z$, then $4 \mid n(e)$; since $n(e) = -dm^2$, for d square-free, we deduce that m is even. Conversely, if m is even we have that $2 \mid n(e) = n_{\mathcal{O},3}(-z, 2x, 2y, 2z)$ and we deduce $2 \mid z$. If $\mathcal{O}$ is an even order, we have that $\mathrm{tr}(v_4) = 0$ and the condition is equivalent to m even. Indeed, if $\mathcal{O}$ is an even order and $d \equiv 1 \bmod 4$, we have that $e = (0, 2x, 2y, 2z)_{\mathcal{B}}$; thus, $4 \mid n(e) = -dm^2$. Now, since d is square-free, we deduce that m is even. Thus, in any event, $\omega \in \mathcal{O}$, with $n(\omega) = m^2(1-d)/4$ and $\mathrm{tr}(\omega) = m$, and because of this, $\varphi(m(1+\sqrt{d})/2) := \omega$ defines an embedding $\varphi \in \mathcal{E}(\mathcal{O}, \Lambda)$. This completes the proof of statement (i).

In particular, if $\mathcal{O}$ is an even order, m is odd and $d \equiv 1 \bmod 4$, then we have $\mathcal{R}(n_{\mathbb{Z}+2\mathcal{O},3}, -D_\Lambda; \mathbb{Z}) = \varnothing$. Note that, under the same conditions, we also have that $\mathcal{E}(\mathcal{O}, \Lambda) = \varnothing$, since in an even order there is no possible image of $m(1+\sqrt{d})/2$.

To prove (ii), let $\varphi \in \mathcal{E}(\mathcal{O}, \Lambda)$ be an embedding of Λ into $\mathcal{O}$. Consider the element $\omega \in \mathcal{O}$ such that $\omega = \varphi(mw) \in H \setminus \mathbb{Q}$ and put P the matrix of the coordinates of 1 and ω with respect to the fixed basis in $\mathcal{O}$. It is clear that P has the coefficients in $\mathbb{Z}$. Since $n(\alpha) = n_{\Lambda,2}(\alpha) = n_{\mathcal{O},4}\big(\varphi(\alpha)\big)$, we have the matrix equality $P^t A_1(n_{\mathcal{O},4}) P = A_1(n_{\Lambda,2})$. Hence, $n_{\mathcal{O},4}$ represents $n_{\Lambda,2}$ over $\mathbb{Z}$ and put $\varrho(\varphi) = P \in \mathcal{R}(n_{\mathcal{O},4}, n_{\Lambda,2}; \mathbb{Z})$. □

Using the details contained in the previous proof, we give the explicit inverse mapping in the following corollary, to make implementation easier.

COROLLARY 4.27. *Let $\mathcal{O} \subseteq H$ a quaternion order with a fixed normalized basis $\mathcal{B}$. Let $\Lambda(d, m)$ be a quadratic order. Then, the mapping ρ^{-1} is given by*

$$\rho^{-1}\colon\ \mathcal{R}(n_{\mathbb{Z}+2\mathcal{O},3}, -D_\Lambda; \mathbb{Z}) \to \mathcal{E}(\mathcal{O}, \Lambda)$$
$$(x, y, z) \mapsto \varphi,$$

where φ is the embedding defined by $\varphi(mw) = \big((rm - z\,\mathrm{tr}(v_4))/2, x, y, z\big)_{\mathcal{B}}$, $r = 0$ if $d \equiv 2, 3 \bmod 4$, and $r = 1$ if $d \equiv 1 \bmod 4$. Namely,

$$\varphi(\sqrt{d}) = \begin{cases} (-z\,\mathrm{tr}(v_4)/2m, x/m, y/m, z/m)_{\mathcal{B}} & \text{if } d \equiv 2, 3 \bmod 4, \\ (-z\,\mathrm{tr}(v_4)/m, 2x/m, 2y/m, 2z/m)_{\mathcal{B}} & \text{if } d \equiv 1 \bmod 4. \end{cases}$$

PROOF. Consider the basis $\{1, mw\}$ of the quadratic order Λ, with $w = \sqrt{d}$ if $d \equiv 2, 3 \bmod 4$, and $w = (1+\sqrt{d})/2$ if $d \equiv 1 \bmod 4$. In the proof of the previous proposition, the embedding is defined from $(x, y, z) \in \mathcal{R}(n_{\mathbb{Z}+2\mathcal{O},3}, -D_\Lambda; \mathbb{Z})$. In a natural way, this representation is in correspondence with an element $e \in \mathbb{Z} + 2\mathcal{O}$, an image of an element in the quadratic order. Using the relation of that element with $\sqrt{d}$, we deduce the expression of $\varphi(\sqrt{d})$ in terms of the representation, which depends on the value of d in $\mathbb{Z}/4\mathbb{Z}$. □

THEOREM 4.28. *Let $\mathcal{O}$ and Λ be orders in H and $F = \mathbb{Q}(\sqrt{d})$, respectively. Let ρ and ϱ be the mappings defined in Theorem 4.26.*

(i) *The restriction of ρ gives a bijective mapping*
$$\rho^*\colon \mathcal{E}^*(\mathcal{O},\Lambda) \to \mathcal{R}^*(n_{\mathbb{Z}+2\mathcal{O},3}, -D_\Lambda; \mathbb{Z}).$$

(ii) *The restriction of ϱ gives a injective mapping*
$$\varrho^*\colon \mathcal{E}^*(\mathcal{O},\Lambda) \hookrightarrow \mathcal{R}^*(n_{\mathcal{O},4}, n_{\Lambda,2}; \mathbb{Z}).$$

PROOF. Assume Λ is the order of conductor m and fix the basis $\{1, mw\}$, where $\{1, w\}$ is a basis of the ring of integers Λ_F. Also fix a normalized basis $\mathcal{B} = \{1, v_2, v_3, v_4\}$ in the quaternion order $\mathcal{O}$.

Let $\varphi \in \mathcal{E}^*(\mathcal{O},\Lambda)$ be given by $\varphi(mw) = \omega$. Then, by Theorem 4.26, we have that $\rho(\varphi) = (x, y, z) \in \mathcal{R}(n_{\mathbb{Z}+2\mathcal{O},3}, -D_\Lambda; \mathbb{Z})$ and, using Corollary 4.27,
$$\omega = \left(\frac{rm - z\,\mathrm{tr}(v_4)}{2}, x, y, z\right)_{\mathcal{B}}, \quad \text{where } r = \begin{cases} 0 & \text{if } d \equiv 2, 3 \bmod 4, \\ 1 & \text{if } d \equiv 1 \bmod 4. \end{cases}$$

Let $s \in \mathbb{Z}$ be such that $s \mid \gcd(x, y, z)$. Consider an integer l such that $l \equiv (rm - z\,\mathrm{tr}(v_4))/2 \bmod s$ and take the element
$$\frac{\omega - l}{s} = \frac{1}{s}\left(\frac{rm - z\,\mathrm{tr}(v_4)}{2} - l, x, y, z\right)_{\mathcal{B}},$$
which belongs to the order $\mathcal{O}$ since all its coordinates are integers. It is clear that $(\omega - l)/s = \varphi\big((mw - l)/s\big)$. Thus, since φ is an optimal embedding, we have $(mw - l)/s \in \Lambda$, which gives $s = 1$. Hence, $(x, y, z) \in \mathcal{R}^*(n_{\mathbb{Z}+2\mathcal{O},3}, -D_\Lambda; \mathbb{Z})$.

Conversely, let $(x, y, z) \in \mathcal{R}^*(n_{\mathbb{Z}+2\mathcal{O},3}, -D_\Lambda; \mathbb{Z})$. By the bijective mapping in Theorem 4.26, this representation is in correspondence with the embedding $\varphi \in \mathcal{E}(\mathcal{O},\Lambda)$ given by $\varphi(mw) = \big((rm - z\,\mathrm{tr}(v_4))/2, x, y, z\big)_{\mathcal{B}}$, where $r = 0$ if $d \equiv 2, 3 \bmod 4$, and $r = 1$ if $d \equiv 1 \bmod 4$. We want to see that φ is an optimal embedding. Assume $\varphi(k + smw) \in \mathcal{O}$ with $k, s \in \mathbb{Q}$; we need $k, s \in \mathbb{Z}$. By definition of φ, we have that
$$\varphi(k + smw) = \left(k + \frac{s\big(rm - z\,\mathrm{tr}(v_4)\big)}{2}, sa, sb, sc\right)_{\mathcal{B}},$$
where $s = s_1/s_2$ with $\gcd(s_1, s_2) = 1$. Then, we deduce $s_2 \mid \gcd(x, y, z) = 1$; hence $s \in \mathbb{Z}$. Then the numbers
$$k + \frac{s\big(rm - z\,\mathrm{tr}(v_4)\big)}{2} \quad \text{and} \quad \frac{s\big(rm - z\,\mathrm{tr}(v_4)\big)}{2}$$
are also integers, and it follows that $k \in \mathbb{Z}$. Thus, $\varphi \in \mathcal{E}^*(\mathcal{O},\Lambda)$, which concludes the proof of (i).

To prove (ii), consider $\varphi \in \mathcal{E}^*(\mathcal{O},\Lambda)$, given by $\varphi(mw) = \omega$. We define $\varrho(\varphi) = P \in \mathcal{R}(n_{\mathcal{O},4}, n_{\Lambda,2}; \mathbb{Z})$, where P is the matrix given by the coordinates of 1 and ω with respect to the fixed basis of $\mathcal{O}$; that is,
$$\omega = \left(\frac{rm - z\,\mathrm{tr}(v_4)}{2}, x, y, z\right)_{\mathcal{B}}, \qquad P = \begin{pmatrix} 1 & (rm - z\,\mathrm{tr}(v_4))/2 \\ 0 & x \\ 0 & y \\ 0 & z \end{pmatrix},$$
where $r = 0$ if $d \equiv 2, 3 \bmod 4$, and $r = 1$ if $d \equiv 1 \bmod 4$, for some $(x, y, z) \in \mathcal{R}(n_{\mathbb{Z}+2\mathcal{O},3}, -D_\Lambda; \mathbb{Z})$. The greatest common divisor of the 2×2 minors of P is

equal to $\gcd(x,y,z)=1$. But, since $\varphi \in \mathcal{E}^*(\mathcal{O},\Lambda)$, we have $\gcd(x,y,z)=1$, by (i). Thus, $P \in \mathcal{R}^*(n_{\mathcal{O},4}, n_{\Lambda,2};\mathbb{Z})$. □

LEMMA 4.29. *Let $\varphi \in \mathcal{E}(H,F)$. Let $\mathcal{O}$ be a quaternion order in H, and Λ the quadratic order of conductor m in F. Assume $\varphi \in \mathcal{E}(\mathcal{O},\Lambda)$, in such a way that $\rho_\Lambda(\varphi)=(x,y,z)\in \mathcal{R}(n_{\mathbb{Z}+2\mathcal{O},3}, -D_\Lambda;\mathbb{Z})$, where ρ_Λ denotes the bijective mapping between $\mathcal{E}(\mathcal{O},\Lambda)$ and $\mathcal{R}(n_{\mathbb{Z}+2\mathcal{O},3},-D_\Lambda;\mathbb{Z})$. Put $m'=\gcd(x,y,z)$. Then we have $\varphi \in \mathcal{E}^*(\mathcal{O},\Lambda')$, where Λ' is the quadratic order in F of conductor m/m'.*

PROOF. Consider $\omega = \varphi(mw)\in\mathcal{O}$. Using the explicit expression for ω in terms of x,y,z obtained in Corollary 4.27, it is easy to see that $(1/m')\omega\in\mathcal{O}$, since $m'=\gcd(x,y,z)$. Thus, $\varphi\big((m/m')w\big)=(1/m')\omega\in\mathcal{O}$. We deduce that $\varphi\big((m/m')w\big)$ is an integral element in H and $(m/m')w$ is an integral element in F. Since w is a primitive element in the ring of integers Λ_F, it is clear that $m/m'\in\mathbb{Z}$. Hence, we have an embedding $\varphi\in\mathcal{E}(\mathcal{O},\Lambda')$.

We must see that φ is an optimal embedding. The bijective mapping in Theorem 4.26, for the quadratic order Λ', gives

$$\rho_{\Lambda'}(\varphi)=(1/m')(x,y,z)\in\mathcal{R}^*(n_{\mathbb{Z}+2\mathcal{O},3},-D_{\Lambda'};\mathbb{Z}).$$

By Theorem 4.28, we have $\varphi\in\mathcal{E}^*(\mathcal{O},\Lambda')$. □

In the following lemma, we make explicit the ternary normic form $n_{\mathbb{Z}+2\mathcal{O},3}$ for certain families of Eichler orders $\mathcal{O}$.

LEMMA 4.30. *Let $\mathcal{O}_M(1,N)$, $\mathcal{O}_A(2p,N)$ and $\mathcal{O}_B(pq,N)$ be the Eichler orders of level N in the quaternion algebras $\left(\frac{1,-1}{\mathbb{Q}}\right)$, $H_A(p)$ and $H_B(p,q)$, respectively, given in Proposition* 1.60. *Then:*

(i) $n_{\mathbb{Z}+2\mathcal{O}_M(1,N),3}=-4NXY-Z^2$.
(ii) $n_{\mathbb{Z}+2\mathcal{O}_A(2p,N),3}=-4pX^2-4pXZ+4N^2Y^2+4NYZ+(1-2p)Z^2$.
(iii) $n_{\mathbb{Z}+2\mathcal{O}_B(pq,N),3}=-4pN^2X^2-4pNXY+p(q-1)Y^2-qZ^2$.

REMARK 4.31. Consider the matrix quaternion algebra $\mathrm{M}(2,\mathbb{Q})$. We fix the normalized basis $\mathcal{B}$ and $\mathcal{B}'$ for the orders $\mathcal{O}_0(1,N)$ and $\mathbb{Z}+2\mathcal{O}_0(1,N)$, respectively, given by

$$\mathcal{B}=\left\{\begin{pmatrix}1&0\\0&1\end{pmatrix},\begin{pmatrix}0&1\\0&0\end{pmatrix},\begin{pmatrix}0&0\\N&0\end{pmatrix},\begin{pmatrix}0&0\\0&1\end{pmatrix}\right\},$$
$$\mathcal{B}'=\left\{\begin{pmatrix}1&0\\0&1\end{pmatrix},\begin{pmatrix}0&2\\0&0\end{pmatrix},\begin{pmatrix}0&0\\2N&0\end{pmatrix},\begin{pmatrix}-1&0\\0&1\end{pmatrix}\right\}.$$

Then, the ternary normic forms are $n_{\mathbb{Z}+2\mathcal{O}_0(1,N),3}=n_{\mathbb{Z}+2\mathcal{O}_M(1,N),3}=-4NXY-Z^2$.

In Appendix A we include Tables A.21–A.24 listing the ternary normic forms $n_{\mathbb{Z}+2\mathcal{O},3}$ for Eichler orders $\mathcal{O}$ in the small ramified quaternion algebras $H_A(3)$, $H_B(2,5)$, $H_A(7)$ and $H_B(3,5)$.

LEMMA 4.32. *Let H be a nonramified quaternion algebra. Let $\mathcal{O}(1,N)\subseteq\mathcal{O}(1,M)$ be two Eichler orders in H of level N and M, respectively. Suppose that $M\mid N$. If $\gcd(m,\frac{N}{M})=1$, then $\mathcal{E}^*(\mathcal{O}(1,N),\Lambda)\subset\mathcal{E}^*(\mathcal{O}(1,M),\Lambda)$.*

EXAMPLE 4.33. Consider the nonramified quaternion algebra $H=M(2,\mathbb{Q})$ and the orders $\mathcal{O}_0(1,1)=M(2,\mathbb{Z})$ and $\mathcal{O}_0(1,5)$, maximal order and Eichler order of level 5, respectively. Let $F=\mathbb{Q}(\sqrt{-1})$ and consider the quadratic orders:

TABLE 4.1. Embeddings of quadratic orders into the quaternion orders $\mathcal{O}(1,1)$ and $\mathcal{O}(1,5)$.

	φ_1	φ_2	φ_3	φ_4	φ_5	φ_6	φ_7	φ_8	φ_9
$\mathcal{E}(\mathcal{O}(1,1),\Lambda(-1,1))$	•	•	•	–	–	–	–	–	–
$\mathcal{E}^*(\mathcal{O}(1,1),\Lambda(-1,1))$	•	•	•	–	–	–	–	–	–
$\mathcal{E}(\mathcal{O}(1,1),\Lambda(-1,2))$	•	•	•	•	•	–	–	–	–
$\mathcal{E}^*(\mathcal{O}(1,1),\Lambda(-1,2))$	–	–	–	•	•	–	–	–	–
$\mathcal{E}(\mathcal{O}(1,1),\Lambda(-1,5))$	•	•	•	–	–	–	–	•	•
$\mathcal{E}^*(\mathcal{O}(1,1),\Lambda(-1,5))$	–	–	–	–	–	–	–	•	•
$\mathcal{E}(\mathcal{O}(1,1),\Lambda(-1,6))$	•	•	•	•	•	•	•	–	–
$\mathcal{E}^*(\mathcal{O}(1,1),\Lambda(-1,6))$	–	–	–	–	–	•	•	–	–
$\mathcal{E}(\mathcal{O}(1,5),\Lambda(-1,1))$	–	–	•	–	–	–	–	–	–
$\mathcal{E}^*(\mathcal{O}(1,5),\Lambda(-1,1))$	–	–	•	–	–	–	–	–	–
$\mathcal{E}(\mathcal{O}(1,5),\Lambda(-1,2))$	–	–	•	–	•	–	–	–	–
$\mathcal{E}^*(\mathcal{O}(1,5),\Lambda(-1,2))$	–	–	–	–	•	–	–	–	
$\mathcal{E}(\mathcal{O}(1,5),\Lambda(-1,5))$	–	•	•	–	–	–	–	–	•
$\mathcal{E}^*(\mathcal{O}(1,5),\Lambda(-1,5))$	–	•	–	–	–	–	–	–	•
$\mathcal{E}(\mathcal{O}(1,5),\Lambda(-1,6))$	–	–	•	–	•	–	•	–	–
$\mathcal{E}^*(\mathcal{O}(1,5),\Lambda(-1,6))$	–	–	–	–	–	–	•	–	–

$\Lambda(m) = \mathbb{Z}[1, mw]$ for $m = 1, 2, 5, 6$, where $w = \sqrt{-1}$. Computing elements of the sets of representations over $\mathbb{Z}$ of 1, 4, 25 and 36 by the ternary normic forms $n_{\mathbb{Z}+2\mathcal{O}(1,1),3}(X,Y,Z) = -4XY - Z^2$ and $n_{\mathbb{Z}+2\mathcal{O}(1,5),3}(X,Y,Z) = -20XY - Z^2$, we obtain the embeddings $\varphi_s \in \mathcal{E}(H,F)$ given by $\varphi_s(w) = \omega_s,\ 1 \le s \le 9$, where

$$\omega_1 := \begin{pmatrix} 0 & -1 \\ 1 & 0 \end{pmatrix}, \qquad \omega_4 := \begin{pmatrix} 0 & -2 \\ \frac{1}{2} & 0 \end{pmatrix}, \qquad \omega_7 := \begin{pmatrix} \frac{1}{3} & -\frac{4}{3} \\ \frac{5}{6} & -\frac{1}{3} \end{pmatrix},$$

$$\omega_2 := \begin{pmatrix} 1 & -2 \\ 1 & -1 \end{pmatrix}, \qquad \omega_5 := \begin{pmatrix} 2 & -2 \\ \frac{5}{2} & -2 \end{pmatrix}, \qquad \omega_8 := \begin{pmatrix} \frac{1}{5} & -13/5 \\ \frac{2}{5} & -\frac{1}{5} \end{pmatrix},$$

$$\omega_3 := \begin{pmatrix} 2 & -1 \\ 5 & -2 \end{pmatrix}, \qquad \omega_6 := \begin{pmatrix} 0 & -\frac{3}{2} \\ \frac{2}{3} & 0 \end{pmatrix}, \qquad \omega_9 := \begin{pmatrix} 0 & -\frac{1}{5} \\ 5 & 0 \end{pmatrix}.$$

For any such embedding φ_s, we check whether it belongs to the set of embeddings of the quadratic orders into the quaternion orders given above. In Table 4.1 we mark the case of affirmative answer by the symbol •, and the case of negative answer by the symbol –.

EXAMPLE 4.34. Consider the order $\mathcal{O}_{\mathrm{A}}(14,1) = \mathbb{Z}[1,i,j,(1+i+j+ij)/2]$ in the quaternion algebra $H_{\mathrm{A}}(7) = \left(\frac{7,-1}{\mathbb{Q}}\right)$. Let $F = \mathbb{Q}(\sqrt{-1})$ and put $w = \sqrt{-1}$. Consider the quadratic orders: $\Lambda(-1,1)$, $\Lambda(-1,3)$ and $\Lambda(-1,15)$. Consider the embeddings $\varphi_s \in \mathcal{E}(H_{\mathrm{A}}(7),F)$, given by $\varphi_s(w) = \omega_s$, $1 \le s \le 4$, where $\omega_1 := j$, $\omega_2 := 3i + 8j$, $\omega_3 := \frac{1}{3}i + \frac{4}{3}j$, $\omega_4 := (1/15)i + (22/15)j + \frac{2}{5}ij$, obtained from representations of 1, 9, and 225 by the ternary normic form $n_{\mathbb{Z}+2\mathcal{O}_{\mathrm{A}}(14,1),3}(X,Y,Z) = -28X^2 + 4Y^2 - 13Z^2 - 28XZ + 4YZ$.

For any such embedding φ_s, we check whether it belongs to the set of embeddings of the quadratic orders into the quaternion orders given above. In Table 4.2

TABLE 4.2. Embeddings of quadratic orders into the quaternion order $\mathcal{O}(7,1)$.

	φ_1	φ_2	φ_3	φ_4
$\mathcal{E}(\mathcal{O}_{\mathrm{A}}(7,1),\Lambda(-1,1))$	•	•	–	–
$\mathcal{E}^*(\mathcal{O}_{\mathrm{A}}(7,1),\Lambda(-1,1))$	•	•	–	–
$\mathcal{E}(\mathcal{O}_{\mathrm{A}}(7,1),\Lambda(-1,3))$	•	•	•	–
$\mathcal{E}^*(\mathcal{O}_{\mathrm{A}}(7,1),\Lambda(-1,3))$	–	–	•	–
$\mathcal{E}(\mathcal{O}_{\mathrm{A}}(7,1),\Lambda(-1,15))$	•	•	•	•
$\mathcal{E}^*(\mathcal{O}_{\mathrm{A}}(7,1),\Lambda(-1,15))$	–	–	–	•

we mark the case of affirmative answer by the symbol •, and the case of negative answer by the symbol –.

PROPOSITION 4.35. *Let $\mathcal{O}$ be a quaternion order with a fixed normalized basis $\mathcal{B}$. Then, we have an homomorphism of groups*

$$\kappa\colon \operatorname{Nor}(\mathcal{O}) \to \operatorname{GL}(3,\mathbb{Z}),$$

which gives rise to a monomorphism $\kappa : \operatorname{Nor}(\mathcal{O})/\mathbb{Q}^* \hookrightarrow \operatorname{GL}(3,\mathbb{Z})$.

PROOF. We will define the mapping κ. Let $\sigma \in \operatorname{Nor}(\mathcal{O})$. Let $\mathcal{B}'$ be the basis obtained from the basis $\mathcal{B}$ by conjugation with σ, which is also a normalized basis of $\mathcal{O}$, cf. Lemma 1.41. Let $S_{k,4}$ be the base change matrix from the basis $\mathcal{B}$ to the basis $\mathcal{B}'$, where $k=1$ if $\mathcal{O}$ is odd and $k=0$ if $\mathcal{O}$ is even. Since both bases are normalized, the matrix $S_{k,4}$ satisfies the conditions in Lemma 1.40 and determines a unique matrix $S_{k,3} \in \operatorname{GL}(3,\mathbb{Z})$. We define $\kappa(\sigma) := S_{k,3}$.

Let us see that κ is a homomorphism. It is clear that, if we consider $\sigma = 1$, then we have $\mathcal{B}' = \mathcal{B}$; hence, $S_{k,3}$ is the identity matrix. Let $\kappa(\sigma) = S_{k,3}$ and $\kappa(\sigma') = S'_{k,3}$. Consider the element $\sigma\sigma'$. We denote by $S''_{k,4}$ the matrix of the basis change corresponding to the conjugation by the element $\sigma\sigma'$. Then we have $\kappa(\sigma\sigma') = S''_{k,3}$. We deduce the relation $S''_{k,3} = S_{k,3}S'_{k,3}$ from $S''_{k,4} = S_{k,4}S'_{k,4}$; that is,

$$\left(\begin{array}{c|c} 1 & * \\ \hline 0 & S''_{k,3} \end{array}\right) = \left(\begin{array}{c|c} 1 & * \\ \hline 0 & S_{k,3} \end{array}\right)\left(\begin{array}{c|c} 1 & * \\ \hline 0 & S'_{k,3} \end{array}\right),$$

where the entries of $*$ are determined by the rest of the matrix.

Finally, let us see how κ factorizes. Assume that $\kappa(\sigma) = S_{k,3}$ coincides with $\kappa(\sigma') = S'_{k,3}$. Then, we have $S_{k,4} = S'_{k,4}$. Thus, the basis $(\sigma')^{-1}\mathcal{B}\sigma'$ and $\sigma^{-1}\mathcal{B}\sigma$ are the same. Therefore $(\sigma')^{-1}v\sigma' = \sigma^{-1}v\sigma$ for all $v \in H$, and we deduce that $\sigma'\sigma^{-1}$ commutes with all the elements of H. Thus, $\sigma'\sigma^{-1}$ is an invertible element in the center of the quaternion algebra $Z(H) = \mathbb{Q}$. □

PROPOSITION 4.36. *Let $\varphi \in \mathcal{E}^*(\mathcal{O},\Lambda)$ and $\sigma \in \operatorname{Nor}(\mathcal{O})$. Then,*

$$\rho^*(\varphi^\sigma) = \kappa(\sigma)\rho^*(\varphi) \text{ in } \mathcal{R}^*(n_{\mathbb{Z}+2\mathcal{O},3}, -D_\Lambda;\mathbb{Z}).$$

PROOF. Consider $\rho^*(\varphi) = (x,y,z)$ and $\rho^*(\varphi^\sigma) = (x',y',z')$, where (x,y,z), $(x',y',z') \in \mathcal{R}^*(n_{\mathbb{Z}+2\mathcal{O},3}, -D_\Lambda;\mathbb{Z})$. By Theorem 4.26, fixing a basis $\{1, mw\}$ of Λ, we have that $\omega := \varphi(mw)$ is equal to $(\lambda,x,y,z)_{\mathcal{B}}$ and $\omega' := \varphi^\sigma(mw) = (\lambda',x',y',z')_{\mathcal{B}}$, where $\lambda := (rm - z\operatorname{tr}(v_4))/2$ and $\lambda' = (rm - z'\operatorname{tr}(v_4))/2$. By definition, $\varphi^\sigma(\alpha) = \sigma^{-1}\varphi(\alpha)\sigma$, for all $\alpha \in F$. Thus, for $\alpha = mw$, we obtain $\omega' = \sigma^{-1}\omega\sigma$.

We will now make this relation explicit. Consider $\mathcal{B}'$ the basis obtained from $\mathcal{B}$ conjugating by σ. Then, by the above relation, the coordinates of ω' with respect to the basis $\mathcal{B}'$ coincide with the coordinates of ω with respect to the basis $\mathcal{B}$; that is, $\omega' = (\lambda', x', y', z')_{\mathcal{B}} = (\lambda, x, y, z)_{\mathcal{B}'}$. In matrix notation, with $S_{k,4}$ the base change matrix from the basis $\mathcal{B}$ to the basis $\mathcal{B}'$, we have

$$S_{k,4}\begin{pmatrix}\lambda\\ x\\ y\\ z\end{pmatrix} = \begin{pmatrix}\lambda'\\ x'\\ y'\\ z'\end{pmatrix}, \qquad S_{k,4} = \left(\begin{array}{c|c} 1 & * \\ \hline 0 & S_{k,3}\end{array}\right).$$

Hence, we deduce

$$S_{k,3}\begin{pmatrix} x\\ y\\ z\end{pmatrix} = \begin{pmatrix} x'\\ y'\\ z'\end{pmatrix}.$$

Now, by the definition of the mapping κ in the previous proposition, the relation obtained is precisely $\rho^*(\varphi^\sigma) = \kappa(\sigma)\rho^*(\varphi)$. □

However, we have no guarantee that two representations related by a matrix $Q \in \mathrm{GL}(3,\mathbb{Z})$ come from equivalent embeddings. For this, it is necessary to refine the equivalence in the set of representations.

NOTATION 4.37. Given a subgroup G of $\mathrm{Nor}\,\mathcal{O}$, take the subgroup $\kappa(G) \subseteq \mathrm{GL}(3,\mathbb{Z})$ and consider the quotient set $\mathcal{R}^*(n_{\mathbb{Z}+2\mathcal{O},3}, -D_\Lambda; \mathbb{Z})/\kappa(G)$. Its number of elements, if finite, will be denoted by $r^*(n_{\mathbb{Z}+2\mathcal{O},3}, -D_\Lambda; G)$; that is,

$$r^*(n_{\mathbb{Z}+2\mathcal{O},3}, -D_\Lambda; G) := \sharp\,\mathcal{R}^*(n_{\mathbb{Z}+2\mathcal{O},3}, -D_\Lambda; \mathbb{Z})/\kappa(G).$$

THEOREM 4.38. *Let $\mathcal{O}$ be a quaternion order and $\Lambda(d,m)$ a quadratic order. Let G be a group, $\mathcal{O}^* \subseteq G \subseteq \mathrm{Nor}(\mathcal{O})$. Then, there is a bijective mapping*

$$\overline{\rho^*}\colon \mathcal{E}^*(\mathcal{O},\Lambda)/G \to \mathcal{R}^*(n_{\mathbb{Z}+2\mathcal{O},3}, -D_\Lambda; \mathbb{Z})/\kappa(G).$$

PROOF. We use Proposition 4.36 and the fact that the mapping ρ^* is bijective, cf. Theorem 4.28.

Suppose that $\overline{\varphi} = \overline{\varphi'} \in \mathcal{E}^*(\mathcal{O},\Lambda)/G$. Then there exists $\sigma \in G$ such that $\varphi' = \varphi^\sigma$. By Proposition 4.36, we have $\rho^*(\varphi^\sigma) = \kappa(\sigma)\rho^*(\varphi^\sigma)$; hence, we obtain the equality $\overline{\rho^*(\varphi^\sigma)} = \overline{\rho^*(\varphi)}$ in $\mathcal{R}^*(n_{\mathbb{Z}+2\mathcal{O},3}, -D_\Lambda; \mathbb{Z})/\kappa(G)$.

Conversely, let us consider two representations in $\mathcal{R}^*(n_{\mathbb{Z}+2\mathcal{O},3}, -D_\Lambda; \mathbb{Z})$. We can assume they are $\rho^*(\varphi)$ and $\rho^*(\varphi')$, since the mapping ρ^* is surjective. Assume $\overline{\rho^*(\varphi)} = \overline{\rho^*(\varphi')}$; that is, assume there exists $\sigma \in G$ such that $\rho^*(\varphi') = \kappa(\sigma)\rho^*(\varphi)$. Then, we have $\rho^*(\varphi') = \rho^*(\varphi^\sigma)$ and by the injectivity of ρ^*, $\varphi' = \varphi^\sigma$. □

COROLLARY 4.39. *Let $\mathcal{O}$ be a quaternion order in H, $\Lambda(d,m)$ a quadratic order and G a group, $\mathcal{O}^* \subseteq G \subseteq \mathrm{Nor}(\mathcal{O})$. Then, the number of $\kappa(G)$-classes of primitive representations of $-D_\Lambda$ by the ternary form $n_{\mathbb{Z}+2\mathcal{O},3}$ is $r^*(n_{\mathbb{Z}+2\mathcal{O},3}, -D_\Lambda; G) = \nu(\mathcal{O},\Lambda;G)$.*

4.4. Binary forms associated to quaternion orders

In this section we will consider binary quadratic forms associated to the elements in an indefinite quaternion algebra.

DEFINITION 4.40. Let H be an indefinite quaternion algebra and Φ a fixed representation of H into $\mathrm{M}(2,\mathbb{R})$. For each element $\omega \in H$, the binary quadratic form $f_{\Phi(\omega)}$ is the binary form associated to the element $\Phi(\omega)$, according to Definition 2.11.

To simplify the expression of the results, we will fix the embedding Φ. For an indefinite quaternion algebra $H = \left(\frac{a,b}{\mathbb{Q}}\right)$, assume $a > 0$. From now on, fix the embedding $\Phi\colon H \hookrightarrow \mathrm{M}(2,\mathbb{R})$ given at Proposition 1.19. Thus, to every quaternion algebra we attach a set of binary quadratic forms with real coefficients; namely, coefficients in the quadratic field $\mathbb{Q}(\sqrt{a})$. Thus, given $\omega = x + yi + zj + tij \in H$, we have

$$\Phi(\omega) = \begin{pmatrix} x + y\sqrt{a} & z + t\sqrt{a} \\ b(z - t\sqrt{a}) & x - y\sqrt{a} \end{pmatrix},$$
$$A_1(f_{\Phi(\omega)}) = \begin{pmatrix} b(z - t\sqrt{a}) & -y\sqrt{a} \\ -y\sqrt{a} & -z - t\sqrt{a} \end{pmatrix},$$

and we have the binary quadratic form

$$f_{\Phi(\omega)} = (b(z - t\sqrt{a}), -2y\sqrt{a}, -z - t\sqrt{a}).$$

LEMMA 4.41. *Let $\omega_1, \omega_2 \in H$ be such that $\omega_2 - \omega_1 \in \mathbb{Q}$. Then, $f_{\Phi(\omega_2)} = f_{\Phi(\omega_1)}$.*

Thus, it is clear that the assignment of binary quadratic forms to the elements of a quaternion algebra is not injective. The above lemma forces us to consider only binary forms attached to pure quaternions. In Proposition 4.44, we will prove that we actually obtain an injective mapping if we restrict ourselves to binary forms attached to pure quaternions.

NOTATION 4.42. We denote by $\mathcal{H}(a,b)$ the set of binary quadratic forms $f_{\Phi(\omega)}$ attached to the elements $\omega \in H$. By the above lemma, it coincides with the set of binary forms associated to the pure quaternions; that is,

$$\mathcal{H}(a,b) = \{f_{\Phi(\omega)} : \omega \in H_0\}.$$

Let $\mathcal{O}$ be a quaternion order in H. We denote by $\mathcal{H}(\mathcal{O})$ the subset of $\mathcal{H}(a,b)$ that contains all the binary quadratic forms associated to elements in the order $\mathcal{O}$. By the above lemma, it coincides with the set of binary forms associated to the elements $\omega \in \mathcal{O} \cap H_0$; that is,

$$\mathcal{H}(\mathcal{O}) = \{f_{\Phi(\omega)} : \omega \in \mathcal{O} \cap H_0\}.$$

REMARK 4.43. If $H = \mathrm{M}(2,\mathbb{Q})$, take $\Phi = \mathrm{id}$. For an order $\mathcal{O} \subseteq \mathrm{M}(2,\mathbb{Q})$, we can also define the set of binary quadratic forms attached to $\mathcal{O}$ as $\mathcal{H}(\mathcal{O}) = \{f_\gamma : \gamma \in \mathcal{O}, \mathrm{tr}(\gamma) = 0\}$.

PROPOSITION 4.44. *Let $H = \left(\frac{a,b}{\mathbb{Q}}\right)$ be an indefinite quaternion algebra. Then:*

(i) $\mathcal{H}(a,b) = \{(b(\lambda_2 + \lambda_3\sqrt{a}), \lambda_1\sqrt{a}, -\lambda_2 + \lambda_3\sqrt{a}) : \lambda_1, \lambda_2, \lambda_3 \in \mathbb{Q}\}$
$= \{(b\beta', \alpha, -\beta) : \alpha, \beta \in \mathbb{Q}(\sqrt{a}), \mathrm{tr}(\alpha) = 0\}.$

(ii) *The mapping $f_\Phi\colon H_0 \to \mathcal{H}(a,b)$, given by $f_\Phi(\omega) := f_{\Phi(\omega)}$, is bijective.*

PROOF. For a pure quaternion $\omega = yi + zj + tij \in H_0$, we have the binary quadratic form

$$f_{\Phi(\omega)} = (b(z - t\sqrt{a}), -2y\sqrt{a}, -z - t\sqrt{a}).$$

For $y, z, t \in \mathbb{Q}$, with $\lambda_1 = -2y$, $\lambda_2 = z$ and $\lambda_3 = -t$, we obtain the first expression of $\mathcal{H}(a,b)$, given in (i). If we put $\beta = z + t\sqrt{a}$ and $\alpha = -2y\sqrt{a}$, we obtain the second one.

By using the explicit expression of $f_{\Phi(\omega)}$ above, it is clear that f_Φ is injective. To prove its surjectivity, consider $f \in \mathcal{H}(a,b)$, $f = (b(\lambda_2 + \lambda_3\sqrt{a}), \lambda_1\sqrt{a}, -\lambda_2 + \lambda_3\sqrt{a})$. It is clear that by putting $y = -\lambda_1/2$, $z = \lambda_2$ and $t = -\lambda_3$ we determine a unique element $\omega = (-\lambda_1/2)i + \lambda_2 j + \lambda_3 ij \in H_0$ such that $f_{\Phi(\omega)} = f$. □

PROPOSITION 4.45. *Let $\mathcal{O}$ be an order in an indefinite quaternion algebra $H = \left(\frac{a,b}{\mathbb{Q}}\right)$. Let $m_\mathcal{O}$ be the denominator of the order $\mathcal{O}$, cf. Definition 1.42. Then we have the following assertions:*

(i) *The binary quadratic forms of $\mathcal{H}(\mathcal{O})$ have coefficients in $\mathbb{Z}[1/m_\mathcal{O}, \sqrt{a}]$; more precisely $\mathcal{H}(\mathcal{O})$ is contained in:*

$$\left\{\frac{1}{m_\mathcal{O}}(b(\lambda_2 + \lambda_3\sqrt{a}), -2\lambda_1\sqrt{a}, -\lambda_2 + \lambda_3\sqrt{a}) : \lambda_1, \lambda_2, \lambda_3 \in \mathbb{Z}\right\}$$
$$= \left\{\frac{1}{m_\mathcal{O}}(b\beta', 2\alpha, -\beta) : \alpha, \beta \in \mathbb{Z}[\sqrt{a}], \operatorname{tr}(\alpha) = 0\right\}.$$

(ii) *The bijective mapping f_Φ restricts to a bijective mapping between the sets $\mathcal{O} \cap H_0$ and $\mathcal{H}(\mathcal{O})$.*

PROOF. Let $\omega \in \mathcal{O} \cap H_0$. Using the definition of $m_\mathcal{O}$, we have that $\omega = yi + zj + tij$, for some $y, z, t \in \mathbb{Z}[\frac{1}{m_\mathcal{O}}]$. Then, applying the above proposition we obtain (i).

To prove (ii), use the definition of $\mathcal{H}(\mathcal{O})$ and the fact that f_Φ is a bijective mapping. □

The set of binary quadratic forms $\mathcal{H}(\mathcal{O})$ does not depend on the choice of basis of the order $\mathcal{O}$. But by fixing a normalized basis $\mathcal{B}$ of the order, we can obtain a generic expression of the binary quadratic forms in the set $\mathcal{H}(\mathcal{O})$, in terms of the coordinates of the elements. Thus, for an element $\omega = (x, y, z, t)_\mathcal{B}$, $f_{\Phi(\omega)}$ is expressed in terms of x, y, z, t. When we restrict ourselves to an element $\omega \in \mathcal{O} \cap H_0$, the coefficients of the binary form depend only on three parameters. Namely, if $\mathcal{O}$ is an even order, the binary form can be directly expressed in terms of y, z, t, since $x = 0$; if $\mathcal{O}$ is an odd order, we substitute the relation $t = -2x$ and obtain also an expression in terms of three parameters. We denote by $f_\mathcal{O}$ the generic binary form obtained in this way; it depends on three parameters, and it satisfies $f_\mathcal{O}(\omega) = f_{\Phi(\omega)}$.

From the expression of the binary form associated to a generic element of H_0 (or of $\mathcal{O} \cap H_0$), we can recover the ternary normic form associated to the algebra H (or to the order $\mathcal{O}$), as shown in the following proposition.

PROPOSITION 4.46. *Let $H = \left(\frac{a,b}{\mathbb{Q}}\right)$ an indefinite quaternion algebra and $\mathcal{O}$ an order in H, with a fixed normalized basis. Then:*

(i) *For every $\omega \in H_0$, $n_{H,3}(\omega) = \det_1(f_{\Phi(\omega)})$.*
(ii) *For every $\omega \in \mathcal{O} \cap H_0$, $n_{\mathcal{O},3}(\omega) = \det_1(f_{\Phi(\omega)})$.*

PROOF. For $\omega \in H_0$ we have

$$\Phi(\omega) = \begin{pmatrix} \alpha & \beta \\ b\beta' & \alpha' \end{pmatrix}, \quad \alpha, \beta \in \mathbb{Q}(\sqrt{a}), \quad \operatorname{tr}(\omega) = \alpha + \alpha' = 0.$$

Note that $(\alpha' - \alpha)/2 = -\alpha = \alpha'$. Thus, $f_{\Phi(\omega)} = (b\beta', -2\alpha, -\beta)$. Now it is clear that $\det_1(f_{\Phi(\omega)}) = -b\beta\beta' - \alpha^2 = -b\beta\beta' + \alpha\alpha' = \det(\Phi(\omega)) = n(\omega) = n_{H,3}(\omega)$.

Let $\omega \in \mathcal{O} \cap H_0$. By expressing ω with respect to the fixed normalized basis of $\mathcal{O}$, we have $n_{\mathcal{O},3}(\omega) = n(\omega)$. Now, since $\omega \in H_0$, we have $n(\omega) = \det_1(f_{\Phi(\omega)})$, which proves (ii). □

COROLLARY 4.47. *Let $H = \left(\frac{a,b}{\mathbb{Q}}\right)$ be an indefinite quaternion algebra and $\mathcal{O}$ an order in H. Fix $\delta \in \mathbb{Z}$. The bijective mapping f_Φ yields the following bijective mappings between sets of quaternions and sets of binary quadratic forms:*

(i) $\{\omega \in H_0 : n(\omega) = \delta\} \leftrightarrow \{f \in \mathcal{H}(a,b) : \det_1(f) = \delta\}$.
(ii) $\{\omega \in \mathcal{O} \cap H_0 : n(\omega) = \delta\} \leftrightarrow \{f \in \mathcal{H}(\mathcal{O}) : \det_1(f) = \delta\}$.

LEMMA 4.48. *Let $\mathcal{O}$ be an order in an indefinite quaternion $\mathbb{Q}$-algebra H. Fix a maximal order $\mathcal{O}(D,1)$ in H. Then, there exists $\sigma \in H^*$ such that $\mathcal{H}(\sigma^{-1}\mathcal{O}\sigma) \subseteq \mathcal{H}(\mathcal{O}(D,1))$.*

PROPOSITION 4.49. *Let $\mathcal{O}$ be a quaternion order in H. Up to conjugation, we have the following inclusions.*

(i) *If $H = \mathrm{M}(2,\mathbb{Q})$ or $H = \left(\frac{1,-1}{\mathbb{Q}}\right)$, then*

$$\mathcal{H}(\mathcal{O}) \subseteq \mathcal{H}(\mathcal{O}_0(1,1)) = \mathcal{H}(\mathcal{O}_M(1,1)) = \{(a, 2b, c) : a,b,c \in \mathbb{Z}\}.$$

(ii) *If $H = H_{\mathrm{A}}(p)$, then*

$$\mathcal{H}(\mathcal{O}) \subseteq \mathcal{H}(\mathcal{O}_{\mathrm{A}}(2p,1)) = \{(a + b\sqrt{p}, 2c\sqrt{p}, a - b\sqrt{p}) : a,b,c \in \mathbb{Z}\}.$$

(iii) *If $H = H_{\mathrm{B}}(p,q)$, then $\mathcal{H}(\mathcal{O}) \subseteq \mathcal{H}(\mathcal{O}_{\mathrm{B}}(pq,1))$, where $\mathcal{H}(\mathcal{O}_{\mathrm{B}}(pq,1))$ is equal to $\{\frac{1}{2}(q(a+b\sqrt{p}), 2c\sqrt{p}, -a+b\sqrt{p}) : a,b,c \in \mathbb{Z}, 2 \mid a, 2 \mid (b-c)\}$.*

PROOF. By the above lemma it is only necessary, in each case, to compute the set of binary quadratic forms attached to a maximal order.

If $H = \left(\frac{1,-1}{\mathbb{Q}}\right)$, take the order $\mathcal{O}_M(1,1) = \mathbb{Z}[1, (j+ij)/2, (-j+ij)/2, (1-i)/2]$. Let us compute $\mathcal{H}(\mathcal{O}_M(1,1))$. Let $\omega \in \mathcal{O}_M(1,1) \cap H_0$, $\omega = x + y(j+ij)/2 + z(-j+ij)/2 + t(1-i)/2$, with $x,y,z,t \in \mathbb{Z}$ and $t = -2x$. Then, we have that

$$\Phi(\omega) = \begin{pmatrix} x & y \\ z & -x \end{pmatrix}, \qquad A_1(f_{\Phi(\omega)}) = \begin{pmatrix} z & -x \\ -x & -y \end{pmatrix}.$$

Hence, $\mathcal{H}(\mathcal{O}_M(1,1)) = \{(z, -2x, -y) : x,y,z \in \mathbb{Z}\} = \{(a, 2b, c) : a,b,c \in \mathbb{Z}\}$.

In $\mathrm{M}(2,\mathbb{Q})$, consider the order $\mathcal{O}_0(1,1) := \mathrm{M}(2,\mathbb{Z})$. Note that $\Phi(\mathcal{O}_M(1,1) \cap H_0) = \{\gamma \in \mathcal{O}_0(1,1) : \mathrm{tr}(\gamma) = 0\}$. Thus, $\mathcal{H}(\mathcal{O}_0(1,1)) = \mathcal{H}(\mathcal{O}_M(1,1))$.

To prove (ii), consider $H_{\mathrm{A}}(p) = \left(\frac{p,-1}{\mathbb{Q}}\right)$, where $p \equiv 3 \bmod 4$, and the maximal order $\mathcal{O}_{\mathrm{A}}(2p,1) = \mathbb{Z}[1, i, j, (1+i+j+ij)/2]$. Let $\omega \in \mathcal{O}_{\mathrm{A}}(2p,1) \cap H_0$, $\omega = x + yi + zj + t(1+i+j+ij)/2$, with $x,y,z,t \in \mathbb{Z}$ and $t = -2x$. We have that

$$\Phi(\omega) = \begin{pmatrix} (y-x)\sqrt{p} & (z-x) - x\sqrt{p} \\ -(z-x) - x\sqrt{p} & -(y-x)\sqrt{p} \end{pmatrix}.$$

Hence, $f_{\Phi(\omega)} = (-(z-x) - x\sqrt{p}, 2(x-y)\sqrt{p}, -(z-x) + x\sqrt{p})$. Thus,

$$\mathcal{H}(\mathcal{O}_{\mathrm{A}}(2p,1)) = \{(a + b\sqrt{p},\ 2c\sqrt{p},\ a - b\sqrt{p}) : a,b,c \in \mathbb{Z}\}.$$

In the same way the analogous result is proved for the algebra $H_{\mathrm{B}}(p,q)$. In this case, take the maximal order $\mathcal{O}_{\mathrm{B}}(pq,1) = \mathbb{Z}[1, i, (1+j)/2, (i+ij)/2]$. For $\omega = x + yi + z(1+j)/2 + t(i+ij)/2 \in \mathcal{O}_{\mathrm{B}}(pq,1) \cap H_0$, we have $z = -2x$. Hence,

$$f_{\Phi(\omega)} = \tfrac{1}{2}(q(-2x - t\sqrt{p}), -2(2y+t)\sqrt{p}, 2x - t\sqrt{p}).$$

Take $a = -2x$, $b = -t$, $c = -(2y+t)$ and note that $a, b, c \in \mathbb{Z}$ satisfy $2 \mid a$ and $2 \mid (b-c)$. Thus, we obtain the expression for $\mathcal{H}(\mathcal{O}_{\mathrm{B}}(pq,1))$ given in the statement. □

EXAMPLE 4.50. Consider the order $\mathcal{O}' = \mathbb{Z}[1, 2i, 2j, i+j+ij]$ in the algebra $H_{\mathrm{A}}(p) = \left(\frac{p,-1}{\mathbb{Q}}\right)$. We have

$$\mathcal{H}(\mathcal{O}') = \{(-(2z+t) + t\sqrt{p}, -2(2y+t)\sqrt{p}, -(2z+t) - t\sqrt{p}) : x, y, z \in \mathbb{Z}\}.$$

Note that $\mathcal{O}' \subseteq \mathcal{O}_{\mathrm{A}}(2p, 1)$ and, indeed, $\mathcal{H}(\mathcal{O}') \subseteq \mathcal{H}(\mathcal{O}_{\mathrm{A}}(2p,1))$, as in the above proposition.

Tables A.25–A.28 in Appendix A contain the generic expression for the binary forms in $\mathcal{H}(\mathcal{O})$, for Eichler orders $\mathcal{O}$ in the small ramified quaternion algebras $H_{\mathrm{A}}(3)$, $H_{\mathrm{B}}(2,5)$, $H_{\mathrm{A}}(7)$ and $H_{\mathrm{B}}(3,5)$.

4.5. Classification of binary forms associated to quaternion orders

In this section, we relate the sets of embeddings of quadratic orders into quaternion orders to certain sets of binary quadratic forms. First, we state the corresponding bijective mappings, and we note the relation with the set of representations by ternary forms in previous sections. This allows us to consider an equivalence relation and to introduce a new concept of primitivity for those binary forms.

NOTATION 4.51. Let H be a quaternion $\mathbb{Q}$-algebra. Given an order $\mathcal{O} \subseteq H$, $m_{\mathcal{O}}$ denotes the denominator of the order, cf. Definition 1.42. Recall the notation used for the set of binary quadratic forms attached to a quaternion order $\mathcal{O}$,

$$\mathcal{H}(\mathcal{O}) := \{f_{\Phi(\omega)} : \omega \in \mathcal{O} \cap H_0\}.$$

Given a quaternion order $\mathcal{O}$ and a quadratic order Λ, put

$$\mathcal{H}(\mathcal{O}, \Lambda) := \{f \in \mathcal{H}(\mathcal{O}) : \det_1(f) = -D_\Lambda\}.$$

REMARK 4.52. Note that $\mathcal{H}(\mathcal{O}, \Lambda)$ is a set of definite or indefinite binary forms according to whether Λ is an imaginary quadratic or real quadratic order, respectively.

THEOREM 4.53. *Let $\mathcal{O}$ be a quaternion order and $\Lambda(d,m)$ be a quadratic order. We have bijective mappings:*

$$\mathfrak{f}\colon \mathcal{E}(\mathcal{O}, \Lambda) \to \mathcal{H}(\mathbb{Z} + 2\mathcal{O}, \Lambda)$$
$$\mu\colon \mathcal{R}(n_{\mathbb{Z}+2\mathcal{O},3}, -D_\Lambda; \mathbb{Z}) \to \mathcal{H}(\mathbb{Z} + 2\mathcal{O}, \Lambda),$$

given by $\mathfrak{f}(\varphi) = f_{\Phi(\varphi(\sqrt{D_\Lambda}))}$ and $\mu := \mathfrak{f} \circ \rho^{-1}$, where ρ is the bijective mapping given in Theorem 4.26.

PROOF. Put $\mathcal{O}' := \mathbb{Z} + 2\mathcal{O}$. By Theorem 4.26, to each embedding $\varphi \in \mathcal{E}(\mathcal{O}, \Lambda)$ there corresponds an element $e \in \mathcal{O}'$ with $n(e) = -D_\Lambda$ and $\mathrm{tr}(e) = 0$,

$$e := \begin{cases} \varphi(2m\sqrt{d}) & \text{if } d \equiv 2, 3 \bmod 4, \\ \varphi(m\sqrt{d}) & \text{if } d \equiv 1 \bmod 4. \end{cases}$$

We can express e as $e = \varphi(\sqrt{D_\Lambda})$. By Proposition 4.45, to each element in $\mathcal{O}' \cap H_0$ there corresponds a binary quadratic form in $\mathcal{H}(\mathcal{O}', \Lambda)$. □

DEFINITION 4.54. Consider the subset of binary forms in $\mathcal{H}(\mathbb{Z}+2\mathcal{O},\Lambda)$ defined by

$$\mathcal{H}^*(\mathbb{Z}+2\mathcal{O},\Lambda):=\mathfrak{f}(\mathcal{E}^*(\mathcal{O},\Lambda)).$$

We call $(\mathcal{O},\Lambda)$-primitive binary forms the binary forms belonging to $\mathcal{H}^*(\mathbb{Z}+2\mathcal{O},\Lambda)$.

COROLLARY 4.55. *The previous bijective mappings* $\mathfrak{f}$ *and* μ *restrict to bijective mappings*:

$$\mathfrak{f}^*\colon \mathcal{E}^*(\mathcal{O},\Lambda)\to\mathcal{H}^*(\mathbb{Z}+2\mathcal{O},\Lambda),$$
$$\mu^*\colon \mathcal{R}^*(n_{\mathbb{Z}+2\mathcal{O},3},-D_\Lambda;\mathbb{Z})\to\mathcal{H}^*(\mathbb{Z}+2\mathcal{O},\Lambda).$$

DEFINITION 4.56. Let $\mathcal{O}\subseteq H$ be a quaternion order and $\Phi: H\hookrightarrow \mathrm{M}(2,\mathbb{R})$ a fixed embedding. Consider the mapping:

$$\widetilde{\Phi}\colon \mathrm{Nor}(\mathcal{O})\to\{P\in\mathrm{GL}(2,\mathbb{R}) : \det P=\pm1\}$$
$$\sigma\mapsto\widetilde{\Phi}(\sigma):=\frac{\Phi(\sigma)}{|\,\mathrm{n}(\sigma)|^{1/2}}.$$

Let G be a group, $G\subseteq\mathrm{Nor}(\mathcal{O})$. We call the group $\Gamma_G:=\widetilde{\Phi}(G)$, the image of G by $\widetilde{\Phi}$, group of quaternion transformations defined by G. In particular, if $G\subseteq\mathcal{O}^*$, then $\widetilde{\Phi}(G)=\Phi(G)$. If $\mathcal{O}$ is an Eichler order of level N in an indefinite quaternion $\mathbb{Q}$-algebra of discriminant D, for $G=\mathcal{O}_+^*$ we obtain $\Gamma_G=\Gamma(D,N)$, the group of quaternion transformations defined in Section 2.3.

THEOREM 4.57. *Let* $\mathcal{O}$ *be a quaternion order and* $\Lambda(d,m)$ *a quadratic order. Let* G *be a subgroup of* $\mathrm{Nor}(\mathcal{O})$. *Then, the bijective mappings* $\mathfrak{f}$ *and* μ *defined in Theorem* 4.53 *induce the bijective mappings between the sets of classes*

$$\overline{\mathfrak{f}^*}\colon \mathcal{E}^*(\mathcal{O},\Lambda)/G\to\mathcal{H}^*(\mathbb{Z}+2\mathcal{O},\Lambda)/\Gamma_G$$
$$\overline{\mu^*}\colon \mathcal{R}^*(n_{\mathbb{Z}+2\mathcal{O},3},-D_\Lambda;\mathbb{Z})/\kappa(G)\to\mathcal{H}^*(\mathbb{Z}+2\mathcal{O},\Lambda)/\Gamma_G.$$

PROOF. We must check that $\varphi,\varphi'\in\mathcal{E}^*(\mathcal{O},\Lambda)$ are G-equivalent if and only if $\mathfrak{f}^*(\varphi),\mathfrak{f}^*(\varphi')\in\mathcal{H}^*(\mathbb{Z}+2\mathcal{O},\Lambda)$ are Γ_G-equivalent, since $\mathfrak{f}^*$ is already a bijective mapping. Put $\alpha=\sqrt{D_\Lambda}$.

$$\mathfrak{f}^*(\varphi^\sigma)=f_{\Phi(\varphi^\sigma(\alpha))}=f_{\Phi(\sigma^{-1}\varphi(\alpha)\sigma)}=f_{\Phi(\sigma)^{-1}\Phi(\varphi(\alpha))\Phi(\sigma)}.$$

Applying Lemma 2.12, we see that the above expression is equal to

$$\frac{1}{\det(\Phi(\sigma))}\Phi(\sigma)^t f_{\Phi(\varphi(\alpha))}\Phi(\sigma)=\frac{\Phi(\sigma)^t}{\sqrt{|\,\mathrm{n}(\sigma)|}}\mathfrak{f}_{\Phi(\varphi(\alpha))}\frac{\Phi(\alpha)}{\sqrt{|\,\mathrm{n}(\sigma)|}}.$$

Put $P:=\widetilde{\Phi}(\sigma)$. Then we have $\mathfrak{f}^*(\varphi')=P^t\mathfrak{f}^*(\varphi)P$. Since the mapping $\widetilde{\Phi}$ is surjective over Γ_G, we deduce the bijective mapping $\mathfrak{f}^*$. The result for μ^* follows by composition. □

NOTATION 4.58. Let $\mathcal{O}\subseteq H$ be a quaternion order and $\Lambda\subseteq F$ a quadratic order. Fix a group $G\subseteq\mathrm{Nor}(\mathcal{O})$. Denote by $h(\mathcal{O},\Lambda;G)$, if it is finite, the class number of the set of $(\mathcal{O},\Lambda)$-primitive binary forms: $\mathcal{H}^*(\mathbb{Z}+2\mathcal{O},\Lambda)/\Gamma_G$.

In particular, if $\mathcal{O} = \mathcal{O}(D,N)$, $\Lambda = \Lambda(d,m)$ and $G = \mathcal{O}(D,N)^*$, we also use the following notations:

$$\begin{aligned}\mathcal{H}(D,N,d,m) &:= \mathcal{H}\big(\mathbb{Z}+2\mathcal{O}(D,N),\Lambda(d,m)\big)\\ \mathrm{H}(D,N,d,m) &:= \mathcal{H}^*\big(\mathbb{Z}+2\mathcal{O}(D,N),\Lambda(d,m)\big)/\Gamma_{\mathcal{O}(D,N)^*},\\ \mathrm{h}(D,N,d,m) &:= \sharp\,\mathrm{H}(D,N,d,m).\end{aligned}$$

COROLLARY 4.59. *Let $\mathcal{O} \subseteq H$ be a quaternion order, $\Lambda(d,m)$ a quadratic order and a group $G \subseteq \mathrm{Nor}(\mathcal{O})$. Then,*

$$h(\mathcal{O},\Lambda;G) = \nu(\mathcal{O},\Lambda;G).$$

Next, consider Eichler orders in nonramified, small ramified of type A and of type B quaternion algebras. For these Eichler orders, we can make explicit the above sets of binary forms and the bijective mappings. Before formulating the results, first we state the following general lemma.

LEMMA 4.60. *Let $\mathcal{O}$ be an order in an indefinite quaternion algebra $H = \left(\frac{a,b}{\mathbb{Q}}\right)$. Assume $a > 0$ and fix the embedding $\Phi\colon H \hookrightarrow M\big(2,\mathbb{Q}(\sqrt{a})\big)$. Then,*

$$\mathcal{H}(\mathbb{Z}+2\mathcal{O}) \subseteq \left\{\frac{2}{m_{\mathcal{O}}}\big(b(\lambda_2-\lambda_3\sqrt{a}),-2\lambda_1\sqrt{a},-(\lambda_2+\lambda_3\sqrt{a})\big) : \lambda_1,\lambda_2,\lambda_3 \in \mathbb{Z}\right\}.$$

PROOF. For a pure quaternion $e \in (\mathbb{Z}+2\mathcal{O}) \cap H_0$, we have

$$\Phi(e) = \frac{2}{m_{\mathcal{O}}}\begin{pmatrix}\alpha & \beta\\ b\beta' & \alpha'\end{pmatrix} = \frac{2}{m_{\mathcal{O}}}\begin{pmatrix}\lambda_1\sqrt{a} & \lambda_2+\lambda_3\sqrt{a}\\ b(\lambda_2-\lambda_3\sqrt{a}) & -\lambda_1\sqrt{a}\end{pmatrix},$$

where $\alpha,\beta \in \mathbb{Z}[\sqrt{a}]$ and $\mathrm{tr}(\alpha) = 0$, and $\lambda_1,\lambda_2,\lambda_3 \in \mathbb{Z}$. The corresponding binary quadratic form is

$$f_{\Phi(e)} = \frac{2}{m_{\mathcal{O}}}(b\beta',-2\alpha,-\beta) = \frac{2}{m_{\mathcal{O}}}\big(b(\lambda_2-\lambda_3\sqrt{a}),-2\lambda_1\sqrt{a},-\lambda_2-\lambda_3\sqrt{a}\big). \qquad \square$$

THEOREM 4.61. *Consider the Eichler order $\mathcal{O} = \mathcal{O}_0(1,N)$ of level N in the nonramified algebra $H = \mathrm{M}(2,\mathbb{Q})$. Let $\Lambda = \Lambda(d,m)$ be a quadratic order. Then:*

(i) $\mathcal{H}(\mathbb{Z}+2\mathcal{O}_0(1,N),\Lambda) = \{f = 2(Na,b,c) : a,b,c \in \mathbb{Z}, \det_1(f) = -D_\Lambda\} \leftrightarrow \{f = (Na,b,c) : a,b,c \in \mathbb{Z}, \det_2(f) = -D_\Lambda\}$.

(ii) *The bijective mapping μ_0 in Theorem 4.53 is given by*

$$\begin{aligned}\mu_0\colon \mathcal{R}(n_{\mathbb{Z}+2\mathcal{O}_0(1,N),3},-D_\Lambda;\mathbb{Z}) &\to \mathcal{H}(\mathbb{Z}+2\mathcal{O}_0(1,N),\Lambda)\\ (x,y,z) &\mapsto 2(Ny,-x,-z).\end{aligned}$$

(iii) *The set $\mathcal{H}^*(\mathbb{Z}+2\mathcal{O}_0(1,N),\Lambda)$ of $(\mathcal{O},\Lambda)$-primitive binary forms is equal to*

$$\{2(Na,b,c) \in \mathcal{H}(\mathbb{Z}+2\mathcal{O}_0(1,N),\Lambda) : \gcd(a,b,c) = 1\}.$$

PROOF. We will compute $\mathcal{H}\big(\mathbb{Z}+2\mathcal{O}_0(1,N)\big)$. We have that

$$\big(\mathbb{Z}+2\mathcal{O}_0(1,N)\big) \cap H_0 = \left\{\begin{pmatrix}x & 2y\\ 2Nz & -x\end{pmatrix} : x,y,z \in \mathbb{Z}\right\}.$$

For every $\omega \in \big(\mathbb{Z}+2\mathcal{O}_0(1,N)\big) \cap H_0$, we have that $f_{\Phi(\omega)} = (2Nz,-2x,-2y)$. Thus,

$$\begin{aligned}\mathcal{H}\big(\mathbb{Z}+2\mathcal{O}_0(1,N)\big) &= \{2(Na,b,c) : a,b,c \in \mathbb{Z}\},\\ \mathcal{H}(\mathbb{Z}+2\mathcal{O}_0(1,N),\Lambda) &= \{g = 2(Na,b,c) : a,b,c \in \mathbb{Z}, \det_1(g) = -D_\Lambda\}.\end{aligned}$$

Now, if we consider $f = \frac{1}{2}g$, we have that $\det_2(f) = \det_1(g) = -D_\Lambda$. Thus,

$$\mathcal{H}(1, N, d, m) \simeq \{f = (Na, b, c) : a, b, c \in \mathbb{Z}, \det_2(f) = -D_\Lambda\}.$$

For (ii) let $(x, y, z) \in \mathcal{R}\left(n_{\mathbb{Z}+2\mathcal{O}_0(1,N),3}, -D_\Lambda; \mathbb{Z}\right)$. By the results in the previous section, this representation determines the element $e = \left(\begin{smallmatrix} x & 2y \\ 2Nz & -x \end{smallmatrix}\right) \in \mathbb{Z}+2\mathcal{O}_0(1, N) \cap H_0$, of norm $\mathrm{n}(e) = -D_\Lambda$. The binary form attached to this element is $f_{\Phi(e)} = 2(Ny, -x, -z)$.

To prove (iii), assume $(x, y, z) \in \mathcal{R}^*(n_{\mathbb{Z}+2\mathcal{O}_0(1,N),3}, -D_\Lambda; \mathbb{Z})$; thus $\gcd(x, y, z) = 1$. Then, it is clear that the binary form given by the bijective mapping in (ii), $f = 2(Ny, -x, -z) = 2(Na, b, c)$, satisfies $\gcd(a, b, c) = 1$. □

THEOREM 4.62. *In the quaternion algebra* $H_A(p)$, *consider the Eichler order* $\mathcal{O}_A(2p, N) = \mathbb{Z}[1, i, Nj, (1 + i + j + ij)/2]$ *of level* N, $N \mid (p-1)/2$ *square-free. Let* $\Lambda = \Lambda(d, m)$ *be any quadratic order.*

(i) *The set* $\mathcal{H}(\mathbb{Z} + 2\mathcal{O}_A(2p, N), \Lambda)$ *of binary forms is equal to*

$$\{f = (a + b\sqrt{p}, 2c\sqrt{p}, a - b\sqrt{p}) : \quad a, b, c \in \mathbb{Z}, a \equiv b \equiv c \bmod 2, N \mid (a + b), \det_1(f) = -D_\Lambda\}.$$

(ii) *Let* $\mu_A : \mathcal{R}(n_{\mathbb{Z}+2\mathcal{O}_A(2p,N),3}, -D_\Lambda; \mathbb{Z}) \to \mathcal{H}(\mathbb{Z} + 2\mathcal{O}_A(2p, N), \Lambda)$ *be the bijective mapping of Theorem* 4.53. *Then,* μ_A *is given by*

$$\mu_A(x, y, z) = (-(2Ny + z) + z\sqrt{p}, -2(2x + z)\sqrt{p}, -(2Ny + z) - z\sqrt{p}).$$

(iii) *The set* $\mathcal{H}^*(\mathbb{Z} + 2\mathcal{O}_A(2p, N), \Lambda)$ *of* $(\mathcal{O}_A(2p, N), \Lambda)$*-primitive binary forms is equal to*

$$\left\{(a + b\sqrt{p}, 2c\sqrt{p}, a - b\sqrt{p}) \in \mathcal{H}(\mathbb{Z} + 2\mathcal{O}_A(2p, N), \Lambda) : \gcd\left(\frac{c + b}{2}, \frac{a + b}{2N}, b\right) = 1\right\}.$$

PROOF. First, we compute $\mathcal{H}\big(\mathbb{Z} + 2\mathcal{O}_A(2p, N)\big)$. It is easy to check that the set $\Phi\big(\mathbb{Z} + 2\mathcal{O}_A(2p, N)\big) \cap H_0$ is equal to

$$\left\{\begin{pmatrix} (2y + t)\sqrt{p} & (2Nz + t) + t\sqrt{p} \\ -(2Nz + t) + t\sqrt{p} & -(2y + t)\sqrt{p} \end{pmatrix} : y, z, t \in \mathbb{Z}\right\}.$$

Therefore, the set of binary forms $\mathcal{H}\big(\mathbb{Z} + 2\mathcal{O}_A(2p, N)\big)$ is equal to

$$\{(-(2Nz + t) + t\sqrt{p}, -2(2y + t)\sqrt{p}, -(2Nz + t) - t\sqrt{p}) : y, z, t \in \mathbb{Z}\}.$$

If we put $a = -(2Nz + t)$, $b = t$ and $c = -(2y + t)$, we obtain the divisibility conditions given in (i): namely $a \equiv b \equiv c \bmod 2$ and $N \mid a + b$.

To prove (ii) consider $(x, y, z) \in \mathcal{R}(n_{\mathbb{Z}+2\mathcal{O}_A(2p,N),3}, -D_\Lambda; \mathbb{Z})$. By results from the previous section, this representation is in correspondence with the element $e \in \big(\mathbb{Z} + 2\mathcal{O}_A(2p, N)\big) \cap H_0 \subseteq \mathcal{O}$, with $\mathrm{n}(e) = -D_\Lambda$. Indeed e is given by $e = (0, x, y, z)_{\mathcal{B}'} = (-z, 2x, 2y, 2z)_{\mathcal{B}} = -z + (2x + z)i + (2Ny + z)j + zij$, where $\mathcal{B}$ and $\mathcal{B}'$ are the fixed normalized basis in the orders $\mathcal{O}_A(2p, N)$ and $\mathbb{Z} + 2\mathcal{O}_A(2p, N)$, respectively. Thus the binary form attached to this element is

$$f_{\Phi(e)} = (-(2Ny + z) + z\sqrt{p}, -2(2x + z)\sqrt{p}, -(2Ny + z) - z\sqrt{p}).$$

Now, for (iii), assume that $(x, y, z) \in \mathcal{R}^*(n_{\mathbb{Z}+2\mathcal{O}_A(2p,N),3}, -D_\Lambda; \mathbb{Z})$; that is, assume that we have $\gcd(x, y, z) = 1$. Apply the change of variables $a = -(2Ny + z)$, $b = z$ and $c = -(2x + z)$ to the binary form given by the bijective mapping in (ii);

that is, $x = -(c+b)/2$, $y = -(a+b)/(2N)$ and $z = b$. Thus, a primitive representation is in correspondence with a binary form in $\mathcal{H}(\mathbb{Z} + 2\mathcal{O}_{\mathrm{A}}(2p, N), \Lambda)$ satisfying the condition $\gcd\big((c+b)/2, (a+b/(2N), b\big) = 1$. □

In the set $\mathcal{H}\big(\mathbb{Z}+2\mathcal{O}_{\mathrm{A}}(2p, N)\big)$ of binary forms, we have the condition that a, b, c are of the same parity. Depending on the parity of the value D_Λ, we deduce the following corollary.

COROLLARY 4.63. *Under the same assumptions as in the previous theorem, we have*

(i) $D_\Lambda \equiv 0 \bmod 4$ *if and only if* $\mathcal{H}(\mathbb{Z} + 2\mathcal{O}_{\mathrm{A}}(2p, N), \Lambda)$ *is equal to*

$$\{f = (a + b\sqrt{p}, 2c\sqrt{p}, a - b\sqrt{p}) : a, b, c \text{ even integers}, N \mid (a+b), \det\nolimits_1(f) = -D_\Lambda\}.$$

(ii) $D_\Lambda \equiv 1 \bmod 4$ *if and only if* $\mathcal{H}(\mathbb{Z} + 2\mathcal{O}_{\mathrm{A}}(2p, N), \Lambda)$ *is equal to*

$$\{f = (a + b\sqrt{p}, 2c\sqrt{p}, a - b\sqrt{p}) : a, b, c \text{ odd integers}, N \mid (a+b), \det\nolimits_1(f) = -D_\Lambda\}.$$

THEOREM 4.64. *In the quaternion algebra $H_{\mathrm{B}}(p, q)$, consider the Eichler order $\mathcal{O}_{\mathrm{B}}(pq, N) = \mathbb{Z}[1, Ni, (1+j)/2, (i+ij)/2]$ of level N square-free, $\gcd(N, p) = 1$ and $N \mid (q-1)/4$. Let Λ be any quadratic order.*

(i) *The set $\mathcal{H}(\mathbb{Z} + 2\mathcal{O}_{\mathrm{B}}(2p, N), \Lambda)$ of binary forms is equal to*

$$\{f = (q(a + b\sqrt{p}), 2c\sqrt{p}, -a + b\sqrt{p}) : a, b, c \in \mathbb{Z}, 2N \mid (c-b), \det\nolimits_1(f) = -D_\Lambda\}.$$

(ii) *Let $\mu_{\mathrm{B}} : \mathcal{R}(n_{\mathbb{Z}+2\mathcal{O}_{\mathrm{B}}(pq,N),3}, -D_\Lambda; \mathbb{Z}) \to \mathcal{H}(\mathbb{Z}+2\mathcal{O}_{\mathrm{B}}(pq, N), \Lambda)$ the bijective mapping of Theorem 4.53. Then, μ_{B} is given by*

$$\mu_{\mathrm{B}}(x, y, z) = (q(z - y\sqrt{p}) - 2(2Nx + y)\sqrt{p}, -z - y\sqrt{p}).$$

(iii) *The set $\mathcal{H}^*(\mathbb{Z} + 2\mathcal{O}_{\mathrm{B}}(pq, N), \Lambda)$ of $(\mathcal{O}_{\mathrm{B}}(pq, N), \Lambda)$-primitive binary forms is equal to*

$$\left\{(q(a + b\sqrt{p}), 2c\sqrt{p}, -a + b\sqrt{p}) \in \mathcal{H}(\mathbb{Z} + 2\mathcal{O}_{\mathrm{B}}(pq, N), \Lambda) : a, b, c \in \mathbb{Z}, \gcd\left(a, b, \frac{c-b}{2N}\right) = 1\right\}.$$

PROOF. As in the previous result, we compute $\mathcal{H}\big(\mathbb{Z} + 2\mathcal{O}_{\mathrm{B}}(pq, N)\big)$. We have that $\big(\mathbb{Z} + 2\mathcal{O}_{\mathrm{B}}(pq, N)\big) \cap H_0$ is equal to

$$\left\{\begin{pmatrix} (2Ny + t)\sqrt{p} & z + t\sqrt{p} \\ q(z - t\sqrt{p}) & -(2Ny + t)\sqrt{p} \end{pmatrix} : y, z, t \in \mathbb{Z}\right\}.$$

Thus, the attached set $\mathcal{H}\big(\mathbb{Z} + 2\mathcal{O}_{\mathrm{B}}(pq, N)\big)$ of binary forms is

$$\{(q(z - t\sqrt{p}), -2(2Ny + t)\sqrt{p}, -z + t\sqrt{p}) : y, z, t \in \mathbb{Z}\}.$$

If we put $a = z$, $b = -t$ and $c = -(2Ny + t)$, we have that $a, b, c \in \mathbb{Z}$, with the condition $2N \mid (c - b)$. This proves (i).

Consider $\mathcal{B} = \{1, Ni, (i + ij)/2, (1 + j)/2\}$ a $\mathbb{Z}$-normalized basis of the order $\mathcal{O}_{\mathrm{B}}(pq, N)$; denote by $\mathcal{B}'$ the normalized basis which corresponds to the order $\mathbb{Z} + 2\mathcal{O}_{\mathrm{B}}(pq, N)$. Let $(x, y, z) \in \mathcal{R}(n_{\mathbb{Z}+2\mathcal{O}_{\mathrm{B}}(pq,N),3}, -D_\Lambda; \mathbb{Z})$. This representation is in one to one correspondence with the element $e \in \big(\mathbb{Z} + 2\mathcal{O}_{\mathrm{B}}(pq, N)\big) \cap H_0 \subseteq \mathcal{O}$ with $\mathrm{n}(e) = -D_\Lambda$, $e = (0, x, y, z)_{\mathcal{B}'} = (-z, 2x, 2y, 2z)_{\mathcal{B}} = (2Nx + y)i + zj + yij$. By Proposition 4.45, this element is in one to one correspondence with the binary form

$$f_{\Phi(e)} = (q(z - y\sqrt{p}), -2(2Nx + y)\sqrt{p}, -z - y\sqrt{p}).$$

Assume $(x, y, z) \in \mathcal{R}^*(n_{\mathbb{Z}+2\mathcal{O}_{\mathrm{B}}(pq,N),3}, -D_\Lambda; \mathbb{Z})$; that is, assume $\gcd(x, y, z) = 1$. Take $a = z$, $b = -y$ and $c = -(2Nx+y)$ in the binary form given by the previous bijective mapping. That is, $x = -(c-b)/(2N)$, $y = -b$ and $z = a$. Thus, the condition of primitive representation corresponds to the condition $\gcd\big(a, b, (c-b)/(2N)\big) = 1$, depending on the coefficients of the corresponding binary form. □

Tables A.29–A.32 in Appendix A contain the generic expression for the binary forms in $\mathcal{H}(\mathbb{Z} + 2\mathcal{O})$ for Eichler orders in the small ramified quaternion algebras $H_{\mathrm{A}}(3)$, $H_{\mathrm{B}}(2,5)$, $H_{\mathrm{A}}(7)$ and $H_{\mathrm{B}}(3,5)$.

REMARK 4.65. In Theorems 4.61, 4.62, and 4.64, we explicitly illustrate the sets of binary forms $\mathcal{H}^*(\mathbb{Z} + 2\mathcal{O}(D, N), \Lambda)$ for some families of values of D and N. Note that these binary forms have the coefficients in a fixed quadratic ring from the initial quaternion algebra. Thus, we obtain a notion of primitivity in sets of binary forms over quadratic rings, even if these quadratic rings are not principal ideal domains.

Furthermore, Theorem 4.57 and Corollary 4.59 give a classification of these $\mathcal{O}(D, N)$-primitive binary forms with respect to the action of Γ_G. In particular, for $G = \mathcal{O}_+^*$ we have a classification by the action of the group of quaternion transformations $\Gamma(D, N) \subseteq \mathrm{SL}(2, \mathbb{R})$, and the class numbers $h(D, N, d, m)$ can be computed.

In the following theorem, we present the equalities of the class numbers resulting from the bijective mappings established in this section and the previous ones.

THEOREM 4.66. *Let $\mathcal{O}(D, N)$ be an Eichler order of level N in a quaternion algebra of discriminant D and $\Lambda(d, m) \subseteq \mathbb{Q}(\sqrt{d})$ be the quadratic order with conductor m. Then the following class numbers coincide:*

(i) *$h(D, N, d, m)$, the number of $\Gamma_{\mathcal{O}(D,N)^*}$-classes of $\big(\mathcal{O}(D, N), \Lambda(d, m)\big)$-primitive binary quadratic forms attached to the order $\mathbb{Z} + 2\mathcal{O}(D, N)$ with determinant equal to $-D_{\Lambda(d,m)}$.*

(ii) *$r^*(n_{\mathbb{Z}+2\mathcal{O}(D,N),3}, -D_{\Lambda(d,m)}; \mathcal{O}(D, N)^*)$, the number of $\kappa(\mathcal{O}(D, N)^*)$-classes of primitive representations of $-D_{\Lambda(d,m)}$ by the ternary normic form $n_{\mathbb{Z}+2\mathcal{O}(D,N),3}$.*

(iii) *$\nu(D, N, d, m; \mathcal{O}(D, N)^*)$, the number of $\mathcal{O}(D, N)^*$-classes of optimal embeddings of $\Lambda(d, m)$ into $\mathcal{O}(D, N)$.*

In particular, all of them can be computed using Theorem 4.19.

Appendix A contains tables of class numbers $h(D, N, d, m)$. In Table A.33, we give the values for Eichler orders in nonramified algebras, $D = 1$; note that in the case $N = 1$, we have $h(1, 1, d, m) = h(d, m)$, which is the classical class number of the quadratic order $\Lambda(d, m)$. In Tables A.34–A.37, we are concerned with Eichler orders in small ramified quaternion algebras of discriminant $D = 6$, 10, 14, and 15, respectively.

CHAPTER 5

Hyperbolic Fundamental Domains for Shimura Curves

Let D, N be natural numbers such that $\gcd(D, N) = 1$. Assume that $D \geq 1$ is equal to the product of an even number of different primes. Fix an indefinite quaternion algebra H of discriminant D and consider an Eichler order $\mathcal{O}(D, N)$, of level N. In Chapter 2, we defined the group of quaternion transformations $\Gamma(D, N)$ and the Shimura curve $X(D, N)$, in such a way that the quotient $\Gamma(D, N)\backslash\mathcal{H}$ is in one to one correspondence with a Zariski open set in $X(D, N)(\mathbb{C})$.

In this chapter, we apply the results for embeddings and quadratic forms in Chapter 4 to obtain elements in the group $\Gamma(D, N)$ and points in the curve $X(D, N)$ with the goal of constructing explicit fundamental domains.

In the first section, we review the known results for the modular curves. In the second section, we describe the relations between the group of quaternion transformations $\Gamma(D, N)$ and the embeddings of quadratic orders into the quaternion order $\mathcal{O}(D, N)$. This allows us to give effective results for elliptic transformations and points, especially for the groups $\Gamma(2p, N)$ and $\Gamma(pq, N)$, corresponding to Eichler orders in small ramified algebras of type A or B. In the third section, we note the special role of some hyperbolic transformations and we introduce the concept of principal homothety of $\Gamma(D, N)$, which plays a similar role in the construction of fundamental domains to that of the translation in the modular case. We also construct a fundamental domain for the subgroup of transformations fixing infinity. In the fourth section we introduce principal hyperbolic lines and principal symmetries. The fifth section contains the explicit construction of fundamental domains for the Shimura curves $X(6, 1)$, $X(10, 1)$ and $X(15, 1)$.

Recall that Φ denotes the embedding of H into $M\big(2, \mathbb{Q}(\sqrt{a})\big)$ given in Proposition 1.19. If F is any quadratic field, for every embedding $\varphi \in \mathcal{E}(H, F)$, put $\phi := \Phi \circ \varphi$ for the corresponding embedding of F in $M(2, \mathbb{R})$.

5.1. Fundamental domains of modular curves

In the case $D = 1$ of nonramified quaternion algebras, the curve $X(1, N)$ is the modular curve usually denoted by $X_0(N)$. In what follows we review some symmetric fundamental domains for N prime, giving explicitly the vertices, the cycles, some invariants and a presentation of the corresponding group of fractional linear transformations. Detailed proofs can be found in [**Als00**].

THEOREM 5.1. *Let $p > 2$ be a prime number. Then,*

$$\mathcal{D}\big(\Gamma(1, p)\big) = \left\{ z \in \mathcal{H} : |\operatorname{Re}(z)| \leq \frac{1}{2}, \left|z - \frac{k}{p}\right| > \frac{1}{p}, k \in \mathbb{Z}, 0 < |k| \leq \frac{p-1}{2} \right\}$$

is a fundamental domain for $\Gamma(1, p)$ acting on $\mathcal{H}$.

The graphical representation of these domains can be obtained by means of the `Poincare` package. Some examples are included in Section 6.4 dealing with the representation of complex multiplication points.

NOTATION 5.2. For a Shimura curve $X(D,N)$ and a fundamental domain for the action of $\Gamma(D,N)$ given by a hyperbolic polygon, we denote by $n(D,N)$ the number of vertices of the fixed fundamental domain, by $n_k(D,N)$ the number of these vertices which are elliptic points of order k, or cusps if $k=\infty$, by $e_k(D,N)$ the number of cycles (i.e., orbits of vertices) of order $k>1$ and by $V_h(D,N)$ the hyperbolic volume of the domain. A cycle will be called elliptic, respectively parabolic, if its points are elliptic, respectively parabolic. We call accidental vertices those vertices in the domain which are neither elliptic nor parabolic. In this case, we use $n_1(D,N)$ for the number of accidental vertices and $e_1(D,N)$ for the number of their cycles.

THEOREM 5.3. *Let $p>2$ be a prime number. The former fundamental domain $\mathcal{D}(\Gamma(1,p))$ has the following properties.*

(i) *$\mathcal{D}(\Gamma(1,p))$ is a hyperbolic polygon of an even number of vertices and edges equal to $n(1,p)=p+1+n_2(1,p)$.*
(ii) *The vertices are $z_j=(2j-2-p)/(2p)+(\sqrt{3}/(2p))\iota$ for $1\le j\le (p+1)/2-1$, $z_{(p+1)/2}=0$, $z_j=(2j-p)/(2p)+(\sqrt{3}/(2p))\iota$ for $(p+1)/2+1\le j\le p$, and $z_{p+1}=\infty$.*
(iii) *The interior angle in the vertices $z_{(p+1)/2}=0$ and $z_{p+1}=\infty$ is 0; in the vertices z_1 and z_p, it is $\pi/3$, and in the other vertices z_j is $2\pi/3$.*
(iv) *The hyperbolic volume is $V_h(1,p)=(p+1)\pi/3$.*
(v) *$n_\infty(1,p)=e_\infty(1,p)=2$ and the parabolic cycles are $\{z_{(p+1)/2}\}=\{0\}$ and $\{z_{p+1}\}=\{\infty\}$.*
(vi) *If $p\equiv 1 \bmod 4$, then $n_2(1,p)=e_2(1,p)=2$ and the elliptic cycles of order 2 are $\{w_{2,1}\}=\{-k_0/p+(1/p)\iota\}$, $\{w_{2,2}\}=\{+k_0/p+(1/p)\iota\}$, where $k_0\in\mathbb{Z}$, $0<k_0\le(p-1)/2$, $k_0^2\equiv -1 \bmod p$. If $p\equiv 3 \bmod 4$, then $n_2(1,p)=e_2(1,p)=0$.*
(vii) *For $p=3$, $n_3(1,3)=2$, $e_3(1,3)=1$ and the elliptic cycle of order 3 is $\{z_1,z_3\}$.*
(viii) *Assume $p>3$. If $p\equiv 1 \bmod 3$, then $n_3(1,p)=e_3(1,p)=2$ and the elliptic cycles of order 3 are $\{w_{3,1}\}=\{(-2k_1-1)/(2p)+(1/p)\iota\}$, $\{w_{3,2}\}=\{(2k_1+1)/(2p)+(1/p)\iota\}$, where $k_1\in\mathbb{Z}$, $0<k_1\le(p-1)/2$, $k_1^2+k_1+1\equiv 0 \bmod p$. If $p\equiv 2 \bmod 3$, then $n_3(1,p)=e_3(1,p)=0$.*
(ix) *$n_1(1,p)=p-1-n_3(1,p)$ and $e_1(1,p)=(p-2-e_3(1,p))/3$.*

NOTATION 5.4. Let $p>2$ be a prime number. Consider the transformations T and γ_k, $k\in\mathbb{Z}$, $0<|k|\le(p-1)/2$, defined by

$$T=\begin{pmatrix}1&1\\0&1\end{pmatrix},\qquad \gamma_k=\begin{pmatrix}a&b\\p&-k\end{pmatrix}\in\Gamma(1,p),$$

where $a,b\in\mathbb{Z}$ are uniquely determined by the conditions $-ak-bp=1$ and $0<|a|\le(p-1)/2$. Note that $\gamma_{-1}=\gamma_1^{-1}$.

PROPOSITION 5.5. *Let $p>2$ be a prime number. The group of quaternion transformations $\Gamma(1,p)/\pm\mathrm{Id}$ is generated by T, γ_1 and the transformations γ_k such that $|a|\ge|k|$ if $|k|>1$, for $0<|a|\le(p-1)/2$. The relations between these generators are given as follows:*

(i) $\gamma_{k_0}^2 = \gamma_{-k_0}^2 = 1$, *if* $k_0 > 0$, $k_0^2 \equiv -1 \bmod p$;
(ii) $\gamma_{k_1}^3 = \gamma_{-k_1-1}^3 = 1$, *if* $k_1 > 0$, $k_1^2 + k_1 + 1 \equiv 0 \bmod p$;
(iii) *a relation* $\gamma_{t_3}^{\varepsilon_{t_3}} \gamma_{t_2}^{\varepsilon_{t_2}} \gamma_{t_1}^{\varepsilon_{t_1}} = 1$ *for every accidental cycle of* $\mathcal{D}(\Gamma(1,p))$ *of the form* $\{z_{j_1}, z_{j_2}, z_{j_3}\}$ *with* $\gamma_{t_i}^{\varepsilon_{t_i}}(z_{j_i}) = z_{j_{i+1}}$, $z_{j_4} = z_{j_1}$, $\varepsilon_{t_i} = \pm 1$;
(iv) *the relation* $T^{-1}\gamma_{s_2}^{\varepsilon_{s_2}} \gamma_{s_1}^{\varepsilon_{s_1}} \gamma_{(1-p)/2} = 1$, *for* $p > 3$, *coming from the unique accidental cycle of* $\mathcal{D}(\Gamma(1,p))$ *which contains four vertices* $\{z_1, z_{a_1}, z_{a_2}, z_p\}$, *with* $\gamma_{(1-p)/2}(z_1) = z_{a_1}$, $\gamma_{s_1}(z_{a_1}) = z_{a_2}$ *and* $\gamma_{s_2}(z_{a_2}) = z_p$, $\varepsilon_{s_i} = \pm 1$.

5.2. Transformations, embeddings and forms

NOTATION 5.6. In the group of quaternion transformations $\Gamma(D,N)$, we define

$$\Gamma_k(D,N) := \{\gamma \in \Gamma(D,N) : \gamma \text{ elliptic of order } k\},$$
$$\Gamma_\infty(D,N) := \{\gamma \in \Gamma(D,N) : \gamma \text{ parabolic}\}.$$

Let $k > 1$ or $k = \infty$. Consider the following sets of points determined from the action of the group $\Gamma(D,N)$ on $\mathcal{H}$:

$$\mathcal{P}_k(D,N) := \{z \in \mathcal{H} : \gamma(z) = z, \gamma \in \Gamma_k(D,N)\},$$
$$\widetilde{\mathcal{P}}_k(D,N) := \{z \in \mathbb{C} : \gamma(z) = z, \gamma \in \Gamma_k(D,N)\} = \{z \in \mathbb{C} : z \text{ or } \bar{z} \in \mathcal{P}_k(D,N)\},$$
$$\mathrm{P}_k(D,N) := \Gamma(D,N)\backslash\mathcal{P}_k(D,N).$$

5.2.1. Quaternion transformations.

Using the normic forms attached to the quaternion algebra, we obtain a direct proof of the well-known fact that the Riemann surface $\Gamma(D,N)\backslash\mathcal{H}$ is compact for $D > 1$.

PROPOSITION 5.7. *If* $D > 1$, *then the Shimura curve* $X(D,N)$ *has no parabolic points*; *that is,* $\mathrm{P}_\infty(D,N) = \varnothing$.

PROOF. Fix a quaternion algebra $H = \left(\frac{a,b}{\mathbb{Q}}\right)$ of discriminant $D > 1$. Assume that there exists $\omega \in H$ such that $n(\omega) = 1$ and that the matrix $\Phi(\omega) = \begin{pmatrix} x+\sqrt{a}y & z+\sqrt{a}t \\ b(z-\sqrt{a}t) & x-\sqrt{a}y \end{pmatrix}$ is parabolic. Then $\mathrm{tr}(\omega) = \pm 2$, hence we have $x = \pm 1$ and $-ay^2 - bz^2 + abt^2 = 0$. Thus, the ternary normic form $n_{H,3}(Y,Z,T) = -aY^2 - bZ^2 + abT^2$ is a $\mathbb{Q}$-isotropic form; therefore, it is $\mathbb{Q}_p$-isotropic for every p. But, applying Proposition 3.71, we know that for $p \mid D$ the form $n_{H,3}$ is $\mathbb{Q}_p$-anisotropic, which gives a contradiction.

In particular, all groups of quaternion transformations $\Gamma(D,N) \subseteq \Phi(H)$, $D > 1$, lack parabolic elements. Therefore, $\mathrm{P}_\infty(D,N) = \varnothing$. □

From now on, we consider quadratic fields $F = \mathbb{Q}(\sqrt{d})$ such that $\mathcal{E}(H,F) \neq \varnothing$, and quadratic orders $\Lambda(d,m) \subseteq F$ such that $\mathcal{E}(\mathcal{O}(D,N), \Lambda(d,m)) \neq \varnothing$. Next, by using these embeddings and units in $\Lambda(d,m)$, we obtain quaternion transformations in the group $\Gamma(D,N)$.

LEMMA 5.8. *Consider an embedding* $\varphi \in \mathcal{E}(\mathcal{O}(D,N), \Lambda(d,m))$ *and a unit* $u \in \Lambda(d,m)^*$. *Put* $\gamma := \phi(u)$, *where* $\phi = \Phi \circ \varphi$.
(i) *If* $\mathrm{n}(u) = 1$, *then* $\gamma \in \Gamma(D,N)$.
(ii) *If* $\mathrm{n}(u) = -1$, *then* $\gamma \notin \Gamma(D,N)$, *but* $\gamma^2 \in \Gamma(D,N)$.

NOTATION 5.9. Given a quadratic order $\Lambda(d,m)$, let ε be its fundamental unit. Put $\xi := \varepsilon$ if $\mathrm{n}(\varepsilon) = 1$ and $\xi := \varepsilon^2$ if $\mathrm{n}(\varepsilon) = -1$.

COROLLARY 5.10. *Let* $\varphi \in \mathcal{E}(\mathcal{O}(D,N), \Lambda(d,m))$ *be an embedding. Then, for any* $n \in \mathbb{Z}$, $\phi(\xi^n) \in \Gamma(D,N)$.

Conversely, the following results show that every element in a quaternion algebra can be obtained from embeddings of quadratic fields as above.

LEMMA 5.11. *Let* $\omega \in H$, $\omega \notin \mathbb{Q}$. *Then there exists an embedding* φ *of the quadratic field* $F_\omega := \mathbb{Q}(\sqrt{\operatorname{tr}(\omega)^2 - 4\operatorname{n}(\omega)})$ *in* H *such that* $\omega \in \varphi(F_\omega)$.

PROOF. Consider the polynomial $p_\omega(X) = X^2 - \operatorname{tr}(\omega)X + \operatorname{n}(\omega) \in \mathbb{Q}(X)$, which defines the quadratic field $F_\omega = \mathbb{Q}(\sqrt{d})$, with $d = \operatorname{tr}(\omega)^2 - 4\operatorname{n}(\omega)$. The element $\alpha = \operatorname{tr}(\omega)/2 + \frac{1}{2}\sqrt{d} \in F_\omega$ satisfies $\operatorname{tr}(\alpha) = \operatorname{tr}(\omega)$ and $\operatorname{n}(\alpha) = \operatorname{n}(\omega)$. Therefore, $\varphi(\alpha) := \omega$ defines an embedding $\varphi \in \mathcal{E}(H, F_\omega)$. □

THEOREM 5.12. *Let* $\gamma \in \Gamma(D,N)$, $D > 1$. *Then the following assertions hold:*

(i) *There exists a quadratic order* $\Lambda(d,m)$, *a number* $n \in \mathbb{Z} - \{0\}$ *and an optimal embedding* $\varphi \in \mathcal{E}^*(\mathcal{O}(D,N), \Lambda(d,m))$ *such that* $\Phi(\varphi(\varepsilon^n)) = \gamma$, *where* ε *is the fundamental unit of* $\Lambda(d,m)$.
(ii) *The transformation* γ *is elliptic if and only if* $d < 0$.
(iii) *The transformation* γ *is hyperbolic if and only if* $d > 0$.

PROOF. By construction of $\Gamma(D,N)$, there exists $\omega \in \mathcal{O}(D,N)_+^*$ such that $\gamma = \Phi(\omega)$. By Lemma 5.11, we have that $\omega \in \varphi(F_\omega)$ for some embedding $\varphi \in \mathcal{E}(H, F_\omega)$, where $F_\omega = \mathbb{Q}(\sqrt{d})$, $d = \operatorname{tr}(\omega)^2 - 4$. Namely, if $\alpha = \operatorname{tr}(\gamma)/2 + \frac{1}{2}\sqrt{d}$, then $\omega = \varphi(\alpha)$ is a unit, since $\operatorname{n}(\alpha) = \operatorname{n}(\omega) = \det(\gamma) = 1$. Thus, there exists $n \in \mathbb{Z} - \{0\}$ such that $\alpha = \varepsilon^n$; n must be even if $\operatorname{n}(\varepsilon) = -1$. If we consider the quadratic order $\mathbb{Z}[\alpha]$, it is clear that φ is an embedding of this order into $\mathcal{O}(D,N)$. Now take the quadratic order $\varphi^{-1}(\mathcal{O}(D,N)) \cap F_\omega$, which equals to $\Lambda(d,m)$, for some m. Then φ is an optimal embedding of $\Lambda(d,m)$ into $\mathcal{O}(D,N)$. This proves (i).

To prove (ii) and (iii), consider $F_\omega = \mathbb{Q}(\sqrt{\operatorname{tr}(\omega)^2 - 4})$, with $n(\omega) = 1$. It is clear that F_ω is an imaginary quadratic field if and only if $|\operatorname{tr}(\omega)| < 2$; this is equivalent to the fact that γ is elliptic. Since there are no parabolic transformations, (iii) is also proved. □

5.2.2. Elliptic points of $X(D,N)$. By Theorem 5.12, the elliptic transformations in the group $\Gamma(D,N)$ are obtained from units of imaginary quadratic orders by means of embeddings of these orders into the quaternion order $\mathcal{O}(D,N)$. Using the results about the equivalence between embeddings and representations of ternary quadratic forms, we can reduce the problem of finding explicit elliptic elements of $\Gamma(D,N)$ and explicit elliptic points of the Shimura curve $X(D,N)$ to the study of integral representations of integers by ternary quadratic forms. We will also obtain analogous results by interpreting the elliptic points as the associated points to binary quadratic forms.

LEMMA 5.13. *Let* $\gamma \in \Gamma(D,N)$ *define an elliptic transformation such that* $\gamma = \Phi(\omega)$, *for* $\omega \in \mathcal{O}_+^*(D,N)$. *Then:*

(i) γ *is elliptic of order* 2 *if and only if* $F_\omega = \mathbb{Q}(\sqrt{-1})$.
(ii) γ *is elliptic of order* 3 *if and only if* $F_\omega = \mathbb{Q}(\sqrt{-3})$.
(iii) *The order of* γ *can only be* 2 *or* 3.

PROOF. By Theorem 5.12, every element $\gamma \in \Gamma(D,N)$ is $\gamma = \Phi(\omega)$ for some ω associated to a fundamental unit in the quadratic field F_ω, and F_ω is an imaginary field since γ is elliptic. The unique imaginary quadratic fields containing nontrivial units are $\mathbb{Q}(\sqrt{-1})$ and $\mathbb{Q}(\sqrt{-3})$. In fact, these are the two only possibilities for the field $F_\omega = \mathbb{Q}(\sqrt{\operatorname{tr}(\gamma)^2 - 4})$, taking into account that, for an elliptic transformation γ with integer trace, $|\operatorname{tr}(\gamma)|$ only takes the values 0 and 1. By Lemma 2.10, the values of the trace of γ determine the order of the transformation. Note also that $\Phi(-1) = -\operatorname{Id} \in \Gamma(D,N)$. This proves (i) and (ii).

Since there are no other possibilities for the trace, (iii) is also proved. □

REMARK 5.14. By Theorem 5.12, we have that the elliptic transformations are determined from embeddings of units of the quadratic orders contained in the fields $\mathbb{Q}(\sqrt{-1})$ and $\mathbb{Q}(\sqrt{-3})$ into quaternion orders. Now, the units $\neq \pm 1$ of the quadratic fields $\mathbb{Q}(\sqrt{-1})$ and $\mathbb{Q}(\sqrt{-3})$ belong to the quadratic maximal orders only. Hence, the embeddings are necessarily optimal.

Thus, the existence of elliptic points of order 2 in the curve $X(D,N)$ is equivalent to the existence of optimal embeddings of the ring of integers of $\mathbb{Q}(\sqrt{-1})$ in the quaternion order $\mathcal{O}(D,N)$. We have the same relation between the existence of the elliptic points of order 3 in the curve $X(D,N)$ and the existence of optimal embeddings of the ring of integers of the quadratic field $\mathbb{Q}(\sqrt{-3})$ in the quaternion order $\mathcal{O}(D,N)$. Note the analogy between the explicit formulas which characterize both facts (cf. Proposition 2.29, Theorem 4.19).

Next, we characterize the quaternion elliptic transformations in terms of representations by ternary normic forms.

THEOREM 5.15. *Let $X(D,N)$ be the Shimura curve corresponding to the group $\Gamma(D,N)$ defined by the Eichler order $\mathcal{O}(D,N)$. To obtain the ternary normic form $n_{\mathbb{Z}+2\mathcal{O}(D,N),3}$, fix a normalized basis $\mathcal{B} = \{v_1, v_2, v_3, v_4\}$ for $\mathcal{O}(D,N)$. Let $\gamma \in \Gamma(D,N)$. Then:*

(i) *γ is an elliptic transformation of order 2 if and only if there exists (x,y,z) in $\mathcal{R}^*(n_{\mathbb{Z}+2\mathcal{O}(D,N),3}, 4; \mathbb{Z})$ such that*
$$\gamma = \Phi\left(-\frac{z\operatorname{tr}(v_4)}{2}, x, y, z\right)_{\mathcal{B}}.$$

(ii) *γ is an elliptic transformation of order 3 if and only if there exists (x,y,z) in $\mathcal{R}^*(n_{\mathbb{Z}+2\mathcal{O}(D,N),3}, 3; \mathbb{Z})$ such that either*
$$\gamma = \Phi\left(\frac{1-z}{2}, x, y, z\right)_{\mathcal{B}} \quad \textit{or} \quad \gamma = \Phi\left(\frac{-1-z\operatorname{tr}(v_4)}{2}, x, y, z\right)_{\mathcal{B}}.$$

PROOF. By Theorem 5.12 and Lemma 5.13, every elliptic transformation comes from a unit in an imaginary quadratic field. We use the bijection between the embeddings and the representations of ternary quadratic forms given in Theorem 4.26 and the explicit expression given in Corollary 4.27.

By Lemma 5.13 and Theorem 5.12, we have that an elliptic transformation γ of order 2 must be $\gamma = \Phi(\varphi(u))$, where u is a nontrivial unit in a quadratic order $\Lambda(-1,m)$ and $\varphi \in \mathcal{E}(\mathcal{O}(D,N), \Lambda(-1,m))$. Now, the nontrivial units of the quadratic field $F = \mathbb{Q}(\sqrt{-1})$ are $\pm\iota$, both belonging to the quadratic maximal order $\Lambda_F = \Lambda(-1,1)$. Thus the set of embeddings $\mathcal{E}(\mathcal{O}(D,N), \Lambda_F)$, which is equal to $\mathcal{E}^*(\mathcal{O}(D,N), \Lambda_F)$ in this case, is in one to one correspondence with the set of representations $\mathcal{R}^*(n_{\mathbb{Z}+2\mathcal{O}(D,N),3}, 4; \mathbb{Z})$, with $D_{\Lambda_F} = -4$. Corollary 4.27 gives explicit expressions of the embeddings in terms of these sets of representations.

Note that the two units ι and $-\iota$ give the same transformation. Thus, we obtain $\varphi(\iota) = (-z/2, x, y, z)_{\mathcal{B}}$, for some representation $(x, y, z) \in \mathcal{R}^*(n_{\mathbb{Z}+2\mathcal{O}(D,N),3}, 4; \mathbb{Z})$.

Analogously, in the case of order 3, we consider the units of the quadratic field $F = \mathbb{Q}(\sqrt{-3})$. The units are 1, ε, ε^2, -1, $-\varepsilon$ and $-\varepsilon^2$, where $\varepsilon = (1+\sqrt{-3})/2$, and they are contained in the maximal quadratic order $\Lambda_F = \Lambda(-3, 1)$. To obtain transformations it is enough to consider the units ε and ε^2.

Therefore, a transformation $\gamma \in \Gamma_3(D, N)$ is $\gamma = \Phi(\varphi(\varepsilon))$ or $\gamma = \Phi(\varphi(\varepsilon^2))$, $\varphi \in \mathcal{E}^*(\mathcal{O}(D, N), \Lambda_F)$, again by Theorem 5.12 and Lemma 5.13. Corollary 4.27 gives the explicit expression of $\varphi(\sqrt{-3})$ in terms of $(x, y, z) \in \mathcal{R}^*(n_{\mathbb{Z}+2\mathcal{O}(D,N),3}, 3; \mathbb{Z})$. From this expression we obtain the following results, which complete the proof of (ii):

$$\varphi(\varepsilon) = \left(\frac{1-z}{2}, x, y, z\right)_{\mathcal{B}}, \qquad \varphi(\varepsilon^2) = \left(\frac{-1-z}{2}, x, y, z\right)_{\mathcal{B}}. \qquad \square$$

COROLLARY 5.16. *Let $X(D, N)$ be the Shimura curve attached to the group $\Gamma(D, N)$. Then, the sets of elliptic points $\mathrm{P}_2(D, N)$ and $\mathrm{P}_3(D, N)$ are determined from the sets of representations $\mathcal{R}^*(n_{\mathbb{Z}+2\mathcal{O}(D,N),3}, 4; \mathbb{Z})$ and $\mathcal{R}^*(n_{\mathbb{Z}+2\mathcal{O}(D,N),3}, 3; \mathbb{Z})$, respectively.*

REMARK 5.17. From Theorem 5.15, given $(x, y, z) \in \mathcal{R}^*(n_{\mathbb{Z}+2\mathcal{O},3}, 3; \mathbb{Z})$, we have two possible elliptic transformations of order 3: $\sigma_1 = \varphi(\varepsilon)$ and $\sigma_2 = \varphi(\varepsilon^2) = (\varphi(\varepsilon))^2 = \sigma_1^2$. But, if $\tau \in \mathcal{H}$ is the elliptic point corresponding to σ_1, it is clear that τ is also a fixed point of σ_1^2. Hence, these two elliptic transformations give the same elliptic point.

REMARK 5.18. Using the class number of optimal embeddings of the rings of integers of $\mathbb{Q}(\sqrt{-1})$ and $\mathbb{Q}(\sqrt{-3})$ into $\mathcal{O}(D, N)$, we reobtain the formulas for the number of elliptic points of order 2 and 3 of the Shimura curves $X(D, N)$ given in Chapter 2. This is a particular case of a more general result which will be obtained for the complex multiplication points in Theorem 6.13.

For every Shimura curve $X(D, N)$, by using the explicit order $\mathcal{O}(D, N)$ and Theorem 5.15, we can compute the elliptic transformations and elliptic points explicitly. In particular, this process can be applied to the orders computed in the Tables A.4–A.7, and the corresponding normic forms, contained in the Tables A.21–A.24.

THEOREM 5.19. *Let $X(2p, N)$ be a Shimura curve for $p \equiv 3 \bmod 4$, $N \mid (p-1)/2$ square-free. Fix the quaternion algebra $H_A(p)$, the Eichler order $\mathcal{O}_A(2p, N) = \mathbb{Z}[1, i, Nj, (1+i+j+ij)/2]$ of level N, and the corresponding group of quaternion transformations $\Gamma(2p, N)$ defining $X(2p, N)$. Then:*

(i) *The set of elliptic points of order 2 of $X(2p, N)$ is*

$$\mathrm{P}_2(2p, N) = \Gamma(2p, N) \backslash \widetilde{\mathcal{P}}_2(2p, N) \cap \mathcal{H},$$

where $\widetilde{\mathcal{P}}_2(2p, N)$ is equal to

$$\left\{ \tau = \frac{(2x+z)\sqrt{p} \pm 2\iota}{-(2Ny+z) + z\sqrt{p}} : (x, y, z) \in \mathcal{R}^*(n_{\mathbb{Z}+2\mathcal{O},3}, 4; \mathbb{Z}) \right\}.$$

(ii) *The set of elliptic points of order 3 of $X(2p, N)$ is*

$$\mathrm{P}_3(2p, N) = \Gamma(2p, N) \backslash \widetilde{\mathcal{P}}_3(2p, N) \cap \mathcal{H},$$

where $\widetilde{\mathcal{P}}_3(2p,N)$ *is equal to*

$$\left\{\nu = \frac{(2x+z)\sqrt{p} \pm \sqrt{3}\iota}{-(2Ny+z)+z\sqrt{p}} : (x,y,z) \in \mathcal{R}^*(n_{\mathbb{Z}+2\mathcal{O},3},3;\mathbb{Z})\right\}.$$

(iii) *The sets of corresponding elliptic transformations are*

$$\Gamma_2(2p,N) = \{\gamma_\tau : \tau \in \mathcal{P}_2(2p,N)\} \text{ and } \Gamma_3(2p,N) = \{\gamma_\nu : \nu \in \mathcal{P}_3(2p,N)\},$$

where

$$\gamma_\tau = \frac{1}{2}\begin{pmatrix} (2x+z)\sqrt{p} & (2Ny+z)+z\sqrt{p} \\ -(2Ny+z)+z\sqrt{p} & -(2x+z)\sqrt{p}\end{pmatrix}$$

and $\gamma_\nu = \gamma_\tau \pm \mathrm{Id}$.

PROOF. We determine the elliptic transformations from Theorem 5.15.

Let γ define an elliptic transformation of order 2. Then, there exists a representation $(x,y,z) \in \mathcal{R}^*(n_{\mathbb{Z}+2\mathcal{O}_A(2p,N),3},4;\mathbb{Z})$ which determines an embedding $\varphi \in \mathcal{E}(\mathcal{O}_A(2p,N),\Lambda(-1,1))$, in such a way that $\gamma = \Phi(\varphi(\iota))$. If we apply Theorem 5.15(i), we explicitly obtain $\gamma = \Phi\Big(\frac{1}{2}\big((2x+z)i+(2Ny+z)j+zij\big)\Big)$; that is,

$$\gamma = \frac{1}{2}\begin{pmatrix} (2x+z)\sqrt{p} & (2Ny+z)+z\sqrt{p} \\ -(2Ny+z)+z\sqrt{p} & -(2x+z)\sqrt{p}\end{pmatrix}.$$

Thus, the fixed points of γ are the elliptic points

$$\frac{(2x+z)\sqrt{p} \pm 2\iota}{-(2Ny+z)+z\sqrt{p}}.$$

For the elliptic points of order 3 of $X(2p,N)$, the proof is analogous, using Theorem 5.15(ii). Given γ an elliptic transformation of order 3, we have $\gamma = \Phi(\varphi(\varepsilon))$ or $\gamma = \Phi(\varphi(\varepsilon^2))$, where $\varepsilon = (1+\sqrt{3})/2$. Now

$$\varphi(\varepsilon) = \frac{1}{2}(1+(2x+z)i+(2Ny+z)j+zij),$$
$$\varphi(\varepsilon^2) = \frac{1}{2}(-1+(2x+z)i+(2Ny+z)j+zij),$$

for some representation $(x,y,z) \in \mathcal{R}^*(n_{\mathbb{Z}+2\mathcal{O},3},3;\mathbb{Z})$. If we apply Φ to these equalities, we obtain the expressions of the transformations and the corresponding elliptic points given in the statement. □

Using other arguments, we obtain the following equivalent description, depending on the ternary normic form associated to the quaternion algebra. This generalizes the result in [**Als97**].

THEOREM 5.20. *Let* $\Gamma(2p,N)$ *be the group of transformations defined from the Eichler order* $\mathcal{O}_A(2p,N) = \mathbb{Z}[1,i,Nj,(1+i+j+ij)/2]$ *in the quaternion algebra* $H_A(p)$, $N \mid (p-1)/2$ *square-free. Then, the sets of elliptic points of order* $k = 2,3$ *in the associated Shimura curve* $X(2p,N)$ *are*

$$\mathrm{P}_k(2p,N) = \Gamma(2p,N)\backslash\widetilde{\mathcal{P}}_k(2p,N) \cap \mathcal{H},$$

where

$$\widetilde{\mathcal{P}}_2(2p,N) = \left\{\frac{b\sqrt{p} \pm 2\iota}{-c+d\sqrt{p}} : (b,c,d) \in \mathcal{R}(n_{H,3},4;\mathbb{Z}), b,c,d \text{ even, } 2N \mid c-d\right\},$$

$$\widetilde{\mathcal{P}}_3(2p,N)=\left\{\frac{b\sqrt{p}\pm\sqrt{3}\iota}{-c+d\sqrt{p}}:(b,c,d)\in\mathcal{R}(n_{H,3},3;\mathbb{Z}),b,c,d\ odd,\ 2N\mid c-d\right\}.$$

For an elliptic point $\tau=\tau(b,c,d)\in\widetilde{\mathcal{P}}_2(2p,N)\cup\widetilde{\mathcal{P}}_3(2p,N)$ *given as above, the corresponding elliptic transformation is*

$$\gamma_\tau=\frac{1}{2}\begin{pmatrix}a+b\sqrt{p} & c+d\sqrt{p}\\ -c+d\sqrt{p} & a-b\sqrt{p}\end{pmatrix},$$

where $a=0$ *if* $\tau\in\widetilde{\mathcal{P}}_2(2p,N)$ *and* $a=1$ *if* $\tau\in\widetilde{\mathcal{P}}_3(2p,N)$.

PROOF. Let $\gamma\in\Gamma_e(2p,N)$ define an elliptic transformation. From the explicit description of the group $\Gamma(2p,N)$, cf. 2.25, we have $\gamma=\frac{1}{2}\left(\begin{smallmatrix}a+b\sqrt{p} & c+d\sqrt{p}\\ -c+d\sqrt{p} & a-b\sqrt{p}\end{smallmatrix}\right)$, where a,b,c are integers, $a\equiv b\equiv c\equiv d \bmod 2$, $2N\mid c-d$, $\det(\gamma)=1$ and $|a|<2$. Thus, the elliptic points are $\tau=(b\sqrt{p}\pm\sqrt{a^2-4})/(-c+d\sqrt{p})$, where $a=0,1,-1$. In fact, it is enough to consider the cases $a=0$ and $a=1$. If $a=0$, then the elliptic points are the points $\tau=(b\sqrt{p}\pm 2\iota)/(-c+d\sqrt{p})\in\mathbb{C}$, where b,c,d are integral solutions of the equation $-pb^2+c^2-pd^2=4$, with $b\equiv c\equiv d\equiv 0 \bmod 2$ and $2N\mid c-d$; in particular, b,c,d even. If $a=1$, the elliptic points are the points $\tau=(b\sqrt{p}\pm\sqrt{3}\iota)/(-c+d\sqrt{p})\in\mathbb{C}$, where b,c,d are integral solutions of the equation $-pb^2+c^2-pd^2=3$, with $b\equiv c\equiv d\equiv 1 \bmod 2$ and $2N\mid c-d$; in particular, b,c,d are odd. The above quadratic equation is equivalent to considering the ternary normic form associated to the quaternion algebra $n_{H,3}(Y,Z,T)=-pY^2+Z^2-pT^2$; thus, it is necessary to study the set of representations of 3 and 4 by $n_{H,3}$ over $\mathbb{Z}$, depending on p, with the above conditions of parity and divisibility. □

REMARK 5.21. The elliptic points belong to the straight lines in $\mathbb{C}$ with slopes $\pm 2/(b\sqrt{p})$ or $\pm\sqrt{3}/(b\sqrt{p})$, depending on whether they are elliptic of order 2 or of order 3, respectively.

The following results for small ramified algebras of type B are obtained analogously.

THEOREM 5.22. *Let* $X(pq,N)$ *be a Shimura curve,* $q\equiv 1 \bmod 4$, $\left(\frac{p}{q}\right)=-1$, $N\mid(q-1)/4$ *square-free, and* $\gcd(p,N)=1$. *Fix the quaternion algebra* $H_B(p,q)$, *the Eichler order* $\mathcal{O}_B(pq,N)=\mathbb{Z}[1,Ni,(1+j)/2,(i+ij)/2]$ *of level* N, *and the corresponding group of quaternion transformations* $\Gamma(pq,N)$ *defining* $X(pq,N)$. *Then, every elliptic point of* $X(pq,N)$ *has order* 3 *and*

$$\mathrm{P}_3(pq,N)=\Gamma(pq,N)\backslash\widetilde{\mathcal{P}}_3(pq,N)\cap\mathcal{H},$$

where $\widetilde{\mathcal{P}}_3(pq,N)$ *is the set*

$$\left\{\nu=\frac{(2Nx+y)\sqrt{p}\pm\sqrt{3}\iota}{q(z-y\sqrt{p})}:(x,y,z)\in\mathcal{R}^*(n_{\mathbb{Z}+2\mathcal{O}(pq,N),3},3;\mathbb{Z})\right\}.$$

The set of elliptic transformations is $\Gamma_e(pq,N)=\Gamma_3(pq,N)=\{\gamma_\nu:\nu\in\mathcal{P}_3(pq,N)\}$, *where*

$$\gamma_\nu=\frac{1}{2}\begin{pmatrix}\pm 1+(2Nx+y)\sqrt{p} & z+y\sqrt{p}\\ q(z-y\sqrt{p}) & \pm 1-(2Nx+y)\sqrt{p}\end{pmatrix}.$$

PROOF. Let us see that there are no elliptic points of order 2. First, we obtain that there are no embeddings of $\mathbb{Q}(\sqrt{-1})$ into $H_B(p,q)$. Namely, we have that $\left(\frac{-1}{q}\right)=1$, since $q\equiv 1 \bmod 4$, hence the prime q splits in the quadratic field

$\mathbb{Q}(\sqrt{-1})$. Further, the discriminant of the algebra $H_{\mathrm{B}}(p,q)$ is pq. Thus, if we apply criterion 1.14, it follows that $\mathcal{E}\big(H_{\mathrm{B}}(p,q),\mathbb{Q}(\sqrt{-1})\big)=\varnothing$. Note that this agrees with the formulas given in Section 2.4. Now, by Theorem 5.12 and Lemma 5.13, we deduce that $\Gamma_2(pq,N)=\varnothing$ and $\mathcal{P}_2(pq,N)=\varnothing$.

We now deal with the elliptic transformations of order 3. Let $\gamma\in\Gamma_3(pq,N)$. There exist representations $(x,y,z)\in\mathcal{R}^*(n_{\mathbb{Z}+2\mathcal{O}_{\mathrm{B}}(pq,N)),3},3;\mathbb{Z})$ which determine an embedding $\varphi\in\mathcal{E}^*\big(\mathcal{O}_{\mathrm{B}}(pq,N),\Lambda(-3,1)\big)$, in such a way that $\gamma=\Phi\big(\varphi(u)\big)$, where u is a unit of $\Lambda(-3,1)$. Applying directly Theorem 5.15(ii) to the order $\mathcal{O}_B(pq,N)$, with the normalized basis $\mathcal{B}=\{1,Ni,(i+ij)/2,(1+j)/2\}$, we obtain

$$\gamma=\tfrac{1}{2}\Phi(1+(2Nx+y)i+zj+yi)\quad\text{or}\quad\gamma=\tfrac{1}{2}\Phi(-1+(2Nx+y)i+zj+yij).$$

Thus, the transformations $\gamma\in\Gamma_3(pq,N)$ are given by:

$$\gamma=\frac{1}{2}\begin{pmatrix}\pm1+(2Nx+y)\sqrt{p} & z+y\sqrt{p}\\ q(z-y\sqrt{p}) & \pm1-(2Nx+y)\sqrt{p}\end{pmatrix},$$

from which we obtain the elliptic points $\big((2Nx+y)\sqrt{p}\pm\sqrt{3}\iota\big)/\big(q(z-y\sqrt{p})\big)$. □

As in Theorem 5.20, we have the following result in the case of type B, which characterizes the elliptic points from the ternary normic form of the algebra.

THEOREM 5.23. *Let $\Gamma(pq,N)$ be the group of matrices defined from the Eichler order $\mathcal{O}_B(pq,N)=\mathbb{Z}\{1,Ni,(1+j)/2,(i+ij)/2\}\subseteq H_{\mathrm{B}}(p,q)$, where N is square-free, $N\mid(q-1)/2$, and $\gcd(p,N)=1$. Then, all the elliptic points of $X(pq,N)$ have order 3 and belong to the straight lines of $\mathbb{C}$ of slope $\pm\sqrt{3}/(b\sqrt{p})$. Explicitly, $\widetilde{\mathcal{P}}_3(pq,N)$ is the set*

$$\left\{\frac{b\sqrt{p}\pm\sqrt{3}\iota}{q(c-d\sqrt{p})}:(b,c,d)\in\mathcal{R}(n_{H,3},3;\mathbb{Z}),c\text{ odd},\ 2N\mid b-d\right\}.$$

For an elliptic point $\tau=\tau(b,c,d)\in\widetilde{\mathcal{P}}_3(pq,N)$, the corresponding elliptic transformation is

$$\gamma_\tau=\frac{1}{2}\begin{pmatrix}1+b\sqrt{p} & c+d\sqrt{p}\\ q(c-d\sqrt{p}) & 1-b\sqrt{p}\end{pmatrix}.$$

We have used the equivalence between embeddings and representations of ternary forms to look for quaternion elliptic transformations of $\Gamma(D,N)$ and their fixed points. Alternatively, we may also use the equivalence between the set of optimal embeddings $\mathcal{E}^*(\mathcal{O},\Lambda)$ and the set of primitive binary forms $\mathcal{H}^*(\mathbb{Z}+2\mathcal{O},\Lambda)$, given in Corollary 4.55. In particular, with respect to the fixed points of the transformations, the relation to the binary quadratic forms was already given in Lemma 2.12. This is especially useful for the elliptic points.

THEOREM 5.24. *Let H be a quaternion algebra, $\mathcal{O}(D,N)$ an Eichler order in H, and $\Gamma(D,N)$ the group of quaternion transformations defining a Shimura curve $X(D,N)$. Put $\delta=4$ if $k=2$, and $\delta=3$ if $k=3$. Then,*

$$\mathcal{P}_k(D,N)=\{\tau(f):f\in\mathcal{H}^*\big(\mathbb{Z}+2\mathcal{O}(D,N)\big),\det{}_1(f)=\delta\}.$$

From this result, by using the description of the sets of binary forms given in Theorems 4.62 and 4.64, we can compute the expression of the elliptic points for the Shimura curves $X(D,N)$ corresponding to the small ramified cases of types A and B, respectively. For the Eichler order $\mathcal{O}_{\mathrm{A}}(2p,N)\subseteq H_{\mathrm{A}}(p)$, the result is exactly the same as the one we obtained in Theorem 5.20, relating the condition $\det_1(f)=$

δ with the representations by the ternary normic form of the algebra $n_{H,3}$, cf. Proposition 4.46. For the algebra $H_{\mathrm{B}}(p,q)$ and the Eichler order $\mathcal{O}_{\mathrm{B}}(pq,N)$ cf. Theorem 5.23. In an analogous way it could be also applied to the nonramified case.

5.3. Local conditions at infinity

5.3.1. Principal homotheties of $\Gamma(D,N)$ for $D>1$. By Theorem 5.12, all the hyperbolic transformations in the group $\Gamma(D,N)$ can be obtained via optimal embeddings of orders of real quadratic fields into the order $\mathcal{O}(D,N)$, by means of quadratic units. Some of these quadratic orders play a special role.

Since $D>1$, we can assume $H=\left(\frac{a,b}{\mathbb{Q}}\right)$ is a quaternion algebra with $a>1$, a square-free. Once and for all, we fix the embedding $\Phi\colon H\hookrightarrow \mathrm{M}\big(2,\mathbb{Q}(\sqrt{a})\big)\subseteq \mathrm{M}(2,\mathbb{R})$, given in Proposition 1.19. The units of the real quadratic field $\mathbb{Q}(\sqrt{a})$ will be very important, especially the fundamental unit which will yield special hyperbolic transformations.

NOTATION 5.25. Let a be a square-free positive integer. Let ε be the fundamental unit of $\mathbb{Q}(\sqrt{a})$. Put $\xi=\varepsilon$ if $\mathrm{n}(\varepsilon)=1$ and $\xi=\varepsilon^2$ if $\mathrm{n}(\varepsilon)=-1$. Then, let us denote by h the hyperbolic transformation obtained from ξ; that is,

$$h=\begin{pmatrix}\xi & 0\\ 0 & \xi'\end{pmatrix},$$

where ξ' denotes the Galois conjugate of ξ in $\mathbb{Q}(\sqrt{a})$.

Note that the action of h is given by $h(z)=\xi^2 z$, where $\xi^2\in\mathbb{R}$. This leads to geometric interpretation of h.

LEMMA 5.26. *The hyperbolic transformation h is an homothety with center 0 and ratio ξ^2. Their two fixed real points are 0 and ∞.*

LEMMA 5.27. *Let D,N be natural numbers such that $\gcd(D,N)=1$ and assume that D is equal to the product of an even number of different primes. We consider the group $\Gamma(D,N)\subseteq \mathrm{SL}(2,\mathbb{R})$ defined by an Eichler order $\mathcal{O}(D,N)$. Then, there exists exactly one $s\in\mathbb{N}$ such that*

(a) *$h^s\in\Gamma(D,N)$.*
(b) *If $h^{s'}\in\Gamma(D,N)$, then $s\mid s'$.*

In particular, if $N=1$, then $s=1$.

PROOF. Fix a quaternion algebra $H=\left(\frac{a,b}{\mathbb{Q}}\right)$ of discriminant D and an Eichler order $\mathcal{O}(D,N)\subseteq H$ of level N. Consider the natural embedding $\varphi\colon\mathbb{Q}(\sqrt{a})\hookrightarrow H$, given by $\varphi(\alpha):=\alpha$, in such a way that $h=\Phi\big(\varphi(\xi)\big)$. Consider the quadratic order $\Lambda:=\Big(\varphi^{-1}\big(\mathcal{O}(D,N)\big)\Big)\cap\mathbb{Q}(\sqrt{a})$ in $\mathbb{Q}(\sqrt{a})$. There exists a unique $k\in\mathbb{N}$ such that ε^k is the fundamental unit of the order Λ. If k is even and $\mathrm{n}(\varepsilon)=-1$, then $\mathrm{n}(\varepsilon^k)=\mathrm{n}(\xi^{k/2})=1$ and take $s=k/2$. Otherwise, take $s=k$. Thus, $\mathrm{n}(\xi^s)=1$ and $\varphi(\xi)\in\mathcal{O}(D,N)^*{}_+$. Therefore, $h^s=\Phi\big(\varphi(\xi^s)\big)\in\Gamma(D,N)$. It is clear that this s satisfies the properties (a) and (b) in the statement.

In particular, for the maximal order $\mathcal{O}(D,1)$ it is clear that $s=1$, since in this case the quadratic order Λ is the ring of integers of $\mathbb{Q}(\sqrt{a})$. $\square$

DEFINITION 5.28. Fix the group $\Gamma(D,N)$. The hyperbolic transformation h^s determined in Lemma 5.27 will be called the principal homothety of $\Gamma(D,N)$.

LEMMA 5.29. *Let h^s be the principal homothety of $\Gamma(D,N)$.*

(i) *The points z and $\varepsilon^{2sn}z$ are $\Gamma(D,N)$-equivalent, for all $n \in \mathbb{Z}$.*

(ii) *Let $\gamma \in \Gamma(D,N)_z$. Then, $\gamma^{h^{-sn}} := h^{sn}\gamma h^{-sn} \in \Gamma(D,N)_{\varepsilon^{2sn}z}$, for all $n \in \mathbb{Z}$.*

PROOF. We have $h^{sn} = \left(\begin{smallmatrix} \varepsilon^{sn} & 0 \\ 0 & (\varepsilon^{sn})' \end{smallmatrix}\right)$. Thus, $h^{sn}(z) = \varepsilon^{2sn}z$, for $z \in \mathcal{H}$, since $\mathrm{n}(\varepsilon)^s = 1$. Since s satisfies (a), we deduce (i).

Assume $\gamma \in \Gamma(D,N)$ is a matrix with z as a fixed point. Hence, applying the above computation, we obtain $h^{sn}\gamma h^{-sn}(\xi^{2sn}z) = \varepsilon^{2sn}z$, which proves (ii). □

REMARK 5.30. Given a quaternion algebra H, there are also other real quadratic fields $F \neq \mathbb{Q}(\sqrt{a})$ such that $\mathcal{E}^*(H,F) \neq \varnothing$. For example, $\mathbb{Q}(\sqrt{D})$. For any such F, by fixing a suitable embedding of H into $\mathrm{M}(2,F)$, any $\varphi \in \mathcal{E}^*(H,F)$ would yield another hyperbolic transformation, which could be considered as a principal homothety.

5.3.2. Construction of a fundamental domain at infinity, for $D > 1$. We determine the subgroup $\Gamma(D,N)_\infty$ of the transformations in $\Gamma(D,N)$ fixing infinity, and we construct a fundamental domain.

Let $H = \left(\frac{a,b}{\mathbb{Q}}\right)$ be an indefinite quaternion algebra of discriminant $D > 1$. Assume $a > 0$ and fix the monomorphism $\Phi\colon H \hookrightarrow M(2,\mathbb{R})$ given in Proposition 1.19, which determines an embedding of H into $M\big(2,\mathbb{Q}(\sqrt{a})\big)$. Let ε be the fundamental unit of the real quadratic field $\mathbb{Q}(\sqrt{a})$. Recall that we put $\xi = \varepsilon$ if $\mathrm{n}(\varepsilon) = 1$, and $\xi = \varepsilon^2$ if $\mathrm{n}(\varepsilon) = -1$.

NOTATION 5.31. Given two numbers $r_1, r_2 \in \mathbb{R}^+$, let $S(r_1,r_2)$ denote the hyperbolic strip defined by

$$S(r_1,r_2) := \{z \in \mathcal{H} : r_1 \leq |z| \leq r_2\}.$$

PROPOSITION 5.32. *Let r be any positive real number. Let h^s be the principal homothety of $\Gamma(D,N)$. Then, $\Gamma(D,N)_\infty = \langle h^s \rangle$ and $S(r,\xi^{2s}r)$ is a fundamental domain in $\mathcal{H}$ for the action of $\Gamma(D,N)_\infty$; that is,*

$$\mathcal{D}_r(\Gamma(D,N)_\infty) = \{z \in \mathcal{H} : r \leq |z| \leq \xi^{2s}r\}.$$

PROOF. Let $\gamma \in \Gamma(D,N)_\infty$. Let $\mathcal{O}(D,N)$ be a quaternion order such that $\Gamma(D,N) = \Phi(\mathcal{O}(D,N)^*_+)$. Using the generic expression of the integral elements of $\Phi(H)$, we get:

$$\Phi(x+yi+zj+tij) = \begin{pmatrix} x+y\sqrt{a} & z+t\sqrt{a} \\ b(z-t\sqrt{a}) & x-y\sqrt{a} \end{pmatrix} = \begin{pmatrix} \alpha & \beta \\ b\beta' & \alpha' \end{pmatrix}.$$

Since $\gamma(\infty) = \infty$, we have $\beta' = z - t\sqrt{a} = 0$, hence $z = t = 0$ and $\beta = 0$, because $a \neq 1$. Then

$$\gamma = \begin{pmatrix} u & 0 \\ 0 & u' \end{pmatrix}, \quad \text{with } u \in \mathbb{Q}(\sqrt{a}).$$

Since $\gamma \in \Gamma(D,N)$, we deduce that $\mathrm{tr}(u) \in \mathbb{Z}$, $\mathrm{n}(u) = 1$; hence u is a unit in the ring of integers of $\mathbb{Q}(\sqrt{a})$ and there exists $m \in \mathbb{Z}$ such that $u = \xi^m$, ξ defined from the fundamental unit. Therefore $\gamma = h^m$. Now, if we apply Lemma 5.27, we have that

$s \mid m$ and $\gamma = h^m = (h^s)^n$ for some $n \in \mathbb{Z}$, $s = s(D, N)$. Since $h^s \in \Gamma(D, N)_\infty$, we deduce $\Gamma(D, N)_\infty$ is the infinite cyclic group generated by h^s; namely,

$$\Gamma(D, N)_\infty = \{h^{sn} : n \in \mathbb{Z}\}.$$

By using the geometric interpretation of h, it is clear that h^s is an homothety of center $(0, 0)$ and ratio ξ^{2s}. Thus, a fundamental domain for $\Gamma(D, N)_\infty = \langle h^s \rangle$ is the hyperbolic strip between two circles, centered at 0 and of radius r, $\xi^{2s}r$, denoted by $S(r, \xi^{2s}r)$, for any r. □

We observe that, for a fixed hyperbolic strip $S = S(r, \xi^{2s}r)$, their images under the group $\Gamma(D, N)_\infty$ cover the Poincaré half-plane.

REMARK 5.33. Proposition 5.32 generalizes a well-known fact in the nonramified case, taking the algebra $\mathrm{M}(2, \mathbb{Q})$ and the Eichler order $\mathcal{O}_0(1, N) \subseteq \mathrm{M}(2, \mathbb{Z})$. In fact, the argument of the above proof also works for the nonramified algebra $H = \left(\frac{1,-1}{\mathbb{Q}}\right)$ and the Eichler order $\mathcal{O}_M(1, N) = \mathbb{Z}[1, (j + ij)/2, N(-j + ij)/2, (1 - i)/2] \subseteq H$. Then we have $a = 1$, the embedding Φ also makes sense and we obtain that the group $\Gamma(D, N)_\infty$ is generated by a translation.

REMARK 5.34. In analogy to the modular case, the idea would be to choose a basic hyperbolic strip $S(r_1, r_2)$, playing the same role as the strip $\{z \in \mathcal{H} : |\operatorname{Re}(z)| \leq \frac{1}{2}\}$ in the nonramified case. In the modular case, the usual fundamental domain is invariant by the symmetry with respect to the hyperbolic line determined by the imaginary axis. In this context, for nonmodular fundamental domains it makes sense to determine a hyperbolic line, which will be a semicircle with center 0, in such a way that $\mathcal{D}(\Gamma(D, N)_\infty)$ turns out to be invariant by the hyperbolic symmetry with respect to this line. The key point is to choose this special hyperbolic line.

LEMMA 5.35. *Let $r \in \mathbb{R}^+$. The fundamental domain $\mathcal{D}_r(\Gamma(D, N)_\infty) = S(r, \xi^{2s}r)$ is invariant with respect to the hyperbolic symmetry given by the hyperbolic line $C = C(0, r\xi^s)$.*

PROOF. Consider $\mathcal{D}_r(\Gamma(D, 1)_\infty) = S(r, \xi^{2s}r)$ and let denote by $C_1 = C(0, r)$ and $C_2 = C\big((0, 0), \xi^{2s}r\big)$ the hyperbolic lines in its boundary. The middle point between the points $r\iota$ and the $\xi^{2s}r\iota$, according to the hyperbolic distance δ defined in Section 2.1, is $(0, y)$ such that $\delta(r, y) = \big|\log|y/r|\big| = \big|\log|\xi^{2s}r/y|\big| = \delta(y, \xi^{2s}r)$; that is, $y = \xi^s r$. Note that there are infinitely many hyperbolic lines that pass through the point $\xi^s r\iota$ and have no intersection with the hyperbolic lines C_1 and C_2. Fix the line $C = C(0, \xi^s r)$ and let f be the hyperbolic symmetry with respect to this line. To see that the hyperbolic strip $S(r, \xi^{2s}r)$ is invariant under f, we prove that $f(C_1) = C_2$. Writing f in terms of the center and the radius of C, we have $f(z) = (\xi^s r)^2/\bar{z}$, cf. Lemma 2.3. In particular, $f(r) = \xi^{2s}r$ and $f(r\iota) = \xi^{2s}r\iota$; since f maps lines to lines, we have $f(C_1) = C_2$. □

REMARK 5.36. As a consequence of Lemma 5.35, for a quaternion group $\Gamma(D, N)$ and for a fixed hyperbolic line $C = C(0, r)$, we have three special fundamental domains for $\Gamma(D, N)_\infty$ with good properties with respect to C:

(a) $S(r, \xi^{2s}r)$, having C as the inferior edge.
(b) $S(\xi^{-2s}r, r)$, having C as the superior edge.
(c) $S(\xi^{-s}r, \xi^s r)$, which is invariant by the symmetry with respect to C.

In the modular case, we can fix the imaginary positive axis as a special hyperbolic line C. Then, among the above three fundamental domains with good properties with respect to C, the usual fundamental domain for $\Gamma(1,N)_\infty$ corresponds to case (c). For $N = 1$, this choice is the only one that yields to a fundamental domain satisfying the condition that in each parabolic cycle there is a unique vertex. But in the case $D > 1$, we do not have parabolic points; thus this can not be applied to distinguish a fundamental domain.

Another option, for example, can be to fix a hyperbolic line C containing elliptic points, if they exist. Moreover, we can apply conditions about the symmetry of the elliptic points with respect to the imaginary axis, cf. Lemma 2.20. Thus, at least C contains two elliptic points, equivalent or not. Then, since the elliptic points cannot be in the interior of the fundamental domain, this leads us to a fundamental domain satisfying (a) or (b) with respect to C.

5.4. Principal symmetries of $\Gamma(D, N)$

Besides the principal homothety, there are other elements acting on the Poincaré half-plane with a special meaning in the construction of fundamental domains. We will use the isometric circles defined in Section 2.2 to focus on some transformations given by elements of $\mathrm{SL}(2,\mathbb{R})$.

From the results in Theorem 4.19 about optimal embeddings, we easily obtain the lemma below.

LEMMA 5.37. *Let $\mathcal{O}(D,N)$ be a quaternion order. Assume N is square-free. Then there exists $\omega \in \mathcal{O}(D,N)$ such that $\mathrm{n}(\omega) = DN$ and $\mathrm{tr}(\omega) = 0$.*

PROOF. The quadratic order $\Lambda(-DN,1)$ satisfies $\mathcal{E}^*\big(\mathcal{O}(D,N),\Lambda(-DN,1)\big) \neq \varnothing$. Actually, by applying Theorem 4.19, $\nu(D,N,-DN,1;\mathcal{O}(D,N)^*) = h(-DN,1)$. For any embedding $\varphi \in \mathcal{E}^*\big(\mathcal{O}(D,N),\Lambda(-DN,1)\big)$, put $\omega = \varphi(\sqrt{-DN})$. We have $\mathrm{n}(\omega) = DN$ and $\mathrm{tr}(\omega) = 0$ and $\omega \in \mathcal{O}(D,N)$, since $\sqrt{-DN} \in \Lambda(-DN,1)$. □

REMARK 5.38. If $-DN \equiv 1 \bmod 4$, we can also consider the set of optimal embeddings $\mathcal{E}^*\big(\mathcal{O}(D,N),\Lambda(-DN,2)\big)$ to obtain elements of $\mathcal{O}(D,N)$ with norm DN and trace 0, since $\sqrt{-DN} \in \Lambda(-DN,2)$.

The existence of embeddings allow us to make the following construction. For each embedding $\varphi \in \mathcal{E}^*\big(\mathcal{O}(D,N),\Lambda(-DN,1)\big)$, put $\omega = \varphi(\sqrt{-DN})$. We have $\mathrm{n}(\omega) = DN$ and $\mathrm{tr}(\omega) = 0$ and $\omega \in \mathcal{O}(D,N)$, since $\sqrt{-DN} \in \Lambda(-DN,1)$. Thus, considering the mapping $\widetilde{\Phi}$ defined in Definition 4.56, we have a transformation $\eta := \widetilde{\Phi}(\omega) \in \mathrm{SL}(2,\mathbb{R})$.

DEFINITION 5.39. Take any $\omega \in \mathcal{O}(D,N)$ such that $\mathrm{n}(\omega) = DN$ and $\mathrm{tr}(\omega) = 0$. By considering the mapping $\widetilde{\Phi}$ defined in Definition 4.56 extended to H^*, we have a transformation $\eta := \widetilde{\Phi}(\omega) \in \mathrm{SL}(2,\mathbb{R})$. Assume that η does not fix infinity. Then, we say that the isometric circle C_η attached to η is a principal hyperbolic line for the group $\Gamma(D,N)$.

DEFINITION 5.40. A hyperbolic symmetry is called principal with respect to $\Gamma(D,N)$ if it is the symmetry with respect to a principal hyperbolic line for $\Gamma(D,N)$.

REMARK 5.41. In the same way, we may consider elements with norm d, for any $d \mid DN$, coming from the quadratic orders $\Lambda(-d,1)$ by means of embeddings, to obtain distinguished hyperbolic lines.

Note that the explicit computations of principal lines and principal symmetries are done via explicit computations of embeddings using quadratic forms.

Next we look for properties of transformations of this kind.

LEMMA 5.42. *Let* $\eta = \left(\begin{smallmatrix} a & b \\ c & d \end{smallmatrix}\right)$ *be a transformation such that the isometric circle* C_η *is principal. Then* η *is an elliptic transformation of order* 2 *and* L_η *is the line* $x = a/c$.

PROOF. Since C_η is principal we have $\mathrm{tr}(\eta) = 0$, thus η is elliptic of order 2. Hence $d = -a$. Let z_1, z_2 be the fixed points z_1, z_2 of η, that satisfy the equation $cZ^2 - 2aZ - b = 0$. By properties of the isometric circles (cf. Section 2.2), the line L_η is $x = (z_1 + z_2)/2$ and $(z_1 + z_2)/2 = a/c$ which coincides with the center of the isometric circle. □

We are going to use these principal hyperbolic lines to get some symmetrical conditions on the fundamental domains. In particular, comparing with the modular case, we are especially interested in the cases where L_η is the imaginary axis. By Lemma 5.42, this is achieved if and only if the transformation η has zeros at the diagonal.

5.43. *Small ramified case of type* A. Let $\mathcal{O}(2p, 1) \subseteq H_{\mathrm{A}}(p)$ be the maximal order corresponding to the group $\Gamma(2p, 1)$. As an example, we fix $p = 3$ and consider $\omega = -3j + ij$; it satisfies $\mathrm{n}(\omega) = 6$ and $\mathrm{tr}(\omega) = 0$. Then,

$$\eta = \widetilde{\Phi}(\omega) = \frac{1}{\sqrt{6}} \begin{pmatrix} 0 & -3+\sqrt{3} \\ 3+\sqrt{3} & 0 \end{pmatrix}, \quad C_\eta = C\left(0, \frac{\sqrt{2}}{1+\sqrt{3}}\right).$$

The isometric circle C_η is a principal hyperbolic line and L_η is the imaginary axis. If we put $\omega' = i + 3j$, it also satisfies $\mathrm{n}(\omega') = 6$ and $\mathrm{tr}(\omega') = 0$. Therefore, we have

$$\eta' = \widetilde{\Phi}(\omega') = \frac{1}{\sqrt{6}} \begin{pmatrix} \sqrt{3} & 3 \\ -3 & -\sqrt{3} \end{pmatrix}, \quad C_{\eta'} = C\left(\frac{-1}{\sqrt{3}}, \frac{\sqrt{2}}{\sqrt{3}}\right).$$

The isometric circle $C_{\eta'}$ is a principal hyperbolic line and $L_{\eta'}$ is the line $x = -1/\sqrt{3}$.

5.44. *Small ramified case of type* B. Let $\mathcal{O}_B(pq, 1) \subseteq H_B(p, q)$ be the maximal order corresponding to the group of quaternion transformations $\Gamma(pq, 1)$. In this case, we give the expression of a principal line of $\Gamma(pq, 1)$ in terms of q. Consider $\omega = -ij$, since $\mathrm{n}(-ij) = pq$ and $\mathrm{tr}(-ij) = 0$. Then,

$$\eta = \widetilde{\Phi}(-ij) = \frac{1}{\sqrt{q}} \begin{pmatrix} 0 & -1 \\ q & 0 \end{pmatrix}, \quad C_\eta = C\left(0, \frac{1}{\sqrt{q}}\right).$$

The isometric circle C_η is a principal hyperbolic line and L_η is the imaginary axis. Namely, $\eta = \widetilde{\Phi}(-ij)$, $\mathrm{n}(-ij) = pq$, $\mathrm{tr}(-ij) = 0$.

REMARK 5.45. Note the parallelism with the nonramified case, $\Gamma(1, N) = \Gamma_0(N)$. Consider

$$\omega = \begin{pmatrix} 0 & -1 \\ N & 0 \end{pmatrix} \in \mathcal{O}_0(1, N), \quad \mathrm{n}(\omega) = N, \ \mathrm{tr}(\omega) = 0.$$

Then we obtain

$$\eta = \widetilde{\Phi}(\omega) = \frac{1}{\sqrt{N}} \begin{pmatrix} 0 & -1 \\ N & 0 \end{pmatrix}, \quad C_\eta = C\left(0, \frac{1}{\sqrt{N}}\right).$$

The isometric circle C_η is a principal hyperbolic line and L_η is the imaginary axis.

If N is odd, we can also consider

$$\omega' = \begin{pmatrix} N & -(N+1)/2 \\ 2N & -N \end{pmatrix} \in \mathcal{O}_0(1, N), \quad \mathrm{n}(\omega) = N, \ \mathrm{tr}(\omega) = 0.$$

Then we obtain

$$\eta' = \widetilde{\Phi}(\omega) = \frac{1}{\sqrt{N}} \begin{pmatrix} N & -(N+1)/2 \\ 2N & -N \end{pmatrix}, \quad C_{\eta'} = C\Big(\frac{1}{2}, \frac{1}{2\sqrt{N}}\Big).$$

The isometric circle $C_{\eta'}$ is a principal hyperbolic line and $L_{\eta'}$ is the line $x = \frac{1}{2}$. Changing the signs of the matrix, we obtain η'' with isometric circle $C_{\eta'} = C\big(-\frac{1}{2}, 1/(2\sqrt{N})\big)$ and $L_{\eta'}$ the line $x = -\frac{1}{2}$.

5.5. Construction of fundamental domains ($D > 1$)

5.5.1. General comments. In this section, we perform effective constructions of fundamental domains using results from previous sections. The graphical representations and the tables of constants can be found at the end of this chapter.

Consider the Shimura curve $X(D, N)$, $D > 1$, attached to the group $\Gamma(D, N)$. Assume that the set of elliptic points is not empty, that is, at least one of the values $e_2(D, N)$ and $e_3(D, N)$ is nonzero. We want to find a fundamental domain for $X(D, N)$ which is a hyperbolic polygon having elliptic points as vertices in order to get a fundamental domain for setting the comments in Section 5.3.2 above. Moreover we can apply Proposition 2.32 to know in advance the number of vertices.

Using the formulas in Section 2.4, we compute the constants attached to the Shimura curve $X(D, N)$: $e_2(D, N)$, $e_3(D, N)$, $g(D, N)$ and $V(D, N)$. Then we look for a hyperbolic polygon with exactly $n_e(D, N)$ elliptic vertices (cf. 2.32) and with those constants.

To compute elliptic points explicitly, we use the results for the Eichler orders $\mathcal{O}(D, N)$ given in Chapter 1 and the group of quaternion transformations $\Gamma(D, N)$ given in Section 2.3 for the small ramified algebras of type A or B. We consider the normic forms attached to $\mathcal{O}(D, N)$ and $\mathbb{Z} + 2\mathcal{O}(D, N)$ and study their representations, which allow us to make an effective determination of elliptic transformations, cf. Section 5.2. To choose a suitable set of elliptic points, we look for $\Gamma(D, N)$-equivalences.

First of all, we make an attempt to determine a hyperbolic polygon with $n_e(D, N)$ elliptic vertices, with the suitable number of cycles according to $e_2(D, N)$, $e_3(D, N)$, the edges identified pairwise, and the suitable genus. After choosing a possible hyperbolic polygon, we check if the hyperbolic volume is correct. If this is the case, the polygon is a fundamental domain for the Shimura curve $X(D, N)$. In fact, the polygon is a fundamental domain for a subgroup of $\Gamma(D, N)$ with the same invariants as $\Gamma(D, N)$. Therefore, the subgroup has to be equal to $\Gamma(D, N)$, by Hurwitz's theorem. Moreover, the paired edges in a fundamental domain provide an explicit presentation of the group $\Gamma(D, N)$.

The process to obtain a fundamental domain involves laborious computations, such as the search for representations of integers by normic forms, formulas related to embeddings, etc. In the package `Poincare` we implemented instructions to make the computations more simply.

5.5.2. Fundamental domain for $X(6,1)$. The following theorem yields a fundamental domain for the Shimura curve $X(6,1)$ with the property that all the vertices are elliptic. It also contains the values of the constants, and the equivalence relations between the vertices. Its graphical representation can be found in Figure 5.1.

THEOREM 5.46. *Consider the Shimura curve $X(6,1)$. Then the hyperbolic hexagon of vertices $(v_1, v_2, v_3, v_4, v_5, v_6)$,*

$$v_1 = \frac{-\sqrt{3}+\iota}{2}, \qquad v_2 = \frac{-1+\iota}{1+\sqrt{3}}, \qquad v_3 = (2-\sqrt{3})\iota,$$

$$v_4 = \frac{1+\iota}{1+\sqrt{3}}, \qquad v_5 = \frac{\sqrt{3}+\iota}{2}, \qquad v_6 = \iota,$$

is a fundamental domain for the Shimura curve $X(6,1)$ in the Poincaré half-plane, which we denote by $\mathcal{D}(6,1)$. It has the following properties.

(i) *All the vertices are elliptic and the corresponding elliptic transformations fixing them are:*

$$\gamma_{v_1} = \begin{pmatrix} \sqrt{3} & 2 \\ -2 & -\sqrt{3} \end{pmatrix}, \qquad \gamma_{v_2} = \frac{1}{2}\begin{pmatrix} 1+\sqrt{3} & 3-\sqrt{3} \\ -3-\sqrt{3} & 1-\sqrt{3} \end{pmatrix},$$

$$\gamma_{v_3} = \begin{pmatrix} 0 & -2+\sqrt{3} \\ 2+\sqrt{3} & 0 \end{pmatrix}, \qquad \gamma_{v_4} = \frac{1}{2}\begin{pmatrix} 1+\sqrt{3} & -3+\sqrt{3} \\ 3+\sqrt{3} & 1-\sqrt{3} \end{pmatrix},$$

$$\gamma_{v_5} = \begin{pmatrix} \sqrt{3} & -2 \\ 2 & -\sqrt{3} \end{pmatrix}, \qquad \gamma_{v_6} = \begin{pmatrix} 0 & 1 \\ -1 & 0 \end{pmatrix}.$$

(ii) *The hyperbolic volume of $\mathcal{D}(6,1)$ is $V_h(6,1) = 2\pi/3$ and the genus of the curve $X(6,1)$ is $g(6,1) = 0$.*

(iii) *The number of elliptic vertices of order 2 is $n_2(6,1) = 4$. There are $e_2(6,1) = 2$ elliptic cycles of order 2, namely $\{v_6\}$ and $\{v_1, v_3, v_5\}$. The relations between these vertices are $\gamma_{v_2}(v_3) = v_1$ and $\gamma_{v_4}(v_3) = v_5$.*

(iv) *The number of elliptic vertices of order 3 is $n_3(6,1) = 2$. There are $e_3(6,1) = 2$ elliptic cycles of order 3: $\{v_2\}$ and $\{v_4\}$.*

(v) *The principal homothety of $\Gamma(6,1)$ is*

$$h = \begin{pmatrix} 2+\sqrt{3} & 0 \\ 0 & 2-\sqrt{3} \end{pmatrix}.$$

(vi) *The pairing of the edges is:*

(v_2v_3, v_2v_1) *by the transformation* γ_{v_2},

(v_3v_4, v_5v_4) *by the transformation* γ_{v_4},

(v_5v_6, v_1v_6) *by the transformation* γ_{v_6}.

(vii) *We have the following presentation of the group $\Gamma(6,1)/\pm \mathrm{Id}$:*

$$\langle \gamma_{v_2}, \gamma_{v_4}, \gamma_{v_6} : \gamma_{v_2}^3 = \gamma_{v_4}^3 = \gamma_{v_6}^2 = (\gamma_{v_2}^{-1}\gamma_{v_6}\gamma_{v_4})^2 = 1\rangle.$$

(viii) *Put $r_1 = 7-4\sqrt{3}$, $r_2 = 2-\sqrt{3}$, $r_3 = 1$, $r_4 = 2+\sqrt{3}$. $\mathcal{D}(6,1)$ is included in the hyperbolic strips $S(r_1, r_3)$ and $S(r_2, r_4)$, which are two fundamental domains by the action of $\Gamma(6,1)_\infty$. These hyperbolic strips are symmetric with respect to the principal hyperbolic lines $C(0, r_2)$ and $C(0, r_3)$, respectively.*

(ix) *The edges* v_5v_6 *and* v_6v_1 *are segments of the isometric circle of* γ_{v_6}.
(x) $\mathcal{D}(6,1)$ *is symmetric with respect to the imaginary axis.*

PROOF. The constants attached to the Shimura curve $X(6,1)$ are $e_2(6,1) = 2$, $e_3(6,1) = 2$, $g(6,1) = 0$ and $V_h(6,1) = 2\pi/3$, cf. Section 2.4. Moreover, if it exists, a fundamental domain for $X(6,1)$ in $\mathcal{H}$ whose vertices are all elliptic ones must have $n_e(6,1) = 6$ vertices.

Applying Theorems 5.19 and 5.20, from representations by normic forms, we find elliptic points with the corresponding elliptic transformations fixing them. For example, we have the following elliptic points of order 2:

$$A_1 = \iota, \qquad \gamma_{A_1} = \begin{pmatrix} 0 & 1 \\ -1 & 0 \end{pmatrix},$$

$$A_2 = (2-\sqrt{3})\iota, \qquad \gamma_{A_2} = \begin{pmatrix} 0 & -2+\sqrt{3} \\ 2+\sqrt{3} & 0 \end{pmatrix},$$

$$A_3 = \frac{-\sqrt{3}+\iota}{2}, \qquad \gamma_{A_3} = \begin{pmatrix} \sqrt{3} & 2 \\ -2 & -\sqrt{3} \end{pmatrix},$$

$$A_4 = \frac{\sqrt{3}+\iota}{2}, \qquad \gamma_{A_4} = \begin{pmatrix} \sqrt{3} & -2 \\ 2 & -\sqrt{3} \end{pmatrix},$$

$$A_5 = \frac{-\sqrt{3}+\iota}{4+2\sqrt{3}}, \qquad \gamma_{A_5} = \begin{pmatrix} \sqrt{3} & 4-2\sqrt{3} \\ -4-2\sqrt{3} & -\sqrt{3} \end{pmatrix},$$

$$A_6 = \frac{\sqrt{3}+\iota}{4+2\sqrt{3}}, \qquad \gamma_{A_6} = \begin{pmatrix} \sqrt{3} & -4+2\sqrt{3} \\ 4+2\sqrt{3} & -\sqrt{3} \end{pmatrix}.$$

The following points are elliptic of order 3:

$$B_1 = \frac{-1+\iota}{1+\sqrt{3}}, \qquad \gamma_{B_1} = \frac{1}{2}\begin{pmatrix} 1+\sqrt{3} & 3-\sqrt{3} \\ -3-\sqrt{3} & 1-\sqrt{3} \end{pmatrix},$$

$$B_2 = \frac{1+\iota}{1+\sqrt{3}}, \qquad \gamma_{B_2} = \frac{1}{2}\begin{pmatrix} 1+\sqrt{3} & -3+\sqrt{3} \\ 3+\sqrt{3} & 1-\sqrt{3} \end{pmatrix}.$$

The previous elliptic transformations give us some relations between these points:

$$\gamma_{A_1}(A_1) = A_1, \qquad \gamma_{A_1}(A_4) = A_3,$$
$$\gamma_{B_1}(B_1) = B_1, \qquad \gamma_{B_1}(A_2) = A_3,$$
$$\gamma_{B_2}(B_2) = B_2, \qquad \gamma_{B_2}(A_2) = A_4.$$

We will prove that the polygon with vertices $(A_3, B_1, A_2, B_2, A_4, A_1)$, which we denote by $(v_1, v_2, v_3, v_4, v_5, v_6)$, satisfies the desired conditions.

First, check that the two elliptic vertices of order 3, v_2 and v_4, are not $\Gamma(6,1)$-equivalent; thus, we have two cycles of order 3. Note that the angle of the fixed hyperbolic polygon in each of its vertices is $2\pi/3$; hence, each one is a cycle. In the same way, $\{v_1, v_3, v_5\}$ and $\{v_6\}$ are two cycles of order 2; the angles between these vertices also satisfy the condition for the sum in each cycle.

The hyperbolic volume of this polygon is $2\pi/3$, which coincides with the hyperbolic volume of $\Gamma(6,1)\backslash\mathcal{H}$. Therefore, this hyperbolic polygon is a fundamental domain $\mathcal{D}(6,1)$ for the Shimura curve $X(6,1)$.

The above relations between the vertices give the pairing of the edges. Thus, we have that the transformation γ_{v_2} maps the edge v_1v_2 to v_1v_6; the transformation

γ_{v_4} maps the edge v_3v_4 to v_5v_4 and the transformation γ_{v_6} maps the edge v_2v_3 to v_6v_5. From the pairing of the edges, we see that the genus is 0 and we obtain the presentation of the group given in the statement.

By using the isometric circles, it is clear that the edges v_5v_6 and v_6v_1 are segments of the isometric circle $C_{\gamma_{v_6}}$.

Consider $\varepsilon = 2+\sqrt{3}$, the fundamental unit of $\mathbb{Q}(\sqrt{3})$. Then the hyperbolic strip $S(r,\varepsilon^2 r)$ is a fundamental domain for $\Gamma(6,1)_\infty$.

For $r_3 = |v_6| = 1 = \varepsilon^2 r$, we have $r = r_1 = 7-4\sqrt{3}$. For $r = r_2 = |v_3| = 2-\sqrt{3}$, we have $r_4 = \varepsilon^2 r = 2+\sqrt{3}$. Now the symmetry conditions follow from Lemma 5.35. Moreover $C(0,r_2) = C_{\gamma_{v_3}}$ and $C(0,r_3) = C_{\gamma_{v_6}}$, hence they are principal hyperbolic lines.

Finally, it is clear that $\mathcal{D}(6,1)$ is symmetric with respect to the imaginary axis, which coincides with $L_{\gamma_{v_3}}$ and $L_{\gamma_{v_6}}$. □

5.5.3. Fundamental domain for $X(10,1)$.

THEOREM 5.47. *Consider the Shimura curve $X(10,1)$. Then the hyperbolic hexagon of vertices $(v_1,v_2,v_3,v_4,v_5,v_6)$,*

$$v_1 = \frac{-\sqrt{2}+\sqrt{3}\iota}{5(-1+\sqrt{2})},\qquad v_2 = \frac{-\sqrt{2}+\sqrt{3}\iota}{5(1+\sqrt{2})},\qquad v_3 = \frac{-\sqrt{2}+\sqrt{3}\iota}{5(7+5\sqrt{2})},$$

$$v_4 = \frac{\sqrt{2}+\sqrt{3}\iota}{5(7+5\sqrt{2})},\qquad v_5 = \frac{\sqrt{2}+\sqrt{3}\iota}{5(1+\sqrt{2})},\qquad v_6 = \frac{\sqrt{2}+\sqrt{3}\iota}{5(-1+\sqrt{2})},$$

is a fundamental domain for the Shimura curve $X(10,1)$ in the Poincaré half-plane, which we denote by $\mathcal{D}(10,1)$. It has the following properties.

(i) *All the vertices are elliptic of order 3 and the corresponding elliptic transformations fixing them are:*

$$\gamma_{v_1} = \frac{1}{2}\begin{pmatrix} 1+\sqrt{2} & 1+\sqrt{2} \\ 5(1-\sqrt{2}) & 1-\sqrt{2}\end{pmatrix},\qquad \gamma_{v_2} = \frac{1}{2}\begin{pmatrix} 1+\sqrt{2} & -1+\sqrt{2} \\ -5(1+\sqrt{2}) & 1-\sqrt{2}\end{pmatrix},$$

$$\gamma_{v_3} = \frac{1}{2}\begin{pmatrix} 1+\sqrt{2} & -7+5\sqrt{2} \\ -5(7+5\sqrt{2}) & 1-\sqrt{2}\end{pmatrix},\qquad \gamma_{v_4} = \frac{1}{2}\begin{pmatrix} 1+\sqrt{2} & 7-5\sqrt{2} \\ 5(7+5\sqrt{2}) & 1-\sqrt{2}\end{pmatrix},$$

$$\gamma_{v_5} = \frac{1}{2}\begin{pmatrix} 1+\sqrt{2} & 1-\sqrt{2} \\ 5(1+\sqrt{2}) & 1-\sqrt{2}\end{pmatrix},\qquad \gamma_{v_6} = \frac{1}{2}\begin{pmatrix} 1+\sqrt{2} & -1-\sqrt{2} \\ 5(-1+\sqrt{2}) & 1-\sqrt{2}\end{pmatrix}.$$

(ii) *The hyperbolic volume of $\mathcal{D}(10,1)$ is $V_h(10,1) = \frac{4}{3}\pi$ and the genus of the curve $X(10,1)$ is $g(10,1) = 0$.*

(iii) *The number of elliptic vertices of order 3 is $n_3(10,1) = 6$. There are $e_3(10,1) = 4$ elliptic cycles of order 3, namely $\{v_1,v_3\}$, $\{v_4,v_6\}$, $\{v_2\}$ and $\{v_5\}$. The relations between the vertices are $\gamma_{v_2}(v_3) = v_1$ and $\gamma_{v_5}(v_4) = v_6$.*

(iv) *The principal homothety of $\Gamma(10,1)$ is*

$$h = \begin{pmatrix} 3+2\sqrt{2} & 0 \\ 0 & 3-2\sqrt{2}\end{pmatrix}.$$

(v) *The pairing of the edges is:*

(v_2v_3, v_2v_1) *by the transformation* γ_{v_2},

(v_3v_4, v_1v_6) *by the transformation* h,

(v_4v_5, v_6v_5) *by the transformation* γ_{v_5}.

(vi) *We have the following presentation of the group* $\Gamma(10,1)/\pm\mathrm{Id}$:

$$\langle h, \gamma_{v_2}, \gamma_{v_5} : \gamma_{v_2}^3 = \gamma_{v_5}^3 = (h^{-1}\gamma_{v_2})^3 = (h^{-1}\gamma_{v_5})^3 = 1\rangle.$$

(vii) *The edges* v_3v_4 *and* v_1v_6 *determine a hyperbolic strip* $S(r_1, r_3)$, $r_1 = |v_3|$, $r_3 = |v_1|$ *which is a fundamental domain by the action of* $\Gamma(10,1)_\infty$.

(viii) *The other edges correspond to segments of isometric circles in the following way:* $v_1v_2 \subseteq C_{\gamma_{v_1}^{-1}}$, $v_2v_3 \subseteq C_{\gamma_{v_2}^{-1}}$, $v_4v_5 \subseteq C_{\gamma_{v_5}^{-1}}$ *and* $v_5v_6 \subseteq C_{\gamma_{v_6}^{-1}}$.

(ix) $\mathcal{D}(10,1)$ *is invariant with respect to*

$$w = \frac{1}{\sqrt{5}}\begin{pmatrix} 0 & 1-\sqrt{2} \\ 5(1+\sqrt{2}) & 0 \end{pmatrix},$$

and it is symmetric with respect to the imaginary axis and the hyperbolic line $C_w = C\left(0, 1/(\sqrt{5}(1+\sqrt{2}))\right)$.

REMARK 5.48. The above fundamental domain $\mathcal{D}(10,1)$ is the intersection of the hyperbolic strip $S(r_1, r_3)$ given in (vii) and the exterior of the isometric circles given in (viii).

PROOF. The constants attached to the Shimura curve $X(10,1)$ are $e_2(10,1) = 0$, $e_3(10,1) = 4$, $g(10,1) = 0$ and $V_h(10,1) = \frac{4}{3}\pi$.

By Proposition 2.32, in a fundamental domain such that all vertices are elliptic there must be $n_e(10,1) = 6$ vertices. Let us look for elliptic points of order 3, according to the descriptions given in Theorems 5.22 or 5.23. For example, we obtain the points and the corresponding transformations below. Note that, taking into account the change of the sign, we obtain groups of 4 solutions. Moreover, the points are symmetric with respect to the imaginary axis.

$$B_1 = \frac{-\sqrt{2}+\sqrt{3}\iota}{5(-1+\sqrt{2})}, \qquad \gamma_{B_1} = \frac{1}{2}\begin{pmatrix} 1+\sqrt{2} & 1+\sqrt{2} \\ 5(1-\sqrt{2}) & 1-\sqrt{2} \end{pmatrix},$$

$$B_2 = \frac{-\sqrt{2}+\sqrt{3}\iota}{5(1+\sqrt{2})}, \qquad \gamma_{B_2} = \frac{1}{2}\begin{pmatrix} 1+\sqrt{2} & -1+\sqrt{2} \\ -5(1+\sqrt{2}) & 1-\sqrt{2} \end{pmatrix},$$

$$B_3 = \frac{\sqrt{2}+\sqrt{3}\iota}{5(1+\sqrt{2})}, \qquad \gamma_{B_3} = \frac{1}{2}\begin{pmatrix} 1+\sqrt{2} & 1-\sqrt{2} \\ 5(1+\sqrt{2}) & 1-\sqrt{2} \end{pmatrix},$$

$$B_4 = \frac{\sqrt{2}+\sqrt{3}\iota}{5(-1+\sqrt{2})}, \qquad \gamma_{B_4} = \frac{1}{2}\begin{pmatrix} 1+\sqrt{2} & -1-\sqrt{2} \\ 5(-1+\sqrt{2}) & 1-\sqrt{2} \end{pmatrix},$$

$$B_5 = \frac{-\sqrt{2}+\sqrt{3}\iota}{5(-7+5\sqrt{2})}, \qquad \gamma_{B_5} = \frac{1}{2}\begin{pmatrix} 1+\sqrt{2} & 7+5\sqrt{2} \\ 5(7-5\sqrt{2}) & 1-\sqrt{2} \end{pmatrix},$$

$$B_6 = \frac{-\sqrt{2}+\sqrt{3}\iota}{5(7+5\sqrt{2})}, \qquad \gamma_{B_6} = \frac{1}{2}\begin{pmatrix} 1+\sqrt{2} & -7+5\sqrt{2} \\ -5(7+5\sqrt{2}) & 1-\sqrt{2} \end{pmatrix},$$

$$B_7 = \frac{\sqrt{2}+\sqrt{3}\iota}{5(7+5\sqrt{2})}, \qquad \gamma_{B_7} = \frac{1}{2}\begin{pmatrix} 1+\sqrt{2} & 7-5\sqrt{2} \\ 5(7+5\sqrt{2}) & 1-\sqrt{2} \end{pmatrix},$$

$$B_8 = \frac{\sqrt{2}+\sqrt{3}\iota}{5(-7+5\sqrt{2})}, \qquad \gamma_{B_8} = \frac{1}{2}\begin{pmatrix} 1+\sqrt{2} & -7-5\sqrt{2} \\ 5(-7+5\sqrt{2}) & 1-\sqrt{2} \end{pmatrix}.$$

We look for relations between these elliptic points, using the elliptic transformations. We select the following ones:

$$\gamma_{B_2}(B_2) = B_2, \qquad \gamma_{B_2}(B_6) = B_1,$$
$$\gamma_{B_3}(B_3) = B_3, \qquad \gamma_{B_4}(B_7) = B_4.$$

The hyperbolic transformation fixing the infinity is

$$h = \begin{pmatrix} 3+2\sqrt{2} & 0 \\ 0 & 3-2\sqrt{2} \end{pmatrix} \text{ and satisfies } h(B_6) = B_1, \ h(B_7) = B_4.$$

Hence, we have the sets of equivalent vertices: $\{B_1, B_6\}$, $\{B_4, B_7\}$, $\{B_2\}$ and $\{B_3\}$. We check that the vertices of two different sets are not equivalent; hence we already have 6 elliptic points of order 3 in 4 cycles.

Consider the hyperbolic polygon with vertices:

$$(B_1, B_2, B_6, B_7, B_3, B_4).$$

Now we perform a change of notation and write $(v_1, v_2, v_3, v_4, v_5, v_6)$ to denote the vertices of this hyperbolic polygon. It is easy to check that the above relations give the pairing of the edges stated in (v). The hyperbolic volume is $\frac{4}{3}\pi$. Thus, effectively this hyperbolic polygon is a fundamental domain $\mathcal{D}(10,1)$ for the Shimura curve $X(10,1)$. In Figure 5.2 there is its graphical representation, with the pairing of the edges.

From the pairing of the edges we check that the genus is 0. Moreover, we obtain the following presentation of the group $\Gamma(10,1)/\pm\mathrm{Id}$:

$$\langle h, \gamma_{v_2}, \gamma_{v_5} : \gamma_{v_2}^3 = 1, \gamma_{v_5}^3 = 1, (h^{-1}\gamma_{v_2})^3 = 1, (h^{-1}\gamma_{v_5})^3 = 1\rangle.$$

We note that most of the edges are segments of isometric circles, as described in statement (viii). The edges v_3v_4 and v_1v_6 are contained in the two circles $C(0,r_1)$ and $C(0,r_3)$, where $r_1 = |v_3|$ and $r_3 = |v_1|$. These circles determine the hyperbolic strip $S(r_1,r_3)$ which is a fundamental domain for $\Gamma(10,1)_\infty$, since the transformation h relates these two edges. Let ε_2 the fundamental unit of $\mathbb{Q}(\sqrt{2})$. Check that $\varepsilon_3 r_1 = r_2$ and $\varepsilon_2 r_2 = r_3$, hence, using Lemma 5.35, we have that the hyperbolic strip $S(r_1,r_3)$ is symmetric with respect to $C(0,r_2)$, $r_2 = |v_2|$.

The symmetry of $\mathcal{D}(10,1)$ with respect to the imaginary axis is evident. Moreover, it is easy to check that $\mathcal{D}(10,1)$ is invariant with respect to the transformation w and by the hyperbolic symmetry with respect to C_w. Note that these three facts are related, since the transformation w is the product of the two symmetries, and L_w is the imaginary axis. □

5.5.4. Fundamental domain for $X(15,1)$.

THEOREM 5.49. *Consider the Shimura curve $X(15,1)$. The hyperbolic polygon of vertices $(v_1, v_2, v_3, v_4, v_5, v_6)$,*

$$v_1 = \frac{-2+\iota}{5(2-\sqrt{3})}, \qquad v_2 = \frac{-2+\iota}{5(2+\sqrt{3})}, \qquad v_3 = \frac{2+\iota}{5(2+\sqrt{3})},$$
$$v_4 = \frac{8+\iota}{5(4+\sqrt{3})}, \qquad v_5 = \frac{8+\iota}{5(4-\sqrt{3})}, \qquad v_6 = \frac{2+\iota}{5(2-\sqrt{3})},$$

is a fundamental domain for the Shimura curve $X(15,1)$ in the Poincaré half-plane, which we denote by $\mathcal{D}(15,1)$. It has the following properties.

(i) *All the vertices are elliptic of order* 3 *and the corresponding elliptic transformations fixing them are* $\gamma_{v_i} = \frac{1}{2} g_{v_i}$, *where*

$$g_{v_1} = \begin{pmatrix} 1+2\sqrt{3} & 3+2\sqrt{3} \\ 5(3-2\sqrt{3}) & 1-2\sqrt{3} \end{pmatrix}, \qquad g_{v_2} = \begin{pmatrix} 1+2\sqrt{3} & -3+2\sqrt{3} \\ 5(-3-2\sqrt{3}) & 1-2\sqrt{3} \end{pmatrix},$$

$$g_{v_3} = \begin{pmatrix} 1+2\sqrt{3} & 3-2\sqrt{3} \\ 5(3+2\sqrt{3}) & 1-2\sqrt{3} \end{pmatrix}, \qquad g_{v_4} = \begin{pmatrix} 1+8\sqrt{3} & 3-4\sqrt{3} \\ 5(3+4\sqrt{3}) & 1-8\sqrt{3} \end{pmatrix},$$

$$g_{v_5} = \begin{pmatrix} 1+8\sqrt{3} & -3-4\sqrt{3} \\ 5(-3+4\sqrt{3}) & 1-8\sqrt{3} \end{pmatrix}, \qquad g_{v_6} = \begin{pmatrix} 1+2\sqrt{3} & -3-2\sqrt{3} \\ 5(-3+2\sqrt{3}) & 1-2\sqrt{3} \end{pmatrix}.$$

(ii) *The hyperbolic volume of* $\mathcal{D}(15,1)$ *is* $V_h(15,1) = \frac{8}{3}\pi$ *and the genus of the curve* $X(15,1)$ *is* $g(15,1) = 1$.

(iii) *The principal homothety of* $\Gamma(15,1)$ *is*

$$h = \begin{pmatrix} 2+\sqrt{3} & 0 \\ 0 & 2-\sqrt{3} \end{pmatrix}.$$

(iv) *We have* $n_3(15,1) = 6$ *elliptic vertices of order* 3. *There are* $e_3(15,1) = 2$ *elliptic cycles of order* 3, *namely* $\{v_1, v_2, v_4, v_5\}$ *and* $\{v_3, v_6\}$. *The relations between these vertices are*

$$h(v_2) = v_1, \qquad h(v_3) = v_6,$$
$$\beta(v_1) = v_5, \qquad \beta(v_2) = v_4,$$
$$\gamma(v_6) = v_3, \qquad \gamma(v_5) = v_4,$$

where

$$\beta = \frac{1}{2}\begin{pmatrix} 3 & 1 \\ 5 & 3 \end{pmatrix}, \qquad \gamma = \frac{1}{2}\begin{pmatrix} -4+3\sqrt{3} & -\sqrt{3} \\ 5(\sqrt{3}) & -4-3\sqrt{3} \end{pmatrix}.$$

(v) *The pairing of the edges is*:

(v_1v_2, v_5v_4) *by the transformation* β,
(v_2v_3, v_1v_6) *by the transformation* h,
(v_3v_4, v_6v_5) *by the transformation* γ^{-1}.

(vi) $\Gamma(15,1)/\pm\mathrm{Id} = \langle h, \beta, \gamma : (\gamma h)^3 = (h\beta^{-1}\gamma\beta)^3 = 1\rangle$.

(vii) *Put* $r_1 = \sqrt{5}(2-\sqrt{3})/5$, $r_2 = \sqrt{5}/5$ *and* $r_3 = \sqrt{5}(2+\sqrt{3})/5$. $\mathcal{D}(15,1)$ *is included in the hyperbolic strip* $S(r_1, r_3)$, *which is a fundamental domain by the action of* $\Gamma(15,1)_\infty$, *symmetric with respect to the principal hyperbolic line* $C(0, r_2)$. *The edges* v_2v_3 *and* v_1v_6 *are on the hyperbolic lines determined by* $S(r_1, r_3)$, *which are, moreover, principal lines.*

(viii) *The other edges correspond to segments of isometric circles in the following way:* $v_1v_2 \subseteq C_{\gamma_{v_1}}$, $v_3v_4 \subseteq C_{\gamma_{v_2}}$, $v_4v_5 \subseteq C_{\gamma_{v_5}}$ *and* $v_5v_6 \subseteq C_{\gamma_{v_6}}$. *Moreover* $\mathcal{D}(15,1)$ *is the intersection between* $S(r_1, r_3)$ *and the exterior of these isometric circles.*

(ix) $\mathcal{D}(15,1)$ *is symmetric with respect to the principal hyperbolic line given by* $C(0, \sqrt{5}/5)$.

PROOF. The constants attached to the Shimura curve $X(15,1)$ are $e_2(15,1) = 0$, $e_3(15,1) = 2$, $g(15,1) = 1$ and $V_h(15,1) = \frac{8}{3}\pi$.

In a fundamental domain such that all the vertices are elliptic, there must be $n_e(15,1) = 6$ vertices, by Proposition 2.32. We look for elliptic points of order

3, according to the descriptions given in Theorems 5.22 or 5.23. For example, we obtain the points and the corresponding transformations below. Note that the points are placed symmetrically with respect to the imaginary axis.

$$B_1 = \frac{-2+\iota}{5(2-\sqrt{3})}, \qquad \gamma_{B_1} = \frac{1}{2}\begin{pmatrix} 1+2\sqrt{3} & 3+2\sqrt{3} \\ 5(3-2\sqrt{3}) & 1-2\sqrt{3} \end{pmatrix},$$

$$B_2 = \frac{-2+\iota}{5(2+\sqrt{3})}, \qquad \gamma_{B_2} = \frac{1}{2}\begin{pmatrix} 1+2\sqrt{3} & -3+2\sqrt{3} \\ 5(-3-2\sqrt{3}) & 1-2\sqrt{3} \end{pmatrix},$$

$$B_3 = \frac{2+\iota}{5(2+\sqrt{3})}, \qquad \gamma_{B_3} = \frac{1}{2}\begin{pmatrix} 1+2\sqrt{3} & 3-2\sqrt{3} \\ 5(3+2\sqrt{3}) & 1-2\sqrt{3} \end{pmatrix},$$

$$B_4 = \frac{2+\iota}{5(2-\sqrt{3})}, \qquad \gamma_{B_4} = \frac{1}{2}\begin{pmatrix} 1+2\sqrt{3} & -3-2\sqrt{3} \\ 5(-3+2\sqrt{3}) & 1-2\sqrt{3} \end{pmatrix},$$

$$B_5 = \frac{-8+\iota}{5(4-\sqrt{3})}, \qquad \gamma_{B_5} = \frac{1}{2}\begin{pmatrix} 1+8\sqrt{3} & 3+4\sqrt{3} \\ 5(3-4\sqrt{3} & 1-8\sqrt{3} \end{pmatrix},$$

$$B_6 = \frac{-8+\iota}{5(4+\sqrt{3})}, \qquad \gamma_{B_6} = \frac{1}{2}\begin{pmatrix} 1+8\sqrt{3} & -3+4\sqrt{3} \\ 5(-3-4\sqrt{3} & 1-8\sqrt{3} \end{pmatrix},$$

$$B_7 = \frac{8+\iota}{5(4+\sqrt{3})}, \qquad \gamma_{B_7} = \frac{1}{2}\begin{pmatrix} 1+8\sqrt{3} & 3-4\sqrt{3} \\ 5(3+4\sqrt{3} & 1-8\sqrt{3} \end{pmatrix},$$

$$B_8 = \frac{8+\iota}{5(4-\sqrt{3})}, \qquad \gamma_{B_8} = \frac{1}{2}\begin{pmatrix} 1+8\sqrt{3} & -3-4\sqrt{3} \\ 5(-3+4\sqrt{3} & 1-8\sqrt{3} \end{pmatrix}.$$

We look for relations between these elliptic points. In this case, the elliptic transformations do not give any relation. If we consider the principal homothety h, we have

$$h = \begin{pmatrix} 2+\sqrt{3} & 0 \\ 0 & 2-\sqrt{3} \end{pmatrix}, \quad h(B_1) = B_2,\ h(B_3) = B_4.$$

To find other relations we look for new hyperbolic transformations, applying the results of embeddings of fundamental units of real quadratic orders. Consider the quadratic field $\mathbb{Q}(\sqrt{5})$, splitting $H_B(3,5)$. The fundamental unit of the ring of integers is $\varepsilon_5 = (1+\sqrt{5})/2$, of norm $\mathrm{n}(\varepsilon_5) = -1$. Fixing an embedding φ of the ring of integers of $\mathbb{Q}(\sqrt{5})$ in $\mathcal{O}(15,1)$, we obtain the transformation

$$\beta = \Phi(\varphi(\varepsilon_5^2)) = \frac{1}{2}\begin{pmatrix} 3 & 1 \\ 5 & 3 \end{pmatrix}, \quad \text{satisfying } \beta(B_2) = B_8 \text{ and } \beta(B_1) = B_7.$$

We look directly for relations between the vertices. We find:

$$\gamma = \frac{1}{2}\begin{pmatrix} -4+3\sqrt{3} & -\sqrt{3} \\ 5(\sqrt{3}) & -4-3\sqrt{3} \end{pmatrix}, \quad \text{satisfying } \gamma(B_4) = B_3 \text{ and } \gamma(B_8) = B_7.$$

Thus, we have the sets of equivalent vertices $\{B_1, B_2, B_5, B_6\}$ and $\{B_3, B_4\}$. After checking that the vertices of two different sets are not equivalent, we have 6 elliptic points of order 3 in 2 cycles.

Consider the hyperbolic polygon formed by the 6 elliptic points:

$$(B_1, B_2, B_3, B_7, B_8, B_4).$$

We change the notation and denote its vertices by $(v_1, v_2, v_3, v_4, v_5, v_6)$. The above relations give a pairing of the edges. Its hyperbolic volume is $\frac{8}{3}\pi$. Thus, effectively the hyperbolic polygon $\mathcal{D}(15,1)$ is a fundamental domain for the Shimura curve $X(15,1)$. Its graphical representation is in Figure 5.3, setting the pairing of the edges.

By taking into account the Euler–Poincaré characteristic, we see that the genus is 1. Moreover, we obtain the following presentation of the group $\Gamma(15,1)/\pm\mathrm{Id}$:

$$\langle h, \beta, \gamma : (\gamma h)^3 = 1, (h\beta^{-1}\gamma\beta)^3 = 1\rangle.$$

We note that most of the edges are segments of isometric circles, as described in statement (viii). The edges v_2v_3 and v_1v_6 are in the two circles $C(0,r_1)$ and $C(0,r_3)$, where $r_1 = |v_2|$ and $r_3 = |v_1|$. Those circles determine the hyperbolic strip $S(r_1,r_3)$ which is a fundamental domain for $\Gamma(15,1)_\infty$, since the transformation h relates these two edges. We denote by ε_3 the fundamental unit of $\mathbb{Q}(\sqrt{3})$. Put $r_2 := \varepsilon_3 r_1 = \sqrt{5}/5$; then $\varepsilon_3 r_2 = r_3$, hence $S(r_1,r_3)$ is symmetric with respect to $C(0,r_2)$, using Lemma 5.35. The hyperbolic lines defined by $C(0,r_i)$ are principal lines, since they come from the transformations η_i given by:

$$\eta_1 = \frac{1}{\sqrt{15}}\begin{pmatrix} 0 & 3-2\sqrt{3} \\ 5(3+2\sqrt{3}) & 0 \end{pmatrix}, \qquad \eta_2 = \frac{1}{\sqrt{5}}\begin{pmatrix} 0 & -1 \\ 5 & 0 \end{pmatrix},$$
$$\eta_3 = \frac{1}{\sqrt{15}}\begin{pmatrix} 0 & -3-2\sqrt{3} \\ 5(-3+2\sqrt{3}) & 0 \end{pmatrix}.$$

It is clear that the fundamental domain $\mathcal{D}(15,1)$ is the intersection of the hyperbolic strip and the exterior of the isometric circles that contains the other edges.

Finally, it is easy to check that $\mathcal{D}(15,1)$ is invariant under the hyperbolic symmetry with respect to the principal hyperbolic line $C(0,r_2)$. □

Figure 5.1 represents the fundamental domain for the Shimura curve $X(6,1)$ given in Theorem 5.46. We collect the corresponding data in Table 5.1.

The fundamental domain constructed for the Shimura curve $X(10,1)$, cf. Theorem 5.47, can be seen in Figure 5.2. We also represent the intersection of the isometric circle C_w with the fundamental domain. It is easy to check graphically that the domain is invariant under the transformation w and that C_w is an axis of symmetry of the domain. Table 5.2 contains the explicit points and cycles.

We reproduce in Figure 5.3 the fundamental domain given in Theorem 5.49 for the Shimura curve $X(15,1)$. This graphical representation, also obtained by Michon, can be found in [**Vig80**]. We represent the principal hyperbolic line giving the principal symmetry in the domain. In Table 5.3, we present data about points and cycles.

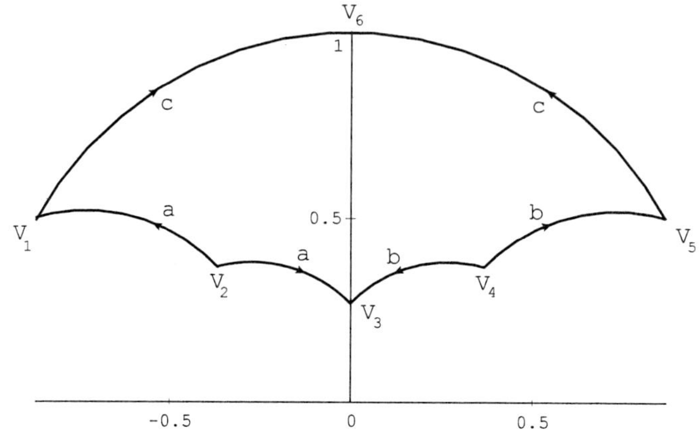

FIGURE 5.1. Fundamental domain for $X(6,1)$.

TABLE 5.1. Elliptic cycles of the Shimura curve $X(6,1)$ and presentation of the group $\Gamma(6,1)/\pm\mathrm{Id}$.

k	$n_k(6,1)$	$e_k(6,1)$	cycles of order k of $\Gamma(6,1)$
2	4	2	$\{v_6\}$ and $\{v_1, v_3, v_5\}$
3	2	2	$\{v_2\}$ and $\{v_4\}$

generators	relations
$\gamma_{v_2}, \gamma_{v_4}, \gamma_{v_6}$	$\gamma_{v_2}^3 = 1, \gamma_{v_4}^3 = 1, \gamma_{v_6}^2 = 1, (\gamma_{v_2}^{-1}\gamma_{v_6}\gamma_{v_4})^2 = 1$

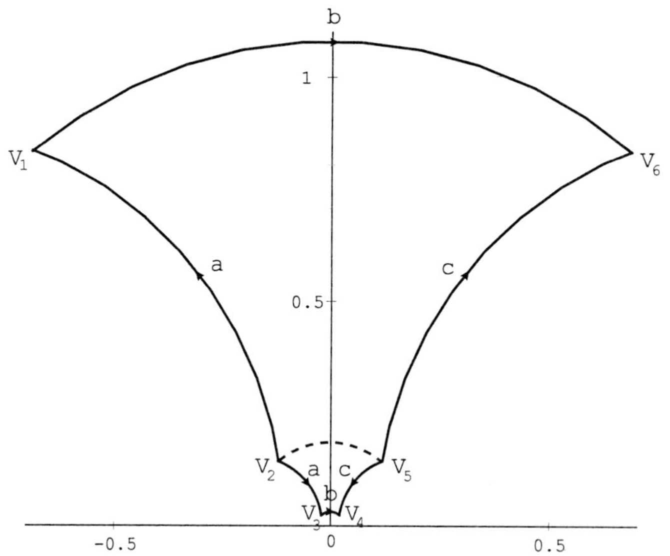

FIGURE 5.2. Fundamental domain for $X(10,1)$.

TABLE 5.2. Elliptic cycles of the Shimura curve $X(10,1)$ and presentation of the group $\Gamma(10,1)/\pm\mathrm{Id}$.

k	$n_k(10,1)$	$e_k(10,1)$	cycles of order k of $\Gamma(10,1)$
2	0	0	
3	6	4	$\{v_2\},\{v_5\},\{v_1,v_3\},\{v_4,v_6\}$

generators	relations
$\gamma_{v_2},\gamma_{v_5},h$	$\gamma_{v_2}^3=1,\gamma_{v_5}^3=1,(h^{-1}\gamma_{v_2})^3=1,(h^{-1}\gamma_{v_5})^3=1$

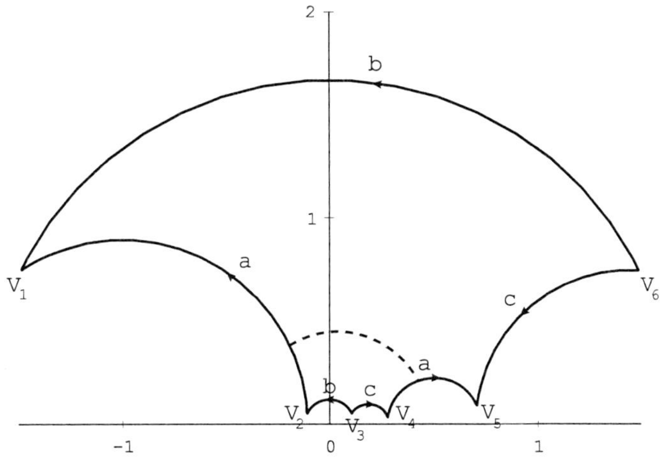

FIGURE 5.3. Fundamental domain for $X(15,1)$.

TABLE 5.3. Elliptic cycles of the Shimura curve $X(15,1)$ and presentation of the group $\Gamma(15,1)/\pm\mathrm{Id}$.

k	$n_k(15,1)$	$e_k(15,1)$	cycles of order k of $\Gamma(15,1)$
2	0	0	
3	6	2	$\{v_1,v_2,v_4,v_5\},\{v_3,v_6\}$

generators	relations
h,k,γ	$(\gamma h)^3=1,(hk^{-1}\gamma k)^3=1$

CHAPTER 6

Complex Multiplication Points in Shimura Curves

In this chapter we study complex multiplication points of Shimura curves $X(D, N)$ using results on the hyperbolic uniformization and the relation between embeddings and quadratic forms from earlier chapters. In Section 6.1, we give definitions and results relating those points with the representations of the ternary forms and with the binary forms of Chapter 4, and we count the number of complex multiplication points for an imaginary quadratic order in Theorem 6.13. In Section 6.2, we introduce the special complex multiplication points, which allow us to restrict ourselves to a finite set of quadratic orders and the corresponding complex points, cf. Corollary 6.20. In Section 6.3, we give results for the calculation of the complex multiplication points in terms of quadratic forms, cf. Theorem 6.21. We deduce results for the nonramified case, and the small ramified cases of type A or B. We collect the graphic results in Sections 6.4 and 6.5. The commands implemented allow us to classify imaginary quadratic orders for which a curve $X(D, N)$ has special complex multiplication points, to obtain the list of quadratic orders with special complex multiplication and to compute the complex multiplication points for a fixed imaginary quadratic order.

As in Proposition 1.19, Φ stands for the fixed embedding of H into $\mathrm{M}(2, \mathbb{R})$.

6.1. Classification of complex multiplication points

To begin with, we present some results in order to give the definition of complex multiplication points.

LEMMA 6.1. *Let $\gamma, \gamma' \in \mathrm{GL}(2, \mathbb{R})$. Then, γ and γ' have the same fixed points if and only if there exist $\lambda, \mu \in \mathbb{R}$, $\lambda \neq 0$, such that $\gamma' = \lambda\gamma + \mu \,\mathrm{Id}$.*

PROOF. Put

$$\gamma = \begin{pmatrix} a & b \\ c & d \end{pmatrix}, \qquad \gamma' = \begin{pmatrix} a' & b' \\ c' & d' \end{pmatrix}$$

and consider the quadratic equations with real coefficients $cZ^2 + (d - a)Z - b = 0$ and $c'Z^2 + (d' - a')Z - b' = 0$, that give rise to the corresponding fixed points. The fixed points are the same if and only if the equation is obtained from the other by multiplication by a nonzero scalar. If $\gamma' = \lambda\gamma + \mu \,\mathrm{Id}$, it is clear that this condition is satisfied. Conversely, if one equation is a multiple of the other, it is enough to put $\lambda := c'/c$ and $\mu := a' - \lambda a$ to obtain the relation $\gamma' = \lambda\gamma + \mu \,\mathrm{Id}$. □

Using that the embeddings preserve the norm and the trace, we obtain the following corollary.

COROLLARY 6.2. *Let F be a quadratic field such that $\mathcal{E}(H, F) \neq \varnothing$. Let $\varphi \in \mathcal{E}(H, F)$. Then, all the transformations $\gamma \neq \mathrm{id}$ in the image $\Phi(\varphi(F^*))$ have the* *same fixed points. If F is an imaginary quadratic field, then they have a unique fixed point in $\mathcal{H}$.*

NOTATION 6.3. Given a quaternion algebra H, let us denote by $\mathcal{J}(H)$ the set of imaginary quadratic fields F such that $\mathcal{E}(H,F) \neq \varnothing$. By Corollary 6.2, we can define a mapping from that set of embeddings to the Poincaré half-plane:

$$\bigcup_{F\in\mathcal{J}(H)} \mathcal{E}(H,F) \to \mathcal{H} \qquad \varphi \mapsto z(\varphi),$$

where $z(\varphi)$ denotes the unique fixed point of $\Phi(\varphi(\alpha))$ in $\mathcal{H}$ for any $\alpha \in F^*$, $\alpha \notin \mathbb{Q}^*$.

DEFINITION 6.4. Let $F \in \mathcal{J}(H)$ and let Λ be an order of F. A point $z \in \Gamma(D,N)\backslash\mathcal{H}$ is called a complex multiplication point (CM) by Λ if there exists an optimal embedding $\varphi \in \mathcal{E}^*(\mathcal{O}(D,N),\Lambda)$ such that $z = z(\varphi)$. In this case, we say also that the point $j_{D,N}(z) \in X(D,N)(\mathbb{C})$ is a complex multiplication point by the order $\Lambda(d,m)$. We denote by $\mathrm{CM}(D,N,d,m)$ the set of points in $\Gamma(D,N)\backslash\mathcal{H}$ of complex multiplication by $\Lambda(d,m)$.

REMARK 6.5. Let $z \in \mathcal{H}$ be the fixed point of a transformation $\gamma = \Phi(\omega) \in \mathrm{GL}(2,\mathbb{R})$, for some $\omega \in H\setminus\mathbb{Q}$. Then, by Lemma 5.11, the quaternion ω determines the quadratic field $F_\omega = \mathbb{Q}(\sqrt{d})$, where $d = \mathrm{tr}(\gamma)^2 - 4\det(\gamma)$, and we have an embedding $\varphi \in \mathcal{E}(H,F_\omega)$ such that $\omega \in \varphi(F_\omega)$. The embedding φ satisfies $z(\varphi) = z$. Note that F_ω is necessarily imaginary, since $\mathrm{Im}(z) > 0$ implies that $\mathrm{tr}(\gamma^2) - 4\det(\gamma) < 0$. This justifies the terminology of complex multiplication.

PROPOSITION 6.6. *Let $\Lambda(d,m)$ be the quadratic order of conductor m in the imaginary quadratic field $F = \mathbb{Q}(\sqrt{d})$. Then, $\mathrm{CM}(D,N,d,m) \neq \varnothing$ if and only if the following equivalent conditions are satisfied:*

(i) $\mathcal{E}^*(\mathcal{O}(D,N),\Lambda(d,m)) \neq \varnothing$.

(ii) $\mathcal{R}^*(n_{\mathbb{Z}+2\mathcal{O}(D,N),3}, -D_{\Lambda(d,m)}; \mathbb{Z}) \neq \varnothing$.

(iii) $\mathcal{H}^*(\mathbb{Z}+2\mathcal{O}(D,N),\Lambda(d,m)) \neq \varnothing$.

If N is a square-free integer, the above conditions are further equivalent to the fact that: $\gcd(D,m) = 1$, $\left(\frac{D_F}{p}\right) \neq 1$ *if* $p \mid D$, *and* $\left(\frac{D_F}{p}\right) \neq -1$ *if* $p \mid N$ *and* $p \nmid m$.

PROOF. The existence of a CM point by $\Lambda(d,m)$ in $X(D,N)$ corresponds to (i), by definition. The equivalence between conditions (i) and (ii) was stated in Theorem 4.26. Conditions (i) and (iii) are equivalent by Theorem 4.57.

If N is square-free, take into account that (i) is equivalent to the fact that the class number $\nu(D,N,d,m;\mathcal{O})$ is nonzero, where $\mathcal{O} = \mathcal{O}(D,N)$. Then, apply the results about the class number given in Corollary 4.20 and the condition of having embeddings of a quadratic field into a quaternion algebra given in Lemma 4.3. $\square$

By Proposition 6.6, for any given Shimura curve $X(D,N)$ there are infinitely many imaginary quadratic orders $\Lambda(d,m)$ such that $\mathrm{CM}(D,N,d,m) \neq \varnothing$. Thus, any Shimura curve has infinitely many CM points. Next we show general properties for CM points.

LEMMA 6.7. *A point of $X(D,N)$ is elliptic if and only if it is a* CM *point by the ring of integers of $\mathbb{Q}(\sqrt{-1})$ or $\mathbb{Q}(\sqrt{-3})$.*

By virtue of Lemma 6.7, we obtain the following corollary with explicit conditions about the existence of elliptic points in the nonramified and small ramified cases of type A and of type B.

COROLLARY 6.8. *Let $X(D,N)$ be a Shimura curve corresponding to a nonramified or small ramified of type* A *or of type* B *quaternion algebra. Suppose N is prime. Then:*

(i) $X(1,N) = X_0(N)$ *has elliptic points if and only if either* $\left(\frac{-1}{N}\right) \neq -1$ *or* $\left(\frac{-3}{N}\right) \neq -1$.

(ii) $X(2p,N)$ *has elliptic points if and only if either* $\left(\frac{-1}{N}\right) \neq -1$, *or* $\left(\frac{-3}{p}\right) \neq 1$ *and* $\left(\frac{-3}{N}\right) \neq -1$.

(iii) $X(pq,N)$ *has elliptic points if and only if* $\left(\frac{-3}{p}\right) \neq 1$, $\left(\frac{-3}{q}\right) \neq 1$, *and* $\left(\frac{-3}{N}\right) \neq -1$.

We have seen that all the elliptic points are CM; but the converse, evidently, is not true. Note that the elliptic transformations $\gamma \in \Gamma(D,N)$ satisfy that $\det(\gamma) = 1$ and $\operatorname{tr}(\gamma) \in \{0,1\}$. For the CM points, we can obtain a general condition.

LEMMA 6.9. *Let $z \in \mathrm{CM}(D,N,d,m)$ be a CM point by $\Lambda(d,m)$. Then, there exists a matrix $\gamma \in \phi(\Lambda(d,m)^*) \subseteq \mathrm{GL}(2,\mathbb{R})$ such that $\gamma(z) = z$ and* $\operatorname{tr}(\gamma) = 0$.

PROOF. Let $F = \mathbb{Q}(\sqrt{d})$. Let $\varphi \in \mathcal{E}^*\big(\mathcal{O}(D,N),\Lambda(d,m)\big)$ such that $z = z(\varphi)$; that is, z is the common fixed point of $\phi(F^*)$. Consider $\gamma = \Phi\big(\varphi(m\sqrt{d})\big)$; it is clear that $\gamma \in \Phi\big(\mathcal{O}(D,N)\big)$ and satisfies $\operatorname{tr}(\gamma) = 0$. □

It is not possible to obtain a similar result for the determinant. Not all the CM points are fixed points of a transformation with a prescribed value of the determinant.

By construction, obtaining the points of $\mathrm{CM}(D,N,d,m)$ is equivalent to obtaining the embeddings of $\Lambda(d,m)$ into the quaternion order $\mathcal{O}(D,N)$. But, note that the complex points of the curve $X(D,N)$ come from $\Gamma(D,N)$-classes of points of $\mathcal{H}$.

Fix an imaginary quadratic order $\Lambda(d,m)$, in such a way that $\mathrm{CM}(D,N,d,m) \neq \varnothing$. Given an optimal embedding $\varphi \in \mathcal{E}^*\big(\mathcal{O}(D,N),\Lambda(d,m)\big)$, consider $z(\varphi) \in \mathcal{H}$ the fixed point of the transformation $\Phi\big(\varphi(\alpha)\big)$, for $\alpha \in \mathbb{Q}(\sqrt{d})$, $\alpha \notin \mathbb{Q}$. Next, we prove the relation between the $\mathcal{O}(D,N)^*_+$-equivalence of embeddings and the $\Gamma(D,N)$-equivalence of the corresponding fixed points giving the explicit relation between the class numbers.

REMARK 6.10. Given $\varphi \in \mathcal{E}(H,F)$, $F = \mathbb{Q}(\sqrt{d})$, put $-\varphi$ the composition of φ with the Galois conjugation in F; that is, $(-\varphi)(\sqrt{d}) := -\varphi(\sqrt{d})$. Note that $-\varphi$ is also an embedding of F into H, since it satisfies $\operatorname{tr}\big((-\varphi)(\sqrt{d})\big) = -\operatorname{tr}(\sqrt{d}) = 0$ and $\mathrm{n}\big((-\varphi)(\sqrt{d})\big) = (-1)^2\,\mathrm{n}(\sqrt{d})$; that is, $-\varphi \in \mathcal{E}(H,F)$. If F is an imaginary field, then φ and $-\varphi$ have the same fixed points. Moreover, if $\varphi \in \mathcal{E}(\mathcal{O},\Lambda)$, we have that $-\varphi \in \mathcal{E}(\mathcal{O},\Lambda)$. The equalities $-\varphi(\Lambda) = \varphi(\Lambda)$ and $-\varphi(F) = \varphi(F)$ assure us that both embeddings are either simultaneously optimal or not.

PROPOSITION 6.11. *Let $\varphi, \varphi' \in \mathcal{E}^*\big(\mathcal{O}(D,N),\Lambda(d,m)\big)$. Then, $z = z(\varphi)$ and $z' = z(\varphi')$ are $\Gamma(D,N)$-equivalent if and only if φ' is $\mathcal{O}(D,N)^*_+$-equivalent to φ or to $-\varphi$.*

PROOF. Assume φ' is $\mathcal{O}(D,N)^*_+$-equivalent to $\pm\varphi$. Since $z(\varphi) = z(-\varphi)$, we can reduce to the case that φ' is $\mathcal{O}(D,N)^*_+$-equivalent to φ. Thus, there exists $\sigma \in \mathcal{O}(D,N)$, with norm $\mathrm{n}(\sigma) = 1$, such that $\varphi'(\alpha) = \sigma^{-1}\varphi(\alpha)\sigma$, for all $\alpha \in \mathbb{Q}(\sqrt{d})$. Since Φ is an embedding, if we put $P := \Phi(\sigma)$, $\gamma := \Phi\big(\varphi(\alpha)\big)$ and $\gamma' := \Phi\big(\varphi'(\alpha)\big)$, the former relation is equivalent to $\gamma' = P^{-1}\gamma P$, with $P \in \Gamma(D,N)$. Now it is clear that if γ fixes z, then γ' fixes $P^{-1}z$. Hence $z' = P^{-1}z$, and z and z' are $\Gamma(D,N)$-equivalent.

Conversely, assume that $z(\varphi) = Pz(\varphi')$, for some transformation $P \in \Gamma(D,N)$. Let $\sigma \in \mathcal{O}_+^*$ be such that $\Phi(\sigma) = P$. Let $0 \neq \alpha \in \mathbb{Q}(\sqrt{d})$, with $\operatorname{tr}(\alpha) = 0$. Consider the transformations $\gamma := \Phi\big(\varphi(\alpha)\big)$ and $\gamma' := \Phi\big(\varphi'(\alpha)\big)$, which fix z and z', respectively. Then,

$$\gamma'' := \Phi\big(\varphi^{\sigma}(\alpha)\big) = \Phi(\sigma)^{-1}\Phi\big(\varphi(\alpha)\big)\Phi(\sigma) = P^{-1}\gamma P$$

is a transformation that also fixes z'. But two transformations γ' and γ'' have the same fixed point if and only if there exist $\lambda, \mu \in \mathbb{R}$, $\lambda \neq 0$ such that $\gamma' = \lambda\gamma'' + \mu \operatorname{Id}$, by Lemma 6.1.

On the one hand, since Φ, φ and φ' preserve the trace, we have that $\operatorname{tr}(\gamma') = \operatorname{tr}(\alpha) = \operatorname{tr}(\gamma'')$. On the other hand, using the relation between γ' and γ'' we have that $\operatorname{tr}(\gamma') = \lambda \operatorname{tr}(\gamma'') + 2\mu$. Since $\operatorname{tr}(\alpha) = 0$, we deduce that $\mu = 0$. About the norm, we have that $\det(\gamma') = \operatorname{n}(\alpha) = \det(\gamma'')$, $\det(\gamma') = \lambda^2 \det(\gamma'')$ and $\operatorname{n}(\alpha) \neq 0$, and it follows that $\lambda^2 = 1$. Hence, $\gamma' = \pm\gamma''$. □

LEMMA 6.12. *Let H be an indefinite quaternion algebra and F an imaginary quadratic field. Let $\varphi \in \mathcal{E}(H,F)$ and consider $-\varphi$. Then, φ and $-\varphi$ are not $\mathcal{O}_+^*$-equivalent, for any quaternion order $\mathcal{O} \subseteq H$.*

PROOF. Fix $\alpha \in F$, $\alpha \neq 0$, and put $\gamma := \Phi\big(\varphi(\alpha)\big)$. We have that $\det(\gamma) = n(\alpha) > 0$, since F is imaginary. Hence, by Lemma 2.14, $P^{-1}\gamma P \neq -\gamma$ for all $P \in \operatorname{GL}(2,\mathbb{R})$. Thus, there exists no $\sigma \in \mathcal{O}(D,N)_+^*$ relating φ and $-\varphi$. □

Now we obtain the following theorem about the number of points of $X(D,N)$ with complex multiplication by an imaginary quadratic order. The class number of $\mathcal{O}(D,N)^*$-equivalent optimal embeddings of the quadratic order $\Lambda(d,m)$ into the quaternion order $\mathcal{O}(D,N)$ is denoted by $\nu(D,N,d,m;\mathcal{O}(D,N)^*)$. The formulas are given in Theorem 4.19.

THEOREM 6.13. *Let $X(D,N)$ be a Shimura curve. Fix an imaginary quadratic order $\Lambda(d,m)$. Then:*

(i) *The number of complex multiplication points of $X(D,N)$ by the order $\Lambda(d,m)$ is finite. Let us denote it by $\operatorname{cm}(D,N,d,m)$.*
(ii) $\operatorname{cm}(D,N,d,m) = \operatorname{h}(D,N,d,m)$.

PROOF. The complex multiplication points of $X(D,N)$ by $\Lambda(d,m)$ are defined from the set of optimal embeddings $\mathcal{E}^*\big(\mathcal{O}(D,N),\Lambda(d,m)\big)$. In Proposition 6.11 we proved that $\mathcal{O}(D,N)_+^*$-equivalent embeddings give the same CM point. Since the class number

$$\nu(D,N,d,m;\mathcal{O}(D,N)_+^*)$$

is finite, this assures us that the number of points in $\operatorname{CM}(D,N,d,m)$ is also finite.

By Corollary 4.59, we have that $\operatorname{h}(D,N,d,m) = \nu(D,N,d,m;\mathcal{O}(D,N)^*)$. Hence, it is enough to prove the equality $\operatorname{cm}(D,N,d,m) = \nu(D,N,d,m;\mathcal{O}(D,N)^*)$.

In Proposition 6.11 we also proved that, up to $\mathcal{O}(D,N)_+^*$-equivalence, the only one possibility for two embeddings to give the same point is that one is obtained from the other by conjugation in the quadratic field. Lemma 6.12 assures us that in this case they are never equivalent. By Remark 6.10, if $\varphi \in \mathcal{E}(\mathcal{O},\Lambda)$, then

$-\varphi \in \mathcal{E}(\mathcal{O}, \Lambda)$. Hence, the number of CM points is exactly $\mathrm{cm}(D, N, d, m) = \frac{1}{2}\nu(D, N, d, m; \mathcal{O}(D, N)_+^*)$.

Finally, the relation $\nu(D, N, d, m; \mathcal{O}(D, N)_+^*) = 2\nu(D, N, d, m; \mathcal{O}(D, N)^*)$ completes the proof, cf. Proposition 4.23. □

We will now be concerned with an application of the complex multiplication points as the points associated to definite binary quadratic forms and the explicit construction of fundamental domains for Shimura curves.

We consider the relation between the points of the Poincaré half-plane and the definite binary forms with real coefficients, cf. Definition 2.11. Then, we define the concept of $\Gamma(D, N)$-reduced binary quadratic form relative to a fixed fundamental domain for the Shimura curve $X(D, N)$. The binary quadratic form involved are those associated to the order $\mathcal{O}(D, N)$.

DEFINITION 6.14. Fix a fundamental domain $\mathcal{D}(D, N)$ for the Shimura curve $X(D, N)$, constructed from the quaternion order $\mathcal{O}(D, N)$ and the group $\Gamma(D, N)$. Given an imaginary quadratic order $\Lambda(d, m)$, consider the set $\mathcal{H}^*(D, N, d, m)$. We say that a definite $(\mathcal{O}(D, N), \Lambda(d, m))$-primitive binary form $f \in \mathcal{H}^*(D, N, d, m)$ is $\Gamma(D, N)$-almost-reduced if $\tau(f) \in \mathcal{D}(D, N)$, where $\tau(f) \in \mathcal{H}$ is defined in 2.11.

Let $f \in \mathcal{H}^*(D, N, d, m)$ be a $\Gamma(D, N)$-almost-reduced form such that $\tau(f)$ belongs to the boundary of the fundamental domain. Then there exists $f' \neq f$, $f' \in \mathcal{H}^*(D, N, d, m)$ such that $\tau(f')$ is also at the boundary of the domain and it is $\Gamma(D, N)$-equivalent to $\tau(f)$. In this case, we consider only one of the two binary forms. Taking into account this choice, we call the $\Gamma(D, N)$-almost-reduced forms $\Gamma(D, N)$-reduced forms.

From the relation between the number of definite primitive $\Gamma(D, N)$-reduced binary quadratic forms f in $\mathcal{H}^*(D, N, d, m)$ and the number of inequivalent points $\tau(f)$, we obtain the result below.

THEOREM 6.15. *Consider the set of primitive binary forms* $\mathcal{H}^*(D, N, d, m)$.

(i) *The number of* $\Gamma(D, N)$*-reduced forms in* $\mathcal{H}^*(D, N, d, m)$ *is finite and equal to* $2\,\mathrm{h}(D, N, d, m)$.
(ii) *The number of positive definite* $\Gamma(D, N)$*-reduced binary forms contained in* $\mathcal{H}^*(D, N, d, m)$ *is equal to* $\mathrm{h}(D, N, d, m)$.

PROOF. It is clear that to each $\Gamma(D, N)$-reduced form $f \in \mathcal{H}^*(D, N, d, m)$ we can assign a complex multiplication point $\tau(f)$ by the quadratic order $\Lambda = \Lambda(d, m)$, in such a way that $\det_2(f) = -D_\Lambda = -D_F m^2$. However, this correspondence is not injective. But in fact the obstruction is that $f, g \in \mathcal{H}^*(D, N, d, m)$ satisfy $\det_2(f) = \det_2(g)$ and $\tau(f) = \tau(g)$ if and only if $g = -f$. Hence, the number of $\Gamma(D, N)$-reduced binary forms of $\mathcal{H}^*(D, N, d, m)$, with fixed even determinant, is twice the number of complex multiplication points. Thus, the proof of (i) is completed applying the relation $\mathrm{cm}(D, N, d, m) = \mathrm{h}(D, N, d, m)$, given in Theorem 6.13.

Taking into account that in $\mathcal{H}^*(D, N, d, m)$ positive and negative definite forms are included, statement (ii) follows. □

Thus, this is a good justification for the notation $\mathrm{h}(D, N, d, m)$ used in Section 4.5, since it coincides with the standard notation for binary quadratic forms with integer coefficients.

In particular, the fundamental domain $\mathcal{D}(1,N)$ given in Theorem 5.1 allows us to define the concept of $\Gamma(1,N)$-reduced form for the positive definite binary forms $f=(Na,b,c)$ with a,b,c coprime integers and N prime.

In the same way, the knowledge of explicit fundamental domains for some Shimura curves allows us to define this concept in a wider context. Thus, for $D=6$, $D=10$, $D=15$ and $N=1$, we have the concept of $\Gamma(D,N)$-reduced binary form.

6.2. Special complex multiplication points

Restrictions over the value of the determinant lead us to consider restrictions over the quadratic orders. Using the principal hyperbolic lines defined in the previous chapter, we give the definition below.

DEFINITION 6.16. Let $z\in \mathrm{CM}(D,N,d,m)$ corresponding to an embedding $\varphi\in \mathcal{E}^*\big(\mathcal{O}(D,N),\Lambda(d,m)\big)$. We call z special complex multiplication point (SCM) if z is the fixed point of a transformation $\gamma\in\phi(\Lambda(d,m)^*)\subseteq \mathrm{GL}(2,\mathbb{R})$ with determinant equal to $D_{\mathcal{O}(D,N)}=DN$.

Note that in the case $D=N=1$, that is, for $\Gamma(1,1)=\mathrm{SL}(2,\mathbb{Z})$, the special complex multiplication points are fixed points of transformations $\gamma\in\mathrm{SL}(2,\mathbb{Z})$. Thus, we obtain exactly the elliptic points. In Sections 6.4 and 6.5, we give the explicit list and the representation in a fundamental domain of the special complex multiplication points for some curves $X(D,N)$, in the nonramified and small ramified cases.

LEMMA 6.17. *The CM points of $X(D,N)$ by the order $\Lambda(d,m)$ are SCM points if and only if there exists $\alpha\in\Lambda(d,m)$ such that $\mathrm{n}(\alpha)=D_{\mathcal{O}(D,N)}$. In this case,* $\mathrm{SCM}(D,N,d,m)=\mathrm{CM}(D,N,d,m)$.

In the following proposition we collect the characterization of the special complex multiplication points, by using Lemma 6.17 and Proposition 6.6.

PROPOSITION 6.18. *Let $\mathcal{O}=\mathcal{O}(D,N)$ be a quaternion order and $\Lambda=\Lambda(d,m)$ an imaginary quadratic order. Then there are SCM points of $X(D,N)$ by $\Lambda(d,m)$ if and only if $\mathcal{R}^*(n_{\mathbb{Z}+2\mathcal{O},3},-D_\Lambda;\mathbb{Z})\neq\varnothing$ and $\mathcal{R}(n_{\Lambda,2},D_{\mathcal{O}};\mathbb{Z})\neq\varnothing$.*

Actually, with the Shimura curve and the quadratic order fixed, the existence of SCM points depends only on properties of the quadratic order. In the following proposition, we state necessary conditions for the quadratic order which give rise to SCM points. From this, a finiteness result is deduced.

PROPOSITION 6.19. *Consider a Shimura curve $X(D,N)$, with N square-free. If there are SCM points of $X(D,N)$ by $\Lambda(d,m)$, then $-m^2d\le DN$ if $d\equiv 2,3 \bmod 4$ and $-m^2d\le 4DN-1$ if $d\equiv 1\bmod 4$.*

PROOF. We use the characterization of the SCM points given in Proposition 6.18. Let $\mathcal{O}=\mathcal{O}(D,N)$ be a quaternion order corresponding to $X(D,N)$. Put $\Lambda=\Lambda(d,m)$.

The condition for the existence of representations of $D_{\mathcal{O}}$ by the normic form attached to Λ is independent of the basis of Λ used to fix the form $n_{\Lambda,\mathbb{Z}}$. Thus, we may use the reduced normic form given in Proposition 3.58,

$$n_{\Lambda,2}(X,Y)=\begin{cases}X^2-(D_\Lambda/4)Y^2 & \text{if } D_\Lambda\equiv 0 \bmod 4,\\ X^2+XY+(1-D_\Lambda/4)Y^2 & \text{if } D_\Lambda\equiv 1\bmod 4.\end{cases}$$

Since the quadratic order Λ is imaginary, $D_\Lambda < 0$. Hence, the normic form of the order is positive definite, independent of the value of $D_\Lambda \bmod 4$. Note that the solutions (x,y) of $n_{\Lambda,2}(X,Y) = D_{\mathcal{O}}$ must satisfy $y \neq 0$. This is because $D_{\mathcal{O}}$ and N are square-free and $\gcd(D,N) = 1$. Using $y \neq 0$, it is clear that D_Λ only can take a finite number of values. Namely, if $D_\Lambda \equiv 0 \bmod 4$, it is necessary that $-D_\Lambda \leq 4D_{\mathcal{O}}$; if $D_\Lambda \equiv 1 \bmod 4$, it is necessary that $-D_\Lambda \leq 4D_{\mathcal{O}} - 1$. Note that these conditions are not sufficient for the existence of solutions. □

COROLLARY 6.20. *In a Shimura curve $X(D,N)$ with N square-free, there is a finite number of special complex multiplication points. We denote by* $\mathrm{SCM}(D,N)$ *the set of SCM points in the curve $X(D,N)$ and by* $\mathrm{scm}(D,N)$ *its number.*

PROOF. By Lemma 6.17, the number of SCM points by a given quadratic order $\Lambda(d,m)$, if it is different from 0, coincides with the number of CM points. Then, by Theorem 6.13 it is finite and equal to $\mathrm{h}(D,N,d,m)$. Now, since N is square-free, by Proposition 6.19, we have a necessary condition that has to be satisfied by the quadratic orders Λ in order that $X(D,N)$ can have special complex multiplication points by Λ. This condition only allows a finite number of quadratic orders. Thus, the number of special complex multiplication points in $X(D,N)$ is finite. □

In Appendix A we illustrate tables about special complex multiplication points for curves corresponding to nonramified or small ramified quaternion algebras of type A and type B. In Tables A.38–A.41, we give the list of quadratic orders for which the curve $X(D,N)$ has special complex multiplication points and the number of such points, for $D = 1$, 6, 10, 15 up to $N \leq 25$ square-free.

6.3. Computation of complex multiplication points

From the results of Chapter 4 relating the embeddings and some quadratic forms, the computation of the CM points can be formulated in terms of computations with quadratic forms. On the one hand, we use the bijective mapping between embeddings and representations by ternary normic forms of quaternion orders. On the other hand, we use the bijective mapping between embeddings and binary quadratic forms associated to the orders.

THEOREM 6.21. *Let $\Lambda = \Lambda(d,m)$ be a quadratic order, $\mathcal{O}(D,N)$ a quaternion order and fix a normalized basis $\mathcal{B}$ in $\mathcal{O}(D,N)$. Consider the corresponding Shimura curve $X(D,N)$. Then, the set of complex multiplication points of $X(D,N)$ by the quadratic order Λ is*

$$\mathrm{CM}(D,N,d,m) = \Gamma(D,N)\backslash\{z \in \mathcal{H} : \gamma(z) = z, \gamma \in S(D,N,m,d)\},$$

where $S(D,N,m,d)$ is the following set of transformations:

$$\big\{\Phi\big((-kz, 2x, 2y, 2z)_{\mathcal{B}}\big) : (x,y,z) \in \mathcal{R}^*(n_{\mathbb{Z}+2\mathcal{O}(D,N),3}, -D_\Lambda; \mathbb{Z})\big\},$$

where $k = 0$ if $\mathcal{O}(D,N)$ is an even order, and $k = 1$ if it is odd.

Equivalently,

$$\mathrm{CM}(D,N,d,m) = \Gamma(D,N)\backslash\big\{\tau(f) : f \in \mathcal{H}^*\big((\mathbb{Z} + 2\mathcal{O}(D,N)); \Lambda\big)\big\}.$$

PROOF. It suffices to see that the transformations of $S(D,N,m,d)$ correspond to the transformations that come from the optimal embeddings of $\Lambda(d,m)$ into $\mathcal{O}(D,N)$. It is enough to choose a transformation $\gamma = \Phi\big(\varphi(\alpha)\big)$ for each embedding $\varphi \in \mathcal{E}^*\big(\mathcal{O}(D,N), \Lambda(d,m)\big)$, for any $\alpha \in \mathbb{Q}(\sqrt{d})$, $\alpha \notin \mathbb{Q}$. Proposition 4.27 describes the embeddings $\varphi \in \mathcal{E}(\mathcal{O}(D,N),\Lambda(d,m))$ by means of the expression of $\varphi(\sqrt{d})$. It depends on the fixed normalized basis in the quaternion order and the representation of $-D_\Lambda$ by the normic form $n_{\mathbb{Z}+2\mathcal{O}(D,N),3}$, according to the value of d modulo 4.

By Lemma 6.1, a transformation γ can be substituted by $\gamma' := \lambda\gamma + \mu\,\mathrm{Id}$ and the fixed points are the same; that is equivalent to choosing α in a convenient way. Consider the transformations $\Phi(\varphi(m\sqrt{d})) = \Phi(m\varphi(\sqrt{d}))$ if $d \equiv 1 \bmod 4$ and $\Phi(\varphi(2m\sqrt{d})) = \Phi(2m\varphi(\sqrt{d}))$ if $d \equiv 2,3 \bmod 4$. Thus, independently of the value of d, for every $(x,y,z) \in \mathcal{R}^*(n_{\mathbb{Z}+2\mathcal{O}(D,N),3}, -D_\Lambda; \mathbb{Z})$ we can fix the transformation

$$\gamma := \Phi((-kz, 2x, 2y, 2z)_\mathcal{B}).$$

Using the bijective mapping between the set of classes of optimal embeddings and the set of $(\mathcal{O}(D,N),\Lambda(d,m))$-primitive binary quadratic forms, given in Theorem 4.57, we obtain the equivalent formulation in terms of the points associated to binary quadratic forms. □

REMARK 6.22. To compute explicit embeddings, it is clear that both approaches are equivalent, since the equation obtained from the binary form asking for a suitable determinant is exactly the equation corresponding to the representation by the ternary form, cf. Proposition 4.46.

Next we present the results in Theorem 6.21 more explicitly for orders in the nonramified and small ramified quaternion algebras of type A and of type B.

THEOREM 6.23. *Consider the curve* $X(1,N) = X_0(N)$. *Fix the Eichler order* $\mathcal{O}_M(1,N) = \mathbb{Z}[1,(j+ij)/2, N(-j+ij)/2, (1-i)/2] \subseteq \left(\frac{1,-1}{\mathbb{Q}}\right)$. *Fix a quadratic order* $\Lambda = \Lambda(d,m)$. *Then,* $\mathrm{CM}(1,N,d,m)$ *is the set of classes of* $\Gamma(1,N)$*-equivalence of the set of points*

$$\left\{\frac{-z \pm \sqrt{-D_\Lambda}\,\iota}{2Ny} \in \mathcal{H} : (x,y,z) \in \mathcal{R}^*(n_{\mathbb{Z}+2\mathcal{O}_M(1,N),3}, -D_\Lambda; \mathbb{Z})\right\},$$

or equivalently,

$$\left\{\frac{-b \pm \sqrt{-D_\Lambda}\,\iota}{2Na} \in \mathcal{H} : f = (Na,b,c),\ a,b,c \in \mathbb{Z},\ \gcd(a,b,c) = 1,\right\}.$$

PROOF. Using the basis $\mathcal{B} = \{1,(j+ij)/2, N(-j+ij)/2, (1-i)/2\}$ of the order $\mathcal{O}_M(1,N)$, we apply Theorem 6.21.

With a representation $(x,y,z) \in \mathcal{R}^*(n_{\mathbb{Z}+2\mathcal{O}_M(1,N),3}, -D_\Lambda; \mathbb{Z})$ fixed, we obtain the transformation

$$\gamma = \frac{1}{2}\begin{pmatrix} -z & 2x \\ 2Ny & z \end{pmatrix}$$

and it is easy to check that it has as fixed points

$$\frac{-z \pm \sqrt{-D_\Lambda}\,\iota}{2Ny}.$$

We only need to choose the sign to obtain a point in the upper half-plane.

Alternatively, we can make explicit the points $\tau(f)$ attached to the binary forms $f \in \mathcal{H}^*(\mathbb{Z} + 2\mathcal{O}_M(1,N); \Lambda)$, given by Theorem 4.61,

$$f = (Na,b,c), \quad a,b,c \in \mathbb{Z},\ \gcd(a,b,c) = 1, \det_2(f) = -D_\Lambda.$$

The results are exactly the same. □

In Section 6.4, Tables 6.1–6.5 collect the special complex multiplication points for the modular curves $X(1,N)$, $N = 1, 2, 3, 11, 13, 23$ and the corresponding quadratic orders. Using the fundamental domains constructed in Chapter 5, we can also show these complex multiplication points graphically.

For the case of small ramified algebras of type A and type B, we now make explicit the result of Theorem 6.21, in a similar way as the computation of elliptic points was made explicit in Section 5.2.

THEOREM 6.24. *Let $X(2p,N)$ be the Shimura curve associated to the group of quaternion transformations $\Gamma(2p,N)$, for $p \equiv 3 \mod 4$, $N \mid (p-1)/2$ square-free. Fix the Eichler order $\mathcal{O}_{\mathrm{A}}(2p,N) = \mathbb{Z}[1,i,Nj,(1+i+j+ij)/2] \subseteq H_{\mathrm{A}}(p)$, with a normalized basis, from which we obtain the ternary normic form $n_{\mathbb{Z}+2\mathcal{O}_{\mathrm{A}}(2p,N),3}$. Fix a quadratic order $\Lambda = \Lambda(d,m)$. Then, $\mathrm{CM}(D,N,d,m)$ is the set of classes of $\Gamma(D,N)$-equivalence of the set of points*

$$\left\{\frac{(2x+z)\sqrt{p} \pm \sqrt{-D_\Lambda}\,\iota}{-(2Ny+z)+z\sqrt{p}} \in \mathcal{H} : (x,y,z) \in \mathcal{R}^*(n_{\mathbb{Z}+2\mathcal{O}_{\mathrm{A}}(2p,N),3}, -D_\Lambda; \mathbb{Z})\right\}.$$

PROOF. We apply Theorem 6.21 to the basis $\mathcal{B} = \{1,i,Nj,(1+i+j+ij)/2\}$. With a representation $(x,y,z) \in \mathcal{R}^*(n_{\mathbb{Z}+2\mathcal{O},3}, -D_\Lambda; \mathbb{Z})$ fixed, we obtain the transformation

$$\gamma = \frac{1}{2}\begin{pmatrix} (2x+z)\sqrt{p} & (2Ny+z)+z\sqrt{p} \\ -(2Ny+z)+z\sqrt{p} & -(2x+z)\sqrt{p} \end{pmatrix}$$

and it is easy to check that it has as fixed points

$$\frac{(2x+z)\sqrt{p} \pm \sqrt{-D_\Lambda}\,\iota}{-(2Ny+z)+z\sqrt{p}}.$$

It is only necessary to choose the point belonging to the upper half-plane. □

In the same way, in the case of the small ramified algebras of type B, we obtain the following result.

THEOREM 6.25. *Let $X(pq,N)$ be the Shimura curve determined by $\Gamma(pq,N)$, for $q \equiv 1 \bmod 4$, $\left(\frac{p}{q}\right) = -1$, $N \mid q-14$, square-free and $\gcd(p,N) = 1$. Fix the Eichler order $\mathcal{O}_{\mathrm{B}}(pq,N) = \mathbb{Z}[1,Ni,(i+ij)/2,(1+j)/2] \subseteq H_{\mathrm{B}}(p,q)$ and a quadratic order $\Lambda = \Lambda(d,m)$. Then, $\mathrm{CM}(D,N,d,m)$ is the set of classes of $\Gamma(D,N)$-equivalence of the set of points*

$$\left\{\frac{(2Nx+y)\sqrt{p} \pm \sqrt{-D_\Lambda}\,\iota}{q(z-y\sqrt{p})} \in \mathcal{H} : (x,y,z) \in \mathcal{R}^*(n_{\mathbb{Z}+2\mathcal{O}_{\mathrm{B}}(pq,N),3}, -D_\Lambda; \mathbb{Z})\right\}.$$

The results are the same if we use the relation between embeddings and binary quadratic forms, since the condition for the determinant of the binary form is equivalent to the condition for the representation, cf. Proposition 4.46.

In Section 6.5, we present tables of complex multiplication points for the curve $X(6,1)$, which corresponds to the small ramified case of type A, and for the curves $X(10,1)$ and $X(15,1)$, which correspond to the small ramified case of type B. Using the fundamental domains constructed in Chapter 5, we can also show these complex multiplication points graphically.

6.4. Examples in the nonramified case

Next we present some examples of special complex multiplication points, together with their graphical representation by using the fundamental domains given in Section 5.1.

Figures 6.1–6.5 and Tables 6.1–6.5 correspond to the special complex multiplication points of the modular curves $X(1,3)$, $X(1,11)$, $X(1,13)$ and $X(1,23)$ respectively. Note that, in the cases in which we have an even number of complex multiplication points by the same quadratic order, these points are displayed symmetrically with respect to the imaginary axis, due to the symmetry of the domain.

Next we include the detailed computation in one example.

EXAMPLE 6.26. Consider the Shimura curve $X(1,11)$. We will make explicit the computation of some special complex multiplication points of $X(1,11)$.

First we obtain the list of quadratic orders giving rise to special complex multiplication points in $X(1,11)$:

$$\begin{array}{lll} \Lambda(-2,1), & \Lambda(-7,1), & \Lambda(-7,2), \\ \Lambda(-10,1), & \Lambda(-11,1), & \Lambda(-11,2), \\ \Lambda(-19,1), & \Lambda(-35,1), & \Lambda(-43,1). \end{array}$$

As a sample, we will consider the quadratic orders $\Lambda(-11,1)$, $\Lambda(-19,1)$ and $\Lambda(-35,1)$. To obtain the SCM points by these orders, first we compute representations of $-D_\Lambda = 11$, 44, 19, 35 by the ternary quadratic form $n_{\mathbb{Z}+2\mathcal{O},3}(X,Y,Z) = -Z^2 - 4NXY$. From that, we have explicit embeddings and the corresponding transformations, and we compute their fixed points. We write $\{\tau_i\}$, the set of complex multiplication points by every order Λ, and for each τ_i, put γ_i to the transformation fixing τ_i, $\gamma_i(\tau_i) = \tau_i$. For each quadratic order $\Lambda(d,m)$, we know in advance how many SCM points we have in $X(1,11)$, that is, $\mathrm{cm}(1,11,d,m) = \mathrm{h}(1,11,d,m)$. Thus, we must obtain at least this number of points and check if they are $\Gamma(1,11)$-equivalent or not. If we are interested in their representation, we must be sure to choose a representative contained in a given fundamental domain $\mathcal{D}(1,11)$.

Recall that we can also compute complex multiplication points in terms of binary quadratic forms. Then the complex multiplication points of $X(1,11)$ by a fixed quadratic order Λ are computed from the points attached to binary quadratic forms $f = (11a,b,c)$, with $a,b,c \in \mathbb{Z}$ such that $\det_2(f) = -D_\Lambda$. This amounts to finding for which a, b, c the expression $44ac - b^2$ takes the values 11, 44, 19 and 35.

Consider the quadratic field $F = \mathbb{Q}(\sqrt{-11})$ and the quadratic order $\Lambda_F = \Lambda(-11,1)$. We know $\mathrm{cm}(1,11,-11,1) = 1$, since $h\big(\mathbb{Q}(\sqrt{-11})\big) = 1$ and $p = 11$ ramifies in $\mathbb{Q}(\sqrt{-11})$. Thus, there will be an unique SCM point by Λ_F in $X(1,11)$. Applying the above results, we obtain

$$\gamma = \begin{pmatrix} 6 & -3 \\ 11 & -5 \end{pmatrix}, \quad \text{with fixed point } \tau = \frac{11+\sqrt{-11}}{22}.$$

Thus τ is the only special complex multiplication point of $X(1,11)$ by the order $\Lambda(-11,1)$. We can consider other solutions, for example

$$\alpha_1 = \begin{pmatrix} 6 & -1 \\ 33 & -5 \end{pmatrix}, \qquad \alpha_2 = \begin{pmatrix} 6 & 1 \\ -33 & -5 \end{pmatrix}, \qquad \alpha_3 = \begin{pmatrix} 6 & 3 \\ -11 & -5 \end{pmatrix},$$

$$\nu_1 = \frac{11+\sqrt{-11}}{66}, \qquad \nu_2 = \frac{-11+\sqrt{-11}}{66}, \qquad \nu_3 = \frac{-11+\sqrt{-11}}{22}.$$

But it is easy to check that the transformations α_i, $i = 1, 2, 3$ are $\Gamma(1, 11)$-equivalent to γ. Therefore, the corresponding fixed points ν_i are $\Gamma(1, 11)$-equivalent to τ.

For $F = \mathbb{Q}(\sqrt{-19})$ and $\Lambda_F = \Lambda(-19, 1)$, we have that $\mathrm{cm}(\mathcal{O}(1, 11), \Lambda_F) = 2$, since $h(\mathbb{Q}(\sqrt{-19})) = 1$ and $p = 11$ splits in $\mathbb{Q}(\sqrt{-19})$. In this case, we obtain the following transformations and fixed points

$$\gamma_1 = \begin{pmatrix} 3 & -1 \\ 11 & -2 \end{pmatrix}, \qquad \gamma_2 = \begin{pmatrix} -2 & -1 \\ 11 & 3 \end{pmatrix},$$

$$\tau_1 = \frac{5 + \sqrt{-19}}{22}, \qquad \tau_2 = \frac{-5 + \sqrt{-19}}{22}.$$

The fixed points τ_1 and τ_2 are not $\Gamma(1, 11)$-equivalent. Then, they are representatives of the special complex multiplication points of $X(1, 11)$ by the order $\Lambda(-19, 1)$.

Finally, we check that there are optimal embeddings of the ring of integers of $F = \mathbb{Q}(\sqrt{-35})$ into the order $\mathcal{O}_0(11)$. Indeed, $\mathrm{cm}(\mathcal{O}(1, 11), \Lambda_F) = 4$, since $h(\mathbb{Q}(\sqrt{-35})) = 2$ and $p = 11$ splits in $\mathbb{Q}(\sqrt{-35})$. From the representations we obtain the transformations

$$\beta_1 = \begin{pmatrix} 10 & -3 \\ 33 & -9 \end{pmatrix}, \quad \beta_2 = \begin{pmatrix} 2 & 1 \\ -11 & 1 \end{pmatrix}, \quad \beta_3 = \begin{pmatrix} 2 & -1 \\ 11 & -1 \end{pmatrix}, \quad \beta_4 = \begin{pmatrix} 10 & -3 \\ 33 & -9 \end{pmatrix}.$$

It is easy to see that the corresponding fixed points σ_i are not $\Gamma(1, 11)$-equivalent. Therefore they yield a system of representatives of the complex multiplication points we are looking for:

$$\sigma_1 = \frac{-19 + \sqrt{-35}}{66}, \quad \sigma_2 = \frac{-3 + \sqrt{-35}}{22}, \quad \sigma_3 = \frac{3 + \sqrt{-35}}{22}, \quad \sigma_4 = \frac{19 + \sqrt{-35}}{66}.$$

Thus, the points σ_1, σ_2, σ_3 and σ_4 are the special complex multiplication points of $X(1, 11)$ by Λ_F, $F = \mathbb{Q}(\sqrt{-35})$.

The graphical representation of all the special complex multiplication points of $X(1, 11)$ are shown in Figure 6.3 using the fundamental domain given in Section 5.1.

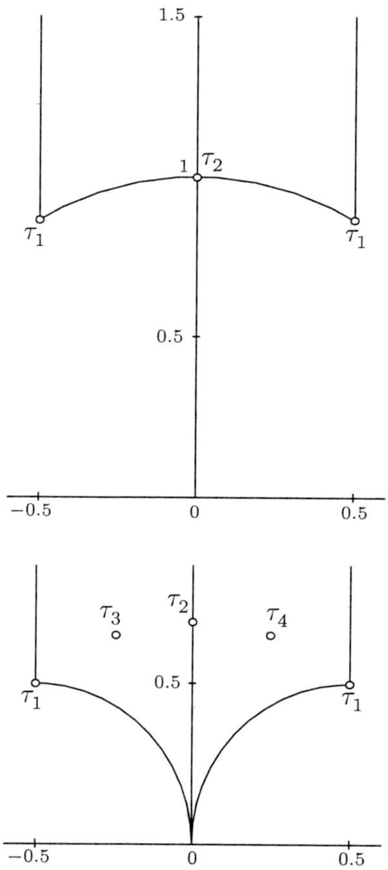

FIGURE 6.1. Representation of the special complex multiplication points SCM(1, 1) and SCM(1, 2) in fundamental domains for $X(1,1)$ and $X(1,2)$, respectively.

TABLE 6.1. The special complex multiplication points of $X(1,2)$ and the corresponding quadratic orders.

(d,m)	n	$\mathrm{CM}(1,2,d,m)$
$(-1,1)$	1	$\left\{\tau_1 = \frac{-1+\iota}{2} \sim \frac{1+\iota}{2}\right\}$
$(-2,1)$	1	$\left\{\tau_2 = \frac{\sqrt{2}\iota}{2}\right\}$
$(-7,1)$	2	$\left\{\tau_3 = \frac{-1+\sqrt{7}\iota}{4}, \tau_4 = \frac{1+\sqrt{7}\iota}{4}\right\}$

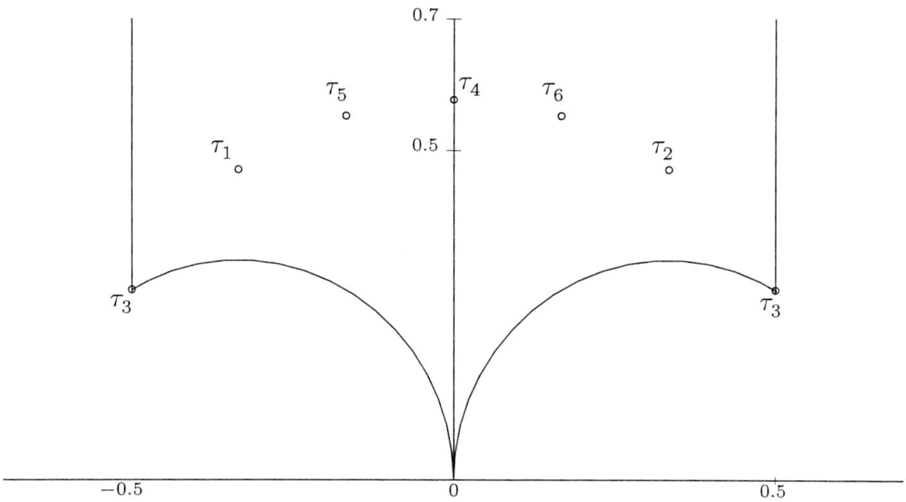

FIGURE 6.2. Representation of the special complex multiplication points SCM(1, 3) in a fundamental domain for $X(1,3)$.

TABLE 6.2. The special complex multiplication points of $X(1,3)$ and the corresponding quadratic orders.

(d,m)	n	$\mathrm{CM}(1,3,d,m)$
$(-2,1)$	2	$\left\{\tau_1 = \frac{-1+\sqrt{2}\iota}{3}, \tau_2 = \frac{1+\sqrt{2}\iota}{3}\right\}$
$(-3,1)$	1	$\left\{\tau_3 = \frac{-3+\sqrt{3}\iota}{6} \sim \frac{3+\sqrt{3}\iota}{6}\right\}$
$(-3,2)$	1	$\left\{\tau_4 = \frac{\sqrt{3}\iota}{3}\right\}$
$(-11,1)$	2	$\left\{\tau_5 = \frac{-1+\sqrt{11}\iota}{6}, \tau_6 = \frac{1+\sqrt{11}\iota}{6}\right\}$

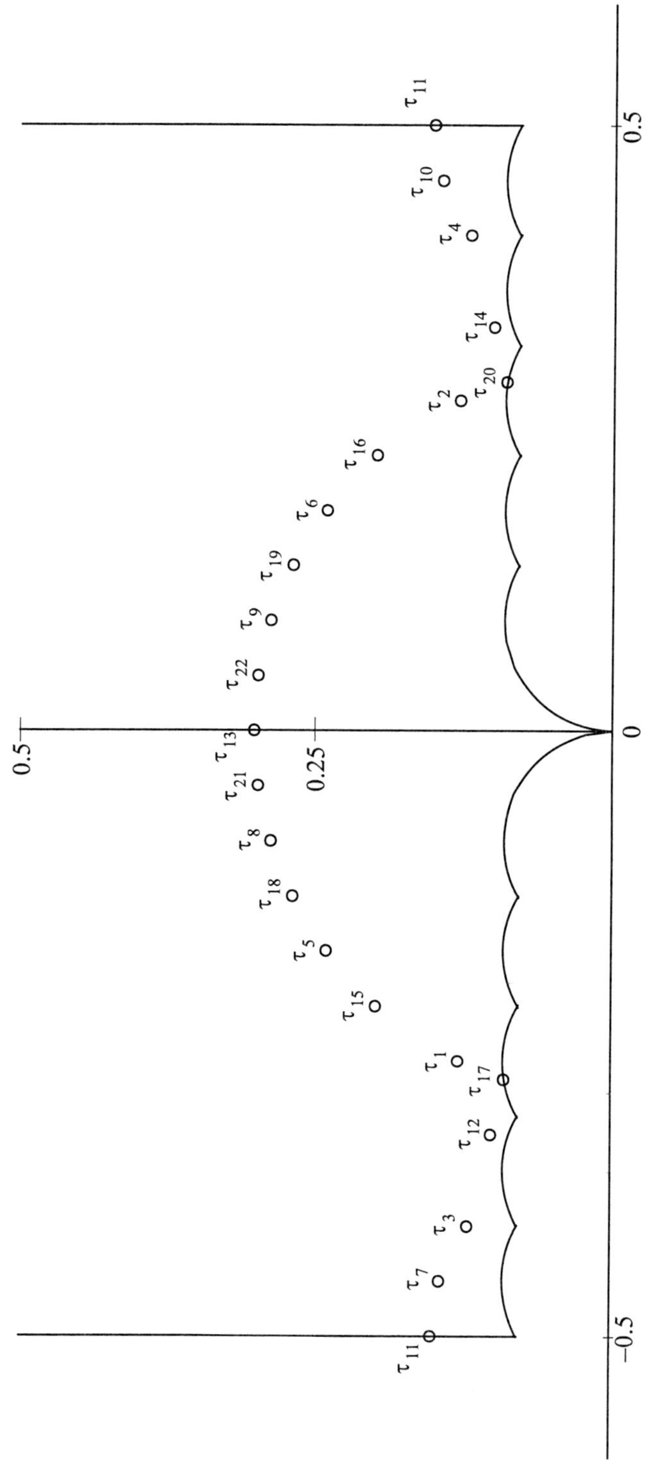

FIGURE 6.3. Representation of the special complex multiplication points SCM(1, 11) in a fundamental domain for $X(1, 11)$.

TABLE 6.3. The special complex multiplication points of $X(1,11)$ and the corresponding quadratic orders.

(d,m)	n	$\mathrm{CM}(1,11,d,m)$
$(-2,1)$	2	$\left\{\tau_1 = \frac{-3+\sqrt{2}\iota}{11}, \tau_2 = \frac{3+\sqrt{2}\iota}{11}\right\}$
$(-7,1)$	2	$\left\{\tau_3 = \frac{-9+\sqrt{7}\iota}{22}, \tau_4 = \frac{9+\sqrt{7}\iota}{22}\right\}$
$(-7,2)$	2	$\left\{\tau_5 = \frac{-2+\sqrt{7}\iota}{11}, \tau_6 = \frac{2+\sqrt{7}\iota}{11}\right\}$
$(-10,1)$	4	$\left\{\tau_7 = \frac{-10+\sqrt{10}\iota}{22}, \tau_8 = \frac{-1+\sqrt{10}\iota}{11}, \tau_9 = \frac{1+\sqrt{10}\iota}{11}, \tau_{10} = \frac{10+\sqrt{10}\iota}{22}\right\}$
$(-11,1)$	1	$\left\{\tau_{11} = \frac{-11+\sqrt{11}\iota}{22} \sim \frac{11+\sqrt{11}\iota}{22}\right\}$
$(-11,2)$	3	$\left\{\tau_{12} = \frac{-11+\sqrt{11}\iota}{33}, \tau_{13} = \frac{\sqrt{11}\iota}{11}, \tau_{14} = \frac{11+\sqrt{11}\iota}{33}\right\}$
$(-19,1)$	2	$\left\{\tau_{15} = \frac{-5+\sqrt{19}\iota}{22}, \tau_{16} = \frac{5+\sqrt{19}\iota}{22}\right\}$
$(-35,1)$	4	$\left\{\tau_{17} = \frac{-19+\sqrt{35}\iota}{66}, \tau_{18} = \frac{-3+\sqrt{35}\iota}{22}, \tau_{19} = \frac{3+\sqrt{35}\iota}{22}, \tau_{20} = \frac{19+\sqrt{35}\iota}{66}\right\}$
$(-43,1)$	2	$\left\{\tau_{21} = \frac{-1+\sqrt{43}\iota}{22}, \tau_{22} = \frac{1+\sqrt{43}\iota}{22}\right\}$

TABLE 6.4. The special complex multiplication points of $X(1,13)$ and the corresponding quadratic orders.

(d,m)	n	$\mathrm{CM}(1,13,d,m)$
$(-1,1)$	2	$\left\{\tau_1 = \frac{-5+\iota}{13}, \tau_2 = \frac{5+\iota}{13}\right\}$
$(-1,2)$	2	$\left\{\tau_3 = \frac{-3+2\iota}{13}, \tau_4 = \frac{3+2\iota}{13}\right\}$
$(-1,3)$	4	$\left\{\tau_5 = \frac{-11+3\iota}{26}, \tau_6 = \frac{-2+3\iota}{13}, \tau_7 = \frac{2+3\iota}{13}, \tau_8 = \frac{11+3\iota}{26}\right\}$
$(-3,1)$	2	$\left\{\tau_9 = \frac{-7+\sqrt{3}\iota}{26}, \tau_{10} = \frac{7+\sqrt{3}\iota}{26}\right\}$
$(-3,2)$	2	$\left\{\tau_{11} = \frac{-6+\sqrt{3}\iota}{13}, \tau_{12} = \frac{6+\sqrt{3}\iota}{13}\right\}$
$(-3,3)$	2	$\left\{\tau_{13} = \frac{-5+3\sqrt{3}\iota}{26}, \tau_{14} = \frac{5+3\sqrt{3}\iota}{26}\right\}$
$(-3,4)$	4	$\left\{\tau_{15} = \frac{-12+2\sqrt{3}\iota}{39}, \tau_{16} = \frac{-1+2\sqrt{3}\iota}{13}, \tau_{17} = \frac{1+2\sqrt{3}\iota}{13}, \tau_{18} = \frac{12+2\sqrt{3}\iota}{39}\right\}$
$(-13,1)$	2	$\left\{\tau_{19} = \frac{\sqrt{13}\iota}{13}, \tau_{20} = \frac{-13+\sqrt{13}\iota}{26} \sim \frac{13+\sqrt{13}\iota}{26}\right\}$
$(-43,1)$	2	$\left\{\tau_{21} = \frac{-3+\sqrt{43}\iota}{26}, \tau_{22} = \frac{3+\sqrt{43}\iota}{26}\right\}$
$(-51,1)$	4	$\left\{\tau_{23} = \frac{-27+\sqrt{51}\iota}{78}, \tau_{24} = \frac{-1+\sqrt{51}\iota}{26}, \tau_{25} = \frac{1+\sqrt{51}\iota}{26}, \tau_{26} = \frac{27+\sqrt{51}\iota}{78}\right\}$

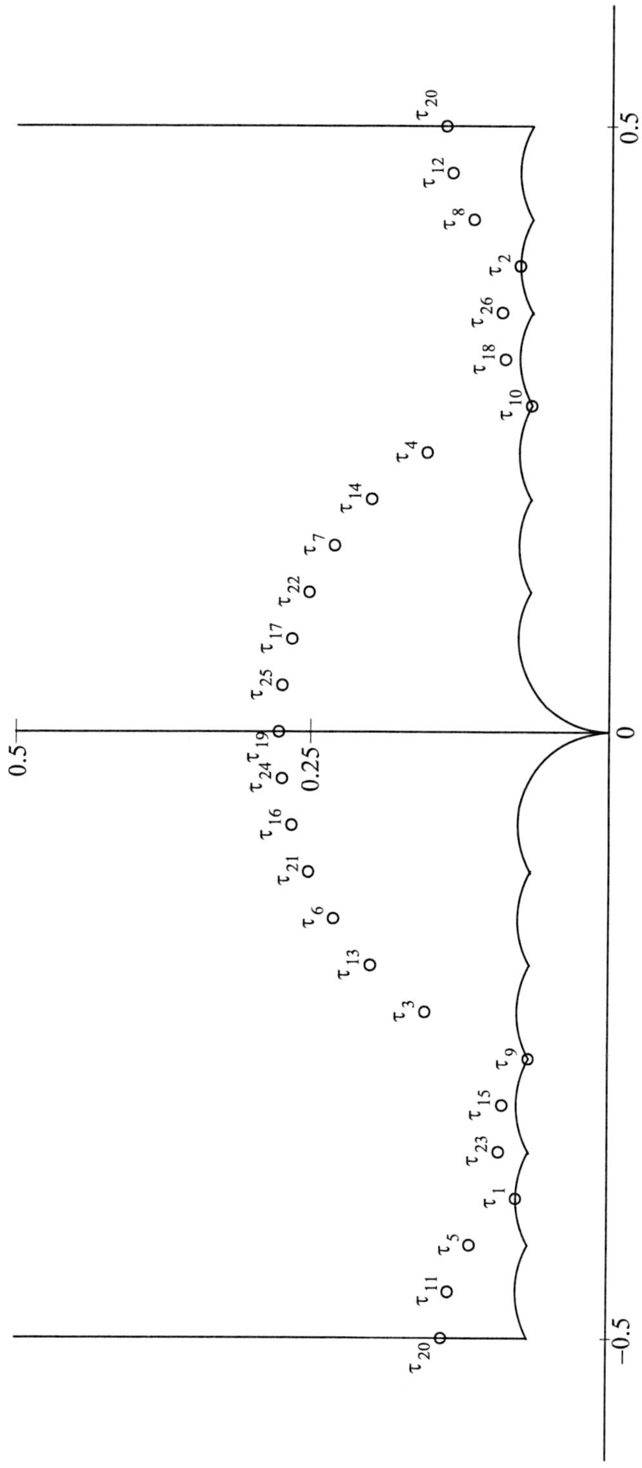

FIGURE 6.4. Representation of the special complex multiplication points SCM(1, 13) in a fundamental domain for $X(1,13)$.

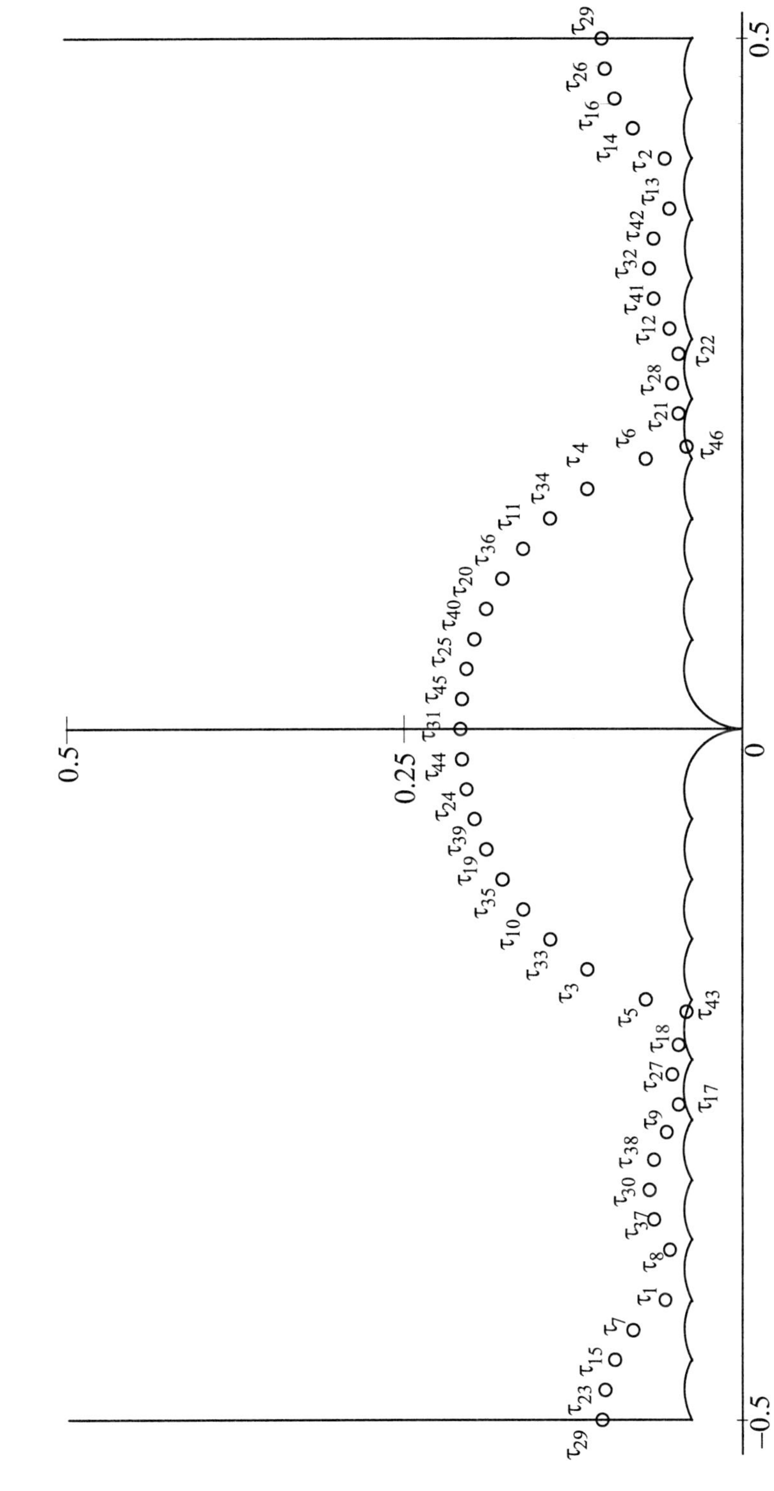

FIGURE 6.5. Representation of the special complex multiplication points SCM(1, 23) in a fundamental domain for $X(1, 23)$.

TABLE 6.5. The special complex multiplication points of $X(1,23)$ and the corresponding quadratic orders.

(d,m)	n	$\mathrm{CM}(1,23,d,m)$
$(-7,1)$	2	$\left\{\tau_1 = \frac{-19+\sqrt{7}\iota}{46}, \tau_2 = \frac{19+\sqrt{7}\iota}{46}\right\}$
$(-7,2)$	2	$\left\{\tau_3 = \frac{-4+\sqrt{7}\iota}{23}, \tau_4 = \frac{4+\sqrt{7}\iota}{23}\right\}$
$(-11,1)$	2	$\left\{\tau_5 = \frac{-9+\sqrt{11}\iota}{46}, \tau_6 = \frac{9+\sqrt{11}\iota}{46}\right\}$
$(-14,1)$	8	$\left\{\tau_7 = \frac{-20+\sqrt{14}\iota}{46}, \tau_8 = \frac{-26+\sqrt{14}\iota}{69}, \tau_9 = \frac{-20+\sqrt{14}\iota}{69}, \tau_{10} = \frac{-3+\sqrt{14}\iota}{23},\right.$ $\left.\tau_{11} = \frac{3+\sqrt{14}\iota}{23}, \tau_{12} = \frac{20+\sqrt{14}\iota}{69}, \tau_{13} = \frac{26+\sqrt{14}\iota}{69}, \tau_{14} = \frac{20+\sqrt{14}\iota}{46}\right\}$
$(-19,1)$	2	$\left\{\tau_{15} = \frac{-21+\sqrt{19}\iota}{46}, \tau_{16} = \frac{21+\sqrt{19}\iota}{46}\right\}$
$(-19,2)$	6	$\left\{\tau_{17} = \frac{-25+\sqrt{19}\iota}{92}, \tau_{18} = \frac{-21+\sqrt{19}\iota}{92}, \tau_{19} = \frac{-2+\sqrt{19}\iota}{23},\right.$ $\left.\tau_{20} = \frac{2+\sqrt{19}\iota}{23}, \tau_{21} = \frac{21+\sqrt{19}\iota}{92}, \tau_{22} = \frac{25+\sqrt{19}\iota}{92}\right\}$
$(-22,1)$	4	$\left\{\tau_{23} = \frac{-22+\sqrt{22}\iota}{46}, \tau_{24} = \frac{-1+\sqrt{22}\iota}{23}, \tau_{25} = \frac{1+\sqrt{22}\iota}{23}, \tau_{26} = \frac{22+\sqrt{22}\iota}{46}\right\}$
$(-23,1)$	3	$\left\{\tau_{27} = \frac{-23+\sqrt{23}\iota}{92}, \tau_{28} = \frac{23+\sqrt{23}\iota}{92}, \tau_{29} = \frac{-23+\sqrt{23}\iota}{46} \sim \frac{23+\sqrt{23}\iota}{46}\right\}$
$(-23,2)$	3	$\left\{\tau_{30} = \frac{-23+\sqrt{23}\iota}{69}, \tau_{31} = \frac{\sqrt{23}\iota}{23}, \tau_{32} = \frac{23+\sqrt{23}\iota}{69}\right\}$
$(-43,1)$	2	$\left\{\tau_{33} = \frac{-7+\sqrt{43}\iota}{46}, \tau_{34} = \frac{7+\sqrt{43}\iota}{46}\right\}$
$(-67,1)$	2	$\left\{\tau_{35} = \frac{-5+\sqrt{67}\iota}{46}, \tau_{36} = \frac{5+\sqrt{67}\iota}{46}\right\}$
$(-83,1)$	6	$\left\{\tau_{37} = \frac{-49+\sqrt{83}\iota}{138}, \tau_{38} = \frac{-43+\sqrt{83}\iota}{138}, \tau_{39} = \frac{-3+\sqrt{83}\iota}{46},\right.$ $\left.\tau_{40} = \frac{3+\sqrt{83}\iota}{46}, \tau_{41} = \frac{43+\sqrt{83}\iota}{138}, \tau_{42} = \frac{49+\sqrt{83}\iota}{138}\right\}$
$(-91,1)$	4	$\left\{\tau_{43} = \frac{-47+\sqrt{91}\iota}{230}, \tau_{44} = \frac{-1+\sqrt{91}\iota}{46}, \tau_{45} = \frac{1+\sqrt{91}\iota}{46}, \tau_{46} = \frac{47+\sqrt{91}\iota}{230}\right\}$

6.5. Examples in the small ramified case

Figures 6.6–6.8 and Tables 6.6–6.8 show the special complex multiplication points of the Shimura curves $X(6,1)$, $X(10,1)$ and $X(15,1)$, respectively.

TABLE 6.6. The special complex multiplication points of $X(6,1)$ and the unique corresponding quadratic order.

(d,m)	n	$\mathrm{CM}(6,1,d,m)$
$(-6,1)$	2	$\left\{\tau_1 = \frac{(\sqrt{6}-\sqrt{2})\iota}{2}, \tau_2 = \frac{-\sqrt{3}+\sqrt{6}\iota}{3} \sim \frac{\sqrt{3}+\sqrt{6}\iota}{3}\right\}$

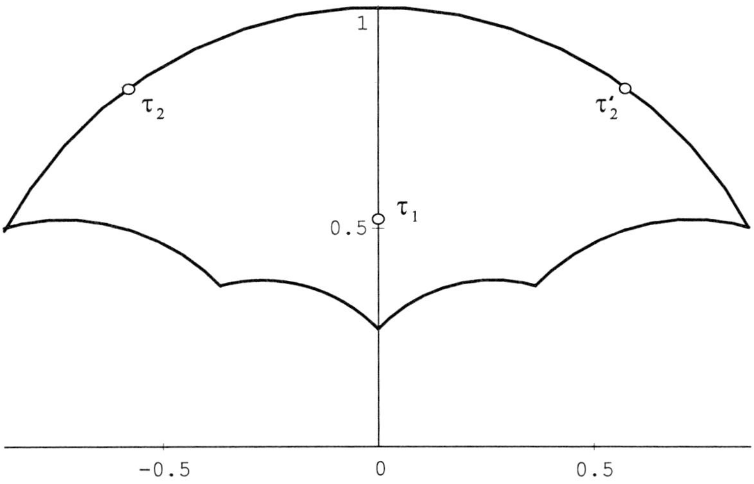

FIGURE 6.6. Representation of the special complex multiplication points SCM(6, 1) in a fundamental domain for $X(6, 1)$.

TABLE 6.7. The special complex multiplication points of $X(10, 1)$ and the unique corresponding quadratic order.

(d, m)	n	$\mathrm{CM}(10, 1, d, m)$
$(-10, 1)$	2	$\left\{\tau_1 = \frac{(3\sqrt{5}-2\sqrt{10})\iota}{5}, \tau_2 = \frac{\sqrt{5}\iota}{5}\right\}$

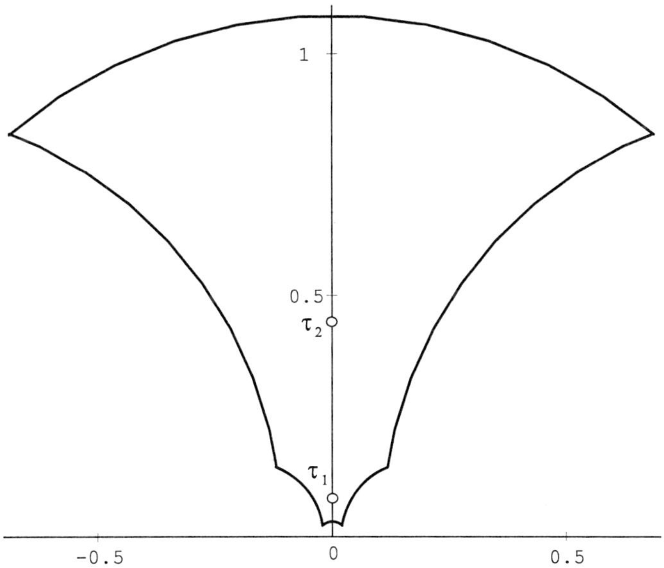

FIGURE 6.7. Representation of the special complex multiplication points SCM(10, 1) in a fundamental domain for $X(10, 1)$.

TABLE 6.8. The special complex multiplication points of $X(15,1)$ and the corresponding quadratic orders.

(d,m)	n	$\mathrm{CM}(15,1,d,m)$
$(-15,1)$	2	$\left\{\tau_1 = \frac{(2\sqrt{5}-\sqrt{15})\iota}{5} \sim \frac{(2\sqrt{5}+\sqrt{15})\iota}{5},\right.$ $\left.\tau_2 = \frac{15-5\sqrt{3}+(3\sqrt{5}-\sqrt{15})\iota}{30} \sim \frac{15+5\sqrt{3}+(3\sqrt{5}+\sqrt{15})\iota}{30}\right\}$
$(-15,2)$	2	$\left\{\tau_3 = \frac{\sqrt{5}\iota}{5}, \tau_4 = \frac{5+2\sqrt{5}\iota}{15}\right\}$

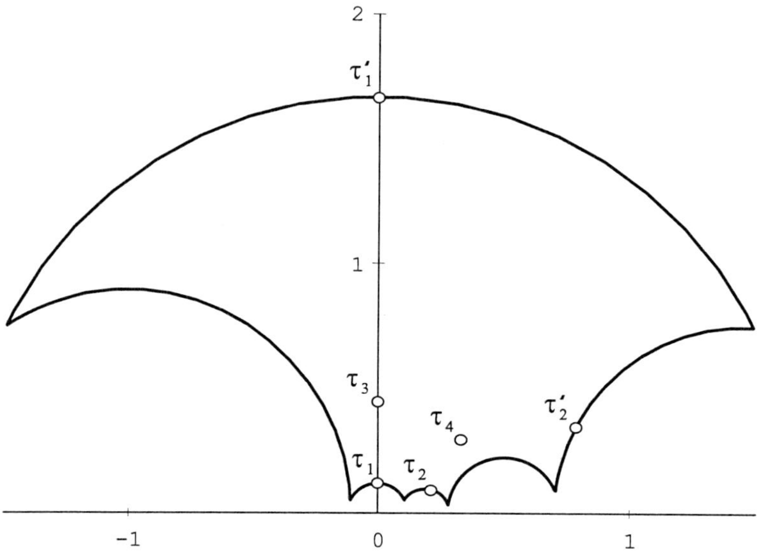

FIGURE 6.8. Representation of the special complex multiplication points SCM(15, 1) in a fundamental domain for $X(15,1)$.

CHAPTER 7

The `Poincare` Package

In this chapter we present the instructions of the `Poincare` package. It contains the implementation in *Maple* V of the algorithms described in the preceding chapters. In Section 7.1, we give the main technical characteristics. Section 7.2 contains some comments on the instructions implemented, ordered by topics corresponding to the chapters of this monograph. In Section 7.3, we give a complete list of the instructions, with input and output arguments.

7.1. Main characteristics

The package is implemented in *Maple* V *Release* 4.00a. It contains about 200 instructions, referring to the manipulation of quaternion algebras and orders, quadratic forms, constants and points of Shimura curves, embeddings, complex multiplication points, etc.

To use the package, the `Poincare.m` file should to be saved in the folder `/MapleV4/Lib/` and the library must be loaded as usual, using the instruction:

```
> with(Poincare);
```

The instructions in the package can be used with the usual syntax of Maple.

The package contains an initialization routine to avoid conflicts of variables with respect to protected variables for quaternion algebras.

In the design of the package we used basic instructions of the *Maple* V and also some instructions from standard libraries such as `linalg`, `numtheory` and `geometry`. It is not necessary to load these libraries, but they must be in the folder of libraries. The `geometry` library allows us to define as objects the points, the lines and the circles, which makes the performance of some computations easier, but it does not contain specific instructions about hyperbolic geometry.

Following standard notations, cf. Chapter 1, the variables i, j, k have a special meaning in quaternion algebras as they are used in the canonical basis. To avoid conflicts, these variables are protected, in such a way that no value can be assigned to them. The parameters a and b also have a special meaning, and they are also protected. To fix a quaternion algebra the values a, b can be assigned by means of the instruction `defQuatAlg`. For security, while loading the `Poincare` package, an initialization routine deletes the previous values of these parameters, if necessary.

A quaternion ω is a linear combination over $\mathbb{Q}$ of the variables i, j, k. A quaternion order $\mathcal{O}$ is determined by a list of four quaternions, $[v_1, v_2, v_3, v_4]$, in such a way that $\mathcal{O} = \mathbb{Z}[v_1, v_2, v_3, v_4]$. We implemented an instruction to check that effectively a list of four quaternions defines an order, `isOrder`. Most of the instructions relatives to orders include an optional argument, `true`, to run this command. This checking process requires, obviously, that the quaternion algebra has been fixed, that is, the parameters a and b must have assigned numerical values.

Given the diversity of areas of application of the package, we chose the names of the instructions in such a way that they show the kind of objects they are dealing with or the area where they are included. Thus, the names of the instructions include: `H`, for quaternion algebras; `F`, for quadratic fields; `Or`, for orders; `f`, for quadratic forms; `M`, in reference to matrices; `X`, for the Shimura curve; `X1`, for the modular curve, which corresponds to fixing $D = 1$; `Hyp`, for hyperbolic geometry, and in particular `L`, `Pol` or `S`, according to whether they refer to a line, a polygon or a hyperbolic strip respectively; `Hom` for homographic transformation; `Emb` for embeddings; `IC`, for the isometric circles. We reserve letters `A` and `B` to denote the specific instructions for the small ramified cases of type A and of type B respectively. We add `q` at the beginning for the instructions which already have a known meaning in other contexts, as the norm and the trace, to note their specificity in relation to the quaternions. The use of capital letters in the interior of a name simplifies the identification of the different parts of the name corresponding to some shortened words. We tried not to overuse capital letters, and the adjectives are written, mostly, in lower case.

7.2. Description of the instructions

7.2.1. Quaternion algebras and quaternion orders. In this section we briefly describe algorithms in the `Poincare` package in relation with quaternion algebras and quaternion orders. We always refer to quaternion algebras over $\mathbb{Q}$.

Basically the instructions are organized in three blocks: instructions relative to operations in the quaternion algebra (note the non-commutativity of the product), instructions for quaternion algebras, and instructions for quaternion orders.

All the instructions relatives to quaternion algebras are expressed in terms of the canonical basis $\{1, i, j, k\}$, $k := ij$, in $H = \left(\frac{a,b}{\mathbb{Q}}\right)$. Thus, the variables i, j, k, a and b are protected and no value can be assigned to them while the `Poincare` package is active. This is done with a routine automatically loaded when the package is initialized and previous values are erased. Some instructions can be used with a and b as parameters; some others need a fixed quaternion algebra, which is done with the command `defQuatAlg`.

We define a quaternion from the *Maple* multivariate polynomials, using the variables i, j, k. The instruction `qcoeffs` gives the coefficients of a quaternion with respect to the basis $\{1, i, j, k\}$ and `Mcoor` gives the coefficients of a list of quaternions, written as columns in a matrix. We point out the logical functions `type/quaternion` and `type/purequaternion`, which check the type of the input element; they take part in the construction of many other instructions.

The sum of quaternions and the product by a scalar coincide with those already defined in *Maple* V for the polynomials. It is only necessary to define the product of quaternions, non commutative, for which we use the instruction `qmul`. For convenience, we also introduce the abbreviation `&q`, as symbol of the quaternion product, emulating the above instruction. We define the conjugate, the norm, the trace and the inverse of a quaternion with the instructions `qbar`, `qnorm`, `qtrace` and `qinv`, respectively.

We can deal with a second block of instructions relatives to quaternion algebras as objects. The discriminant of an algebra is computed by the instruction `DiscH`, which uses the instruction `Hilbert` to compute Hilbert symbols (not included in *Maple* V R4). In relation to the morphisms of algebras, we have the instruction `embH` for the embedding of a quaternion algebra into the real matrix algebra, following Proposition 1.19, and the more general command `embHg`, allowing parameters. For inner automorphisms see the command `qconj`.

The Propositions 1.22 and 1.25 give algorithms for computing quaternion algebras with prefixed characteristics. The instructions `canH` and `canHdisc` are the implementations of these algorithms. Thus, given two prime numbers p and q, `canH` gives a pair (a, b), canonical in the sense of Proposition 1.22, such that $\left(\frac{a,b}{\mathbb{Q}}\right) \simeq \left(\frac{p,q}{\mathbb{Q}}\right)$. This allows us to classify those quaternion algebras in nonramified, small ramified of type A and small ramified of type B with the instruction `typeH`. Given different primes p and q, `canHdisc` gives a pair (a, b) determining a quaternion algebra of discriminant $D = pq$.

In the third block we collect the instructions relative to the orders. For all these instructions, an order will be introduced by means of a basis, given as a list of four quaternions. But, not every list l of four quaternions is necessarily a $\mathbb{Z}$-basis of an order. The logical function `isOrder` checks if l generates an order or not, and it states the reason in case of negative answer, by means of an optional parameter. To avoid unnecessary checkings, the rest of the instructions relative to orders, having as input a list l of four quaternions, assume that the list determines an order. If this is not true the output will give a mistake or an answer that does not make sense. To avoid these mistakes, all these instructions can use an optional argument `true` to check previously if the list defines an order.

In reference to the bases of an order, we point out the following instructions: `HermiteOr`, to simplify the basis of the order; `isnBasisOr`, logical function to check if a basis is normalized or not; `nBasisOr`, which gives a normalized basis of the order; `McbOr`, to obtain the matrix of change of basis between two bases of the order; `coorOr`, which gives the coordinates of an element with respect to a basis, and `OrM`, which converts a matrix into the list of quaternions determined by its columns, with respect to the basis $\{1, i, j, k\}$. We also compute some constants attached to the order. The instructions `DiscOr`, `ParOr` and `denOr` give the discriminant, the parity and the denominator of the order respectively. The logical function `isMaxOrder` identifies if the order is maximal or not. In the case of Eichler orders, `NivOr` gives the level.

Given two orders we also implemented algorithms checking inclusion and equality. These are the logical functions `IncOr` and `EqOr`, respectively. We obtain the index between two orders with the command `IndOr` (with the optional argument `true` to check previously the inclusion). The instruction `IndMaxOr` gives the index of the order in a maximal order. We also draw attention to the instruction `IntOr`, which implements an algorithm to compute the intersection of two orders. The command `ConjOr` computes the conjugate of an order by a quaternion. Finally, the instruction `IntConjOr` gives the intersection of an order with a conjugate order. Note that applying `IntOr` and `IntConjOr` to maximal orders, Eichler orders could be obtained.

7.2.2. Introduction to Shimura curves. In relation to Chapter 2, in the `Poincare` package we implement instructions related to hyperbolic geometry, fractional linear transformations, quaternion groups and constants associated to Shimura curves.

In particular, some commands about hyperbolic geometry are implemented. Given two points in the Poincaré half-plane, `eqHypL`, `defHypL` and `plotHypL` answer the equation, the definition as a geometric object and the graphical representation of the hyperbolic line linking both points, respectively. The command `hypDist` gives the hyperbolic distance between the two points. To compute the angle determined by three ordered points in the Poincaré half-plane, we have the instructions `hypAngE` and `hypAng`; they compute its value in decimals and an approximation in terms of π, respectively. Given n points in the complex half plane defining a hyperbolic polygon, we have the analogous instructions `eqHypPol`, `defHypPol` and `plotHypPol`, giving the equations of the lines determined by the edges, the definition as a geometric object and the graphical representation of the hyperbolic polygon, respectively. To compute the hyperbolic volume of a polygon we have the command `volHypPol`, which uses the approximation of the angles in terms of π computed by `hypAng`.

We also present some instructions referring to the transformations with real coefficients, to simplify the use of the package. Thus, from the four entries of a matrix, or the matrix, the commands `Hom` and `HomM` define the transformation as a function of *Maple* V. To compute the fixed points we have `fixPHom` and `fixPHomM`, which differ in terms of input. With `multHom` we obtain the multiplicator related with the geometrical interpretation. The instructions `bfHom` and `bfHomM` give the binary quadratic form associated to a transformation. The logical function `typeHom`, with the arguments `elliptic`, `parabolic` or `hyperbolic`, classifies the transformation. We also note the command `cirInv`, to define the inversion with respect to a circle, as a function.

There are some instructions to handle the isometric circles associated to the transformations. Namely, we have the instructions `cenIC` and `radIC` to obtain the center and the radius of the isometric circle, respectively. To define the isometric circle as a geometrical object use the command `defIC`. The instruction `symLIC` gives the symmetry line, cf. Lemma 2.6.

Given an Eichler order $\mathcal{O}(D,N)$ by means of a $\mathbb{Z}$-basis, the instruction `embOr` gives the generic expression of a matrix in the corresponding group of quaternion transformations $\Gamma(D,N)$, depending on $x,y,z,t \in \mathbb{Z}$, with the exception of the condition of determinant equals to 1.

To compute constants attached to a Shimura curve $X(D,N)$ we have the following instructions, on the inputs D and N: `VolhX`, `VolX`, for the hyperbolic volume and the normalized volume; `e2X`, `e3X` and `einfX` for the number of elliptic and parabolic cycles; `genusX` for the genus, and `neX` for the number of vertices of a fundamental domain for $X(D,N)$ with all vertices elliptic, assuming that this exists. We also implement the Euler function `Euler`, as an accessory instruction.

7.2.3. Quaternion algebras and quadratic forms. Using the definitions and the results contained in Chapter 3, we introduced instructions to simplify the computation of the constants and the quadratic forms associated to a given quadratic form.

To use a quadratic form f as argument, we use the associated matrix $A_1(f)$. With the instruction `expf` we recover the expression of the quadratic form. Some of the instructions work more in general for forms of real coefficients or even with parameters; note that it makes no sense to use parameters to compute the greater common divisor, or when square roots need to be computed, or when some values need to be integers, as for example to compute the reciprocal form. To give sense to all those instructions, we restrict ourselves to quadratic forms over $\mathbb{Z}$. To simplify, in some commands, we restrict ourselves to $n \leq 4$ variables.

First, command `A2f` gives the even matrix and it allows us to deal with matrices with rational integer coefficients. We have the following instructions to compute the constants attached to a quadratic form: `det1f`, `det2f`, `discf`, `contf`, `d1f` and `d2f`, according to definitions and notations in former sections. To compute the level of a quadratic form we implement the instruction `nivf`.

In relation with the invariants with respect to equivalence relations, we have the instructions: `discpf`, giving the discriminant in $\mathbb{Q}_p$; `signf`, for the signature; `diagf`, to make f diagonal over $\mathbb{Q}$; `Sf`, to obtain the list of primes for which f is $\mathbb{Q}$-anisotropic; `S1f`, which gives the list of critical primes; `HWinvf`, for the Hasse–Witt invariant; and `S2f`, which gives the list of primes for which the Hasse–Witt invariant of f is -1. To look for representations of rational integers by quadratic forms in 2, 3, or 4 variables over $\mathbb{Z}$, we include the command `FindRepf`.

We compute the quadratic forms associated using the commands `adf`, `polf`, `rec1f` and `rec2f`, which give the adjoint form, the polar form, the 1-reciprocal form and the 2-reciprocal form, respectively. We also implement the logical functions `isPrimf`, `isIsotf`, `isK1f` and `isK2f`. Their answer is *true* or *false* according to whether or not the form satisfies the properties of being primitive, isotropic, K_1-form and K_2-form, respectively.

The generalization of the definitions of the reciprocal form associated to f, in the case $|\det_1(f)|$ is a square leads to the commands `d1gf`, `d2gf`, `rec1gf`, `rec2gf`, `isK1gf` and `isK2gf`.

With respect to the forms associated to the quadratic fields, the `Poincare` package contains the instructions: `fundDiscF`, which gives the fundamental discriminant of the field; `bnfF`, which gives the binary normic form associated to the quadratic field; `bnfLatF` and `bnfOrF`, which gives the binary normic form associated to a lattice and to a quadratic order, respectively, and `redbnfOrF`, which gives the reduced binary normic form of a quadratic order.

There are also instructions referred to the normic and trace forms associated to quaternion algebra.

First, we obtain the normic forms attached to a quaternion algebra $H = \left(\frac{a,b}{\mathbb{Q}}\right)$, directly from the arguments a and b, with the instructions `nfH4` and `nfH3`. To simplify the use of the constants associated to these forms, the instructions `det1nfH` and `levnfH` give the determinant and the level of the normic forms attached to the algebra, respectively. We also have instructions to compute forms associated to the normic forms, as `adH3`, `adH4`, `polH3`, `polH4`, `rec1H3`, `rec1H4` and `rec2H4`.

We also implement the correspondence between quadratic forms and quaternion algebras from Section 3.5.4, using commands of quadratic forms. The determination of a quaternion algebra H from prefixed quadratic forms with some properties is implemented in the instructions `Hnf4` and `Hnf3`. The reverse sense corresponds to the instructions `nfH4` and `nfH3`.

In relation to the concepts and results of Section 3.6, we introduce some instructions to find the quadratic forms associated to the quaternion orders. Recall that an order is given for any $\mathbb{Z}$-basis, that is a list of four quaternions. First the optional argument `true` checks if the input defines an order. The normic forms attached to the orders are computed by means of `nfOr4` and `nfOr3`. If the input is not a normalized basis, the same command previously computes a normalized one. Using the trace form, we also can compute the discriminant of an order from the definition wit; it can be used with parameters.

7.2.4. Embeddings and quadratic forms. Next we comment on the algorithms referred to the results in Chapter 4, implemented in the `Poincare` package.

The instruction `hqqOr` gives the class number of optimal embeddings of a quadratic order into a quaternion order. To do this, other instructions are necessary: `hqqOrp`, for the corresponding local class number; `hOrF`, for the class number of a quadratic order; `valp`, to compute the p-adic valuation; and `Psip`, for the arithmetic function.

To make effective the embeddings we also implemented some instructions. With `nBasisZ2Or` we obtain a normalized basis of the order $\mathbb{Z} + 2\mathcal{O}$. The command `FindRepOr` looks for representations in the set $\mathcal{R}(n_{\mathbb{Z}+2\mathcal{O}(D,N)}, -D_{\Lambda(d,m)}; \mathbb{Z})$, from the basis of $\mathcal{O}(D,N)$ and the parameters d and m. It also includes an optional argument to control the search, since the process requires a lot of computations. Given a representation of the former set, the instructions `EmbHRep` and `EmbMRep` give us the corresponding embedding, as an element of H or directly as a matrix, respectively.

The procedure `belongNOr` checks if a quaternion belongs to the normalizer of a given order, to make effective the application κ, implemented in the command `kappa`.

For the computation of the binary form associated to an element in the quaternion algebra and of the binary form corresponding to a generic element of the order, we have the instructions `bfH` and `bfOr`, respectively. Finally, the logical functions `isPrimbfA` and `isPrimbfB` check whether or not the binary form attached to a quaternion satisfies the primitivity condition defined in the sets of binary forms attached to some Eichler orders in the algebras $H_{\mathrm{A}}(p)$ and $H_{\mathrm{B}}(p,q)$, respectively.

7.2.5. Hyperbolic fundamental domains for Shimura curves. We implement instructions to handle the construction of fundamental domains for Shimura curves $X(D,N)$, according to Chapter 5.

First, there are some instructions dealing with the properties and the graphical representation of the fundamental domain of the modular curve $X(1,p)$. The constants are given by the instructions: `einfX1`, `e2X1`, `e3X1`, `volhX1`, `volX1` and `genusX1`.

Details referred to the fundamental domain obtained in Section 5.1 are also available. Thus, `ninfX1`, `n2X1`, `n3X1` and `n1X1` give the cardinal of the set of parabolic, elliptic of order 2, elliptic of order 3 and accidental vertices, respectively. Moreover, the instruction `e1X1`gives the number of accidental cycles. The total number total of vertices is given by `nvX1`. The command `vintX1` gives the explicit list of the vertices obtained as intersection of the isometric circles and the semilines; and the instructions `vpX1`, `ve3X1` and `ve2X1` give the list of the parabolic, elliptic of order 2 and elliptic of order 3 vertices, respectively. The total list of vertices is given by `vtoX1` and `vtcX1`. The cycles are obtained with the command `cyclesX1` and the identified pairs of edges with `pairEdX1`.

In reference to the group $\Gamma(1,p)/\pm\mathrm{Id}$, an explicit presentation by means of generators and relations is given by the instructions `geneX1` and `relX1`.

Finally, the instruction `plotFDX1` gives the graphical representation of this particular fundamental domain $\mathcal{D}(\Gamma(1,p))$.

In the case $D > 1$ there are general instructions to compute the constants attached to the Shimura curve $X(D,N)$, cf. Section 7.2.2. For the computation of the elliptic points and transformations, instructions dealing with embeddings could be used (cf. Section 5.2.2). To make the package more useful, we incorporated explicit instructions. Thus, the instructions `FindRepOrE2` and `FindRepOrE3` look for the necessary representations to determine elliptic points of order 2 and 3, respectively. The instructions `E2HomPRep` and `E3HomPRep` give us the transformation and the corresponding elliptic point from a fixed representation. With the instructions `E2HomP` and `E3HomP`, we join the two algorithms to obtain a list of transformations and elliptic points. In general, in the instructions looking for representations there is an optional parameter to control the depth of the search.

To look for relations between elliptic points given in advance, we use instructions relative to the transformations, from Chapter 2. Moreover, we incorporate the instruction `FindCond`, which gives necessary conditions for two points to be related; it is especially useful to see that two points are not equivalent. To deal with the hyperbolic polygon, use the instructions relative to hyperbolic geometry (cf. Section 7.2.2).

We also implement instructions in relation with the principal homothety and the fundamental domain for the generated group. Thus, `prHom` gives the principal homothety, as a transformation, from the fundamental unit of the corresponding quadratic order. To deal with a hyperbolic strip, which is a fundamental domain for the group generated by the principal homothety, we have the instructions `symcS`, `symiS` and `symsS`, using different criteria (cf. Remark 5.36) from a circle of fixed radius. Finally, the instruction `defsS` defines the circles determining the hyperbolic strip as a geometric object.

To make explicit that all the elements in the quaternion algebra come from embeddings, we have the instruction `Fquat`. From a quaternion $\omega \in H^*$, the instruction gives the quadratic field F_ω such that ω is obtained from an embedding of F_ω into H.

7.2.6. Complex multiplication points in Shimura curves. In this section we briefly comment on the instructions implemented to determine the existence of complex multiplication points in a Shimura curve $X(D,N)$ by a fixed quadratic order, and the instructions to compute its number and the coordinates. Most of the procedures are valid for any Shimura curve in general. In the case of the non-ramified and small ramified algebras of type A and type B, we implement some instructions to simplify the computations, using the results given in the previous sections.

First, note the implementation of the following logical functions: `isCMOrF` and `isCMOrFg`, which check the existence of complex multiplication points of the Shimura curve $X(D,N)$ by a prescribed quadratic order, in the case of N square-free and in the general case. In an analogous way the instruction `isSCMOrF` and `isSCMOrFg` check the existence of special complex multiplication points. The second of these instructions needs the internal use of the logical function `isDNrepOrF`, which checks if the reduced normic form of the quadratic order represents DN.

The number of complex multiplication points by a fixed quadratic order is obtained with the instruction `cmOrF`, from the parameters D, N, d, m.

To explicitly compute complex multiplication points of $\mathcal{H}$ we have the instruction `CMP`, which uses as argument a basis of the quaternion order and the parameters that determine the quadratic order. It also includes an optional argument to control the depth in the search for solutions.

In `SCMOrF` we implement an algorithm to obtain the list of quadratic orders for which a fixed curve $X(D,N)$ has special complex multiplication points. Some variants of this instruction also include the number of points of complex multiplication by those orders, the ideal class number of the quadratic order, etc. The instruction `scm` gives the total number of special complex multiplication points of a Shimura curve.

Like the complex multiplication points that we called specials, the points determined by other similar conditions may be of interest. For example, we may want to consider the quadratic orders whose binary normic form represent a divisor of the discriminant of the quaternion order $D_{\mathcal{O}} = DN$. To simplify those computations we also present a modified version of the above instructions, where we can choose which number is going to be represented by the binary form. Thus, we have the series of modified instructions: `SCMMOrFg`, `scmM`, `isSCMMOrFg`, etc., based on the command `isRrepOrF`.

7.3. List of instructions

`A2f`(M): M, matrix $A_1(f)$ of a quadratic form f.
  Gives the even matrix $A_2(f)$ associated to f.

`adH3`(a,b): $a, b \in \mathbb{Z}$, defining the quaternion $\mathbb{Q}$-algebra $H = \left(\frac{a,b}{\mathbb{Q}}\right)$.
  Gives the matrix of the adjoint form of the ternary normic form associated to the algebra H.

`adH4`(a,b): $a, b \in \mathbb{Z}$, defining the quaternion $\mathbb{Q}$-algebra $H = \left(\frac{a,b}{\mathbb{Q}}\right)$.
  Gives the matrix of the adjoint form of the quaternary normic form associated to the algebra H.

`adf`(M): M, matrix $A_1(f)$ of a quadratic form f.
  Gives the matrix $A_1\big(\mathrm{ad}(f)\big)$ of the adjoint form of f.

`belongNOr`(l,σ): l list of 4 quaternions, $\mathbb{Z}$-basis of an order $\mathcal{O}$; σ a quaternion.
  Logical function. The answer is *true* if $\sigma \in \mathrm{Nor}(\mathcal{O})$ and *false* otherwise. Optional argument `true`.

`bfH`(ω): ω quaternion.
  Gives the matrix $A_1(f_{\Phi(\omega)})$ of the binary quadratic form associated to the quaternion ω.

`bfHom`(a,b,c,d): $a,b,c,d \in \mathbb{R}$, defining the transformation $\gamma = \left(\begin{smallmatrix} a & b \\ c & d \end{smallmatrix}\right) \in \mathrm{SL}(2,\mathbb{R})$.
  Gives the matrix $A_1(f_\gamma)$ of the binary quadratic form associated to the transformation γ.

`bfHomM`(γ): $\gamma \in \mathrm{SL}(2,\mathbb{R})$.
  Gives the matrix $A_1(f_\gamma)$ of the binary quadratic form associated to the transformation γ.

`bfOr`(l): l list of 4 quaternions, $\mathbb{Z}$-basis of an order $\mathcal{O}$.
  Gives the matrix of the binary quadratic form associated to a generic element of the order $\mathcal{O}$, with respect to the basis l. Optional argument `true`.

`bnfF`(d): $d \in \mathbb{Z}$, $d \neq k^2$ for all $k \in \mathbb{Z}$.
  Gives the matrix of the binary normic form associated to the quadratic field $\mathbb{Q}(\sqrt{d})$, with respect to the basis $\{1, \sqrt{d}\}$.

`bnfLatF`(d,x,y): $d \in \mathbb{Z}$, $d \neq k^2$ for all $k \in \mathbb{Z}$; $x, y \in \mathbb{Q}$.
  Gives the matrix of the normic form associated to the lattice in $\mathbb{Q}(\sqrt{d})$ of $\mathbb{Z}$-basis $\{1, x + y\sqrt{d}\}$.

`bnfOrF(`d, m`):` $d \in \mathbb{Z}$, $d \neq k^2$ for all $k \in \mathbb{Z}$; $m \in \mathbb{N}$.
Gives the matrix of the normic form associated to the quadratic order in $\mathbb{Q}(\sqrt{d})$ of conductor m, with respect to the basis $\{1, mw\}$, where $w = \sqrt{d}$ if $d \equiv 2, 3 \bmod 4$ and $w = (1+\sqrt{d})/2$ if $d \equiv 1 \bmod 4$; d' is the square-free part of d.

`canH(`p, q`):` p, q prime numbers.
Gives a list $[a, b]$, where $a, b \in \mathbb{Z}$ are such that $\left(\frac{p,q}{\mathbb{Q}}\right) \simeq \left(\frac{a,b}{\mathbb{Q}}\right)$, according to Proposition 1.22.

`canHdisc(`p, q`):` p, q prime numbers, $p \neq q$.
Gives a list $[a, b]$, where $a, b \in \mathbb{Z}$ are such that the quaternion $\mathbb{Q}$-algebra $H = \left(\frac{a,b}{\mathbb{Q}}\right)$ has discriminant $D_H = pq$, according to Proposition 1.25.

`cenIC(`γ`):` $\gamma \in \mathrm{SL}(2, \mathbb{R})$.
Gives the center of the isometric circle attached to the transformation γ.

`cirInv(`o, r`):` $o \in \mathbb{C}$; $r \in \mathbb{R}$.
Gives the inversion map with respect to the circle of center o and radius r, in the complex plane.

`cmOrF(`D, N, d, m`):` $D \in \mathbb{N}$, product of an even number of different primes; $N \in \mathbb{N}$, $\gcd(D, N) = 1$; $d \in \mathbb{Z}$, $d \neq k^2$ for all $k \in \mathbb{Z}$; $m \in \mathbb{N}$.
Gives the number of points of $X(D, N)$ with complex multiplication by the quadratic order $\Lambda(d, m)$.

`CMP(`l, d, m, r`):` l list of 4 quaternions, $\mathbb{Z}$-basis of an order $\mathcal{O}$; $d \in \mathbb{Z}$, $d \neq k^2$ for all $k \in \mathbb{Z}$; $m \in \mathbb{N}$; $r > 0$ optional parameter (by default its value is 5; we advise $r \leq 10$).
Gives a list of points with complex multiplication by the quadratic order $\Lambda(d, m) \subseteq \mathbb{Q}(\sqrt{d})$ of conductor m, in the Shimura curve determined by the order $\mathcal{O}$, assuming it is an Eichler order in an indefinite quaternion algebra. They are obtained from solutions (x_1, x_2, x_3) of the equation $n_{\mathbb{Z}+2\mathcal{O},3}(X_1, X_2, X_3) = -D_\Lambda$, $|x_i| < r$.

`ConjOr(`l, σ`):` l list of 4 quaternions, $\mathbb{Z}$-basis of an order $\mathcal{O}$; σ invertible quaternion.
Gives a list of 4 quaternions $[v_1, v_2, v_3, v_4]$, $\mathbb{Z}$-basis of the order obtained conjugating the order $\mathcal{O}$ by the element σ; that is, such that $\sigma^{-1}\mathcal{O}\sigma = \mathbb{Z}[v_1, v_2, v_3, v_4]$. Optional argument `true`.

`contf(`M`):` M, matrix $A_1(f)$ of a quadratic form f over $\mathbb{Z}$.
Gives the content of the quadratic form f.

`CoorOr(`ω, l`):` ω a quaternion; l list of 4 quaternions, $\mathbb{Z}$-basis of an order $\mathcal{O}$.
Gives the coordinates of ω with respect to l. Optional argument `true`.

`cyclesX1(`p`):` p prime number.
Gives the cycles of a fundamental domain for the curve $X_0(p)$.

`d1f(`M`):` M, matrix $A_1(f)$ of a quadratic form f over $\mathbb{Z}$, with square determinant.
Gives the value $d_1(f)$, attached to the quadratic form f.

`d2f(`M`):` M, matrix $A_1(f)$ of a quadratic form f over $\mathbb{Z}$, with determinant square.
Gives the value $d_2(f)$, attached to the quadratic form f.

`defHypL(`z_1, z_2`, lin):` $z_1, z_2 \in \mathbb{C}$, $\mathrm{Im}(z_1), \mathrm{Im}(z_2) \geq 0$; `lin` a name.
Defines the hyperbolic line through the two points z_1, z_2, as geometrical object, saved with the name *lin*.

`defHypPol(`l`, lin):` $l = [z_1, \ldots, z_n]$, $z_1 \in \mathbb{C}$, $\mathrm{Im}(z_i) \geq 0$; `lin` a name.

Defines the hyperbolic polygon given by the n points of l, as the collection of the hyperbolic lines giving the edges as geometrical objects saved with the names *lin*1, ..., *linn*.

`defHypS(`r_1`,`r_2`, cir1, cir2):` $r_1, r_2 \in \mathbb{R}^+$; `cir1`, `cir2` names.

Defines the hyperbolic strip given by the circles of center 0 and radius r_1, r_2, defining these circles as geometrical objects, saved with the names *cir*1, *cir*2 respectively.

`defIC(`γ`, cir):` $\gamma \in \mathrm{SL}(2, \mathbb{R})$; `cir` a name.

Defines the isometric circle attached to the transformation γ, as geometrical object, saved with the name *cir*.

`defQuatAlg(`a, b`):` $a, b \in \mathbb{Q}^*$.

Fixes the quaternion $\mathbb{Q}$-algebra $\left(\frac{a,b}{\mathbb{Q}}\right)$.

`denOr(`l`):` l list of 4 quaternions, $\mathbb{Z}$-basis of an order $\mathcal{O}$.

Gives the denominator of the order $\mathcal{O}$, $m_{\mathcal{O}}$.

`det1f(`M`):` M, matrix $A_1(f)$ of a quadratic form f.

Gives the determinant of the quadratic form f, $\det_1(f)$.

`det1nfH(`a`,`b`):` $a, b \in \mathbb{Z}$, defining the quaternion $\mathbb{Q}$-algebra $H = \left(\frac{a,b}{\mathbb{Q}}\right)$.

Gives the determinant of the quaternary and ternary normic forms, attached to the algebra H, $\det_1(n_{H,4}) = \det_1(n_{H,3})$.

`det2f(`M`):` M, matrix $A_1(f)$ of a quadratic form f.

Gives the even determinant $\det_2(f)$ of the quadratic form f.

`diagf(`M`):` M, matrix $A_1(f)$ of a quadratic form f.

Gives the matrix of a diagonal quadratic form $\mathbb{Q}$-equivalent to f.

`discf(`M`):` M, matrix $A_1(f)$ of a quadratic form f over $\mathbb{Z}$.

Gives the $\mathbb{Q}$-discriminant $\mathrm{disc}_{\mathbb{Q}}(f)$ of the quadratic form f.

`DiscH(`a, b`):` $a, b \in \mathbb{Z}$, different from 0.

Gives the discriminant D_H of the quaternion $\mathbb{Q}$-algebra $H = \left(\frac{a,b}{\mathbb{Q}}\right)$.

`DiscOr(`l`):` l list of 4 quaternions, $\mathbb{Z}$-basis of an order $\mathcal{O}$.

Gives the discriminant of the order $\mathcal{O}$, $D_{\mathcal{O}}$. Optional argument `true`. The quaternion algebra must have been fixed by means of the instruction `defQuatAlg`.

`DiscOrg(`l`):` l list of 4 quaternions, $\mathbb{Z}$-basis of an order $\mathcal{O}$.

Gives the discriminant $D_{\mathcal{O}}$ of the order $\mathcal{O}$. Unlike `DiscOr`, this instruction allows parameters, that is unfixed values for the parameters a and b in the quaternion algebra.

`discpf(`M`):` M, matrix $A_1(f)$ of a quadratic form f over $\mathbb{Z}$.

Gives the discriminant $\mathrm{disc}_{\mathbb{Q}_p}(f)$ in $\mathbb{Q}_p$, of the quadratic form f.

`e1X1(`p`):` p prime number.

Gives the number $e_1(1, p)$ of accidental cycles of the fundamental domain for the curve $X(1, p) = X_0(p)$ given in Chapter 5.

`E2HomP(`l`):` l_1 list of 4 quaternions, $\mathbb{Z}$-basis of an Eichler order $\mathcal{O}(D, N)$; $D \in \mathbb{N}$ product of an even number of different primes.

Gives a list of elliptic transformations of order 2 and the corresponding elliptic points of the Shimura curve $X(D, N)$. The optional parameter `r` allows us to control the depth in the search for representations; by default its value is 5; we advise assigning values ≤ 10.

`E2HomPRep(`l_1, l_2`):` l_1 list of 4 quaternions, $\mathbb{Z}$-basis of an Eichler order $\mathcal{O}(D, N)$; $D \in \mathbb{N}$ product of an even number of different primes; $l_2 = [x_1, x_2, x_3]$

a representation in $\mathcal{R}^*(n_{\mathbb{Z}+2\mathcal{O}(D,N),3}, 4; \mathbb{Z})$; that is, $x_i \in \mathbb{Z}$ such that $n_{\mathbb{Z}+2\mathcal{O}(D,N),3}(x_1, x_2, x_3) = 4$. (For example, an element of the list obtained by `FindRepE2`(l_1).)

Gives a list formed by a transformation of order 2 and the corresponding elliptic point in the Shimura curve $X(D,N)$.

`e2X`(D,N): $D \in \mathbb{N}$, product of an even number of different primes; $N \in \mathbb{N}$, $\gcd(D,N) = 1$.

Gives the number $e_2(D,N)$ of elliptic cycles of order 2 in the Shimura curve $X(D,N)$.

`e2X1`(p): p prime number.

Gives the number $e_2(1,p)$ of elliptic cycles of order 2 in the curve $X(1,p) = X_0(p)$.

`E3HomP`(l): l list of 4 quaternions, $\mathbb{Z}$-basis of an Eichler order $\mathcal{O}(D,N)$; $D \in \mathbb{N}$ product of an even number of different primes.

Gives a list of elliptic transformations of order 3 and the corresponding elliptic points in the Shimura curve $X(D,N)$. The optional parameter `r` allows to control the depth in the search of representations; by default its value is 5; we advise assigning values ≤ 10.

`E3HomPRep`(l_1, l_2): l_1 list of 4 quaternions, $\mathbb{Z}$-basis of an Eichler order $\mathcal{O}(D,N)$; $D \in \mathbb{N}$ product of an even number of different primes; $l_2 = [x_1, x_2, x_3]$ a representation in $\mathcal{R}^*(n_{\mathbb{Z}+2\mathcal{O}(D,N),3}, 3; \mathbb{Z})$; that is, $x_i \in \mathbb{Z}$ such that $n_{\mathbb{Z}+2\mathcal{O}(D,N),3}(x_1, x_2, x_3) = 3$. (For example, an element of the list obtained by `FindRepE2`(l_1).)

Gives a list formed by a transformation of order 3 and the corresponding elliptic point in the Shimura curve $X(D,N)$.

`e3X`(D,N): $D \in \mathbb{N}$, product of an even number of different primes; $N \in \mathbb{N}$, $\gcd(D,N) = 1$.

Gives the number $e_3(D,N)$ of elliptic cycles of order 3 of the Shimura curve $X(D,N)$.

`e3X1`(p): p prime number.

Gives the number $e_3(1,p)$ of elliptic cycles of order 3 of the curve $X(1,p) = X_0(p)$.

`einfX`(D,N): $D \in \mathbb{N}$, product of an even number of different primes; $N \in \mathbb{N}$, $\gcd(D,N) = 1$.

Gives the number $e_\infty(D,N)$ of parabolic cycles of the Shimura curve $X(D,N)$.

`einfX1`(p): p prime number.

Gives the number $e_\infty(1,p)$ of parabolic cycles of the curve $X(1,p) = X_0(p)$.

`embH`(ω): ω quaternion.

Gives the matrix $\Phi(\omega) \in \mathrm{M}(2, \mathbb{Q}(\sqrt{a}))$, image of ω by the embedding Φ fixed in Proposition 1.19, assuming $a > 0$. It is necessary to assign values to the parameters a and b in advance, with the instruction `defQuatAlg`.

`embHg`(ω): ω quaternion.

Gives the matrix $\Phi(\omega) \in \mathrm{M}(2, \mathbb{Q}(\sqrt{a}))$, image of ω by the embedding Φ fixed in Proposition 1.19, without checking if $a > 0$. It is not necessary to assign values to the parameters a and b in advance.

`EmbHRep`(l,d,m,l_1): l list of 4 quaternions, $\mathbb{Z}$-basis of an order $\mathcal{O}$; $d\in\mathbb{Z}$, $d\neq k^2$ for all $k\in\mathbb{Z}$; $m\in\mathbb{N}$; l_1 list of three integers, which is a representation of $-D_\Lambda$, by the ternary normic form of the order $\mathbb{Z}+2\mathcal{O}$, where Λ is the order of conductor m in the quadratic field $\mathbb{Q}(\sqrt{d})$.
Gives the quaternion $\varphi(\sqrt{d})$, where φ is the embedding of the quadratic order Λ into the order $\mathcal{O}$ determined by the representation l_1.

`EmbMRep`(l,d,m,l_1): l list of 4 quaternions, $\mathbb{Z}$-basis of an order $\mathcal{O}$; $d\in\mathbb{Z}$, $d\neq k^2$ for all $k\in\mathbb{Z}$; $m\in\mathbb{N}$; l_1 list of three integers which is a representation of $-D_\Lambda$, by the ternary normic form of the order $\mathbb{Z}+2\mathcal{O}$, where Λ is the order of conductor m in the quadratic field $\mathbb{Q}(\sqrt{d})$.
Gives the matrix $\Phi(\varphi(\sqrt{d}))$, where φ is the embedding of the quadratic order Λ into the quaternion order $\mathcal{O}$ determined by the representation l_1.

`embOr`(l): l list of 4 quaternions, $\mathbb{Z}$-basis of an order $\mathcal{O}$.
Gives a matrix in $\mathrm{M}(2,\mathbb{Q}(\sqrt{a}))$ image of a generic element of the order $\mathcal{O}$, by the embedding fixed in Proposition 1.19, if $a>0$. It is necessary to assign values to the parameters a and b in advance, with the instruction `defQuatAlg`. Optional argument `true`.

`eqHypL`(z_1,z_2): $z_1,z_2\in\mathbb{C}$, $\mathrm{Im}(z_i)\geq 0$.
Gives the equation of the hyperbolic line through the two points z_1,z_2.

`eqHypPol`(l): $l=[z_1,\ldots,z_n]$, $z_1\in\mathbb{C}$, $\mathrm{Im}(z_i)\geq 0$.
Gives the equations of the hyperbolic lines that contain the edges of the hyperbolic polygon given by the n points of l.

`EqOr`(l_1,l_2): l_1,l_2 lists of 4 quaternions, $\mathbb{Z}$-bases of the orders $\mathcal{O}_1$, $\mathcal{O}_2$ respectively.
Logical function. The answer is *true* if the two orders are equal, and *false* otherwise. Optional arguments `true`.

`Euler`(n): $n\in\mathbb{N}$.
Gives the value of the Euler function at n, $\varphi(n):=n\prod_{p|n}(1-1/p)$.

`expf`(M): M, matrix $A_1(f)$ of a quadratic form f.
Gives the polynomial expression of the quadratic form f.

`findCond`(z_1,z_2): $z_1,z_2\in\mathbb{C}$.
Gives conditions on $x,y,z,t\in\mathbb{Q}$ so that $\gamma=\begin{pmatrix} x+y\sqrt{a} & z+t\sqrt{a}\\ b(z-t\sqrt{a}) & x-y\sqrt{a}\end{pmatrix}$ satisfies $\gamma(z_1)=z_2$. Specifically, gives a binary form which must represent the determinant of γ. It is applied especially when a and b are the parameters fixing a quaternion algebra.

`findRepf`(M,δ,r): M matrix $A_1(f)$ of a quadratic form f over $\mathbb{Z}$, in $n=2$, 3 or 4 variables; $\delta\in\mathbb{Z}$; $r>0$ optional argument (by default its value is 5; we advise $r\leq 10$).
It looks for representations $(x_1,\ldots,x_n)$ of δ by the quadratic form f, with $|x_i|\leq r$.

`findRepOr`(l,d,m,r): l list of 4 quaternions, $\mathbb{Z}$-basis of an order $\mathcal{O}$, $d\in\mathbb{Z}$, $d\neq 0$; $m\in\mathbb{N}$; $r>0$ optional argument (by default its value is 5; we advise $r\leq 10$).
It looks for representations (x_1,x_2,x_3) of $-D_\Lambda$, with $|x_i|\leq r$, by the ternary normic form of the order $\mathbb{Z}+2\mathcal{O}$, where Λ is the order of conductor m in the quadratic field $\mathbb{Q}(\sqrt{d})$.

`findRepOrE2`(l,r): l list of 4 quaternions, $\mathbb{Z}$-basis of an odd Eichler order of $\mathcal{O}(D,N)$; $r>0$ optional parameter (by default its value is 5; we advise $r\leq 10$).

Gives a list of representations in $\mathcal{R}^*(n_{\mathbb{Z}+2\mathcal{O}(D,N),3}, 4; \mathbb{Z})$ less than r; that is, a list of triples x_1, x_2, x_3 solving $n_{\mathbb{Z}+2\mathcal{O}(D,N),3}(X_1, X_2, X_3) = 4$, with $|x_i| < r$. From these solutions we can determine transformations and elliptic points of order 2 in the Shimura curve $X(D, N)$ using `E2HomPRep`, which implements Theorem 5.15.

`findRepOrE3(`l, r`):` l list of 4 quaternions, $\mathbb{Z}$-basis of an odd Eichler order $\mathcal{O}(D, N)$; $r > 0$ optional parameter (by default its value is 5; we advise $r \leq 10$).
Gives a list of representations in $\mathcal{R}^*(n_{\mathbb{Z}+2\mathcal{O}(D,N),3}, 3; \mathbb{Z})$ less than r; that is, a list of triples x, y, z solving $n_{\mathbb{Z}+2\mathcal{O}(D,N),3}(X, Y, Z) = 3$, with $|x_i| < r$. From these solutions we can determine transformations and elliptic points of order 3 in the Shimura curve $X(D, N)$ using `E2HomPRep`, which implements Theorem 5.15.

`fixPHom(`a, b, c, d`):` a, b, c, d real numbers, defining $\gamma = \left(\begin{smallmatrix} a & b \\ c & d \end{smallmatrix}\right)$.
Gives the list of fixed points of γ contained in $\mathcal{H} \cup \mathbb{R}$.

`fixPHomM(`γ`):` $\gamma \in \mathrm{GL}(2, \mathbb{R})$.
Gives the list of fixed points of γ contained in $\mathcal{H} \cup \mathbb{R}$.

`Fquat(`ω`):` ω invertible quaternion.
Gives the square-free integer d such that ω comes from an embedding of $\mathbb{Q}(\sqrt{d})$ into H.

`fundDiscF(`d`):` $d \in \mathbb{Z}$, $d \neq k^2$ for all $k \in \mathbb{Z}$.
Gives the fundamental discriminant of the quadratic field $\mathbb{Q}(\sqrt{d})$.

`geneX1(`p`):` p prime number.
Gives a list of generators of the group $\Gamma_0(p)/\pm \mathrm{Id}$.

`genusX(`D, N`):` $D \in \mathbb{N}$, product of an even number of different primes; $N \in \mathbb{N}$ such that $\gcd(D, N) = 1$.
Gives the genus $g(D, N)$ of the Shimura curve $X(D, N)$.

`genusX1(p):` p prime number.
Gives the genus $g(1, p)$ of the curve $X(1, p) = X_0(p)$.

`HermiteOr(`l`):` l list of 4 quaternions, $\mathbb{Z}$-basis of an order $\mathcal{O}$.
Gives the Hermite basis of the order $\mathcal{O}$.

`hF(`d`):` $d \in \mathbb{Z}$, $d \neq k^2$ for all $k \in \mathbb{Z}$.
Gives the ideal class number of the field $\mathbb{Q}(\sqrt{d})$, d square free, $|d| < 100$. Otherwise, it works as a parameter.

`Hilbert(`a, b, p`):` a, b enters; $p \geq 2$ prime number.
Gives the Hilbert symbol $(a, b)_p$. For $p \neq 2$, it uses the standard maple instruction for the Legendre symbol `numtheory[L]`.

`Hnf3(`M`):` M matrix of a ternary quadratic form f over $\mathbb{Z}$, with determinant equal to a square, $\det_1(f) = \lambda^2$.
Gives a pair of integers (a, b) such that the ternary normic form associated to the quaternion algebra $H = \left(\frac{a,b}{\mathbb{Q}}\right)$ is $\mathbb{Q}$-equivalent to f.

`Hnf4(`M`):` M matrix of a quaternary quadratic form f over $\mathbb{Z}$, non negative definite, with determinant equal to a square, $\det_1(f) = \lambda^2$.
Gives a pair of integers (a, b) such that the quaternary normic form associated to the quaternion algebra $H = \left(\frac{a,b}{\mathbb{Q}}\right)$ is $\mathbb{Q}$-equivalent to f.

`Hom(`a, b, c, d`):` $a, b, c, d \in \mathbb{R}$, defining a transformation $\gamma = \left(\begin{smallmatrix} a & b \\ c & d \end{smallmatrix}\right) \in \mathrm{SL}(2, \mathbb{R})$.
Gives the transformation γ as a function.

`HomM(`γ`):` $\gamma \in \mathrm{SL}(2, \mathbb{R})$.
Gives the transformation γ as function.

`hOrF`(d, m): $d \in \mathbb{Z}$, $d \neq k^2$ for all $k \in \mathbb{Z}$; $m \in \mathbb{N}$.
Gives the ideal class number of the quadratic order of conductor m in the field $\mathbb{Q}(\sqrt{d})$.

`hqqOr`(D, N, d, m): $D \in \mathbb{N}$ product of an even number of different primes; $N \in \mathbb{N}$, $\gcd(D, N) = 1$; $d \in \mathbb{Z}$, $d \neq k^2$ for all $k \in \mathbb{Z}$; $m \in \mathbb{N}$.
Gives the class number of optimal embeddings of the quadratic order of conductor m of the field $\mathbb{Q}(\sqrt{d})$ into an Eichler order of level N of a quaternion algebra of discriminant D, module $\mathcal{O}^*$.

`hqqOrp`(D, N, d, m, p): $D \in \mathbb{N}$ product of an even number of different primes; $N \in \mathbb{N}$, $\gcd(D, N) = 1$; $d \in \mathbb{Z}$, $d \neq k^2$ for all $k \in \mathbb{Z}$; $m \in \mathbb{N}$; p prime.
Gives the class number of local optimal embeddings of the quadratic order of conductor m of the field $\mathbb{Q}(\sqrt{d})$ localized at p, into an Eichler order of level N of a quaternion algebra of discriminant D localized at p, module $\mathcal{O}_p^*$-equivalence.

`HWinvf`(M, p): M, matrix of a quadratic form f, over $\mathbb{Z}$; p, prime.
Gives the Hasse–Witt invariant of the quadratic form f at $\mathbb{Q}_p$.

`hypAng`(z_1, z_2, z_3): $z_1, z_2, z_3 \in \mathbb{C}$, $\mathrm{Im}(z_i) \geq 0$.
Gives an approximation of the value of the hyperbolic angle determined by the ordered points z_1, z_2, z_3 expressed in terms of π, from the hyperbolic lines determined by the points. The approximation is due to the conversion of the decimal value of the angle, in 10 digits, in a rational multiple of π.

`hypAngE`(z_1, z_2, z_3): $z_1, z_2, z_3 \in \mathbb{C}$, $\mathrm{Im}(z_i) \geq 0$.
Gives the value of the hyperbolic angle determined by the ordered points z_1, z_2, z_3, from the hyperbolic lines determined by the points.

`hypDist`(z_1, z_2): $z_1, z_2 \in \mathbb{C}$, $\mathrm{Im}(z_1), \mathrm{Im}(z_2) \geq 0$.
Gives the hyperbolic distance $\delta(z_1, z_2)$ between the points z_1, z_2.

`IncOr`(l_1, l_2): l_1, l_2 lists of 4 quaternions, $\mathbb{Z}$-bases of orders $\mathcal{O}_1$, $\mathcal{O}_2$, respectively.
Logical function. Gives *true* if the order defined by l_1 is included in the order defined by l_2, and *false* otherwise. Optional arguments `true`.

`IndMaxOr`(l): l list of 4 quaternions, $\mathbb{Z}$-basis of an order $\mathcal{O}$.
Gives the index of $\mathcal{O}$ in any maximal order that contains it. Optional argument `true`.

`IndOr`(l_1, l_2): l_1, l_2 lists of 4 quaternions, $\mathbb{Z}$-bases of orders $\mathcal{O}_1$, $\mathcal{O}_2$, respectively.
Gives the index of the order $\mathcal{O}_1$ in the order $\mathcal{O}_2$, checking previously that $\mathcal{O}_1 \subseteq \mathcal{O}_2$. Optional arguments `true`.

`IntConjOr`(l, σ): l list of 4 quaternions, $\mathbb{Z}$-basis of an order $\mathcal{O}$; σ invertible quaternion.
Gives a list of 4 quaternions, the basis of the intersection order of $\mathcal{O}$ and the conjugate order of $\mathcal{O}$ by σ. Optional argument `true`. Note that applied to a maximal order, an Eichler order is obtained.

`IntOr`(l_1, l_2): l_1, l_2 lists of 4 quaternions, $\mathbb{Z}$-bases of orders $\mathcal{O}_1$ and $\mathcal{O}_2$, respectively.
Gives list of 4 quaternions, the basis of the intersection order of $\mathcal{O}_1$ and $\mathcal{O}_2$. Optional arguments `true`. Note that, applied to maximal orders, Eichler orders are obtained.

`isCMOrF`(D, N, d, m): $D \in \mathbb{N}$, product of an even number of different primes; $N \in \mathbb{N}$, square-free, $\gcd(D, N) = 1$; $d \in \mathbb{Z}$, $d \neq k^2$ for all $k \in \mathbb{Z}$; $m \in \mathbb{N}$.

Logical function. The answer is *true* if the Shimura curve $X(D,N)$ has complex multiplication points by the quadratic order of conductor m in the quadratic field $\mathbb{Q}(\sqrt{d})$; the answer is *false* otherwise.

`isCMOrFg`(D,N,d,m): $D\in\mathbb{N}$, product of an even number of primes; $N\in\mathbb{N}$, not necessarily square-free, $\gcd(D,N)=1$; $d\in\mathbb{Z}$, $d\neq k^2$ for all $k\in\mathbb{Z}$; $m\in\mathbb{N}$.
Logical function. The answer is *true* if the Shimura curve $X(D,N)$ has complex multiplication points by the quadratic order of conductor m of the quadratic field $\mathbb{Q}(\sqrt{d})$; the answer is *false* otherwise.

`isDNrepOrF`(D,N,d,m): $D\in\mathbb{N}$, product of an even number of different primes; $N\in\mathbb{N}$, $\gcd(D,N)=1$; $d\in\mathbb{Z}$, $d\neq k^2$ for all $k\in\mathbb{Z}$; $m\in\mathbb{N}$.
Logical function. The answer is *true* if the reduced binary normic form of the quadratic order of conductor m in the quadratic field $\mathbb{Q}(\sqrt{d})$ represents the value DN; the answer is *false* otherwise.

`isIsotf`(M,p): M, matrix $A_1(f)$ of a quadratic form f, over $\mathbb{Z}$; p, prime.
Logical function. The answer is *true* if f is a $\mathbb{Q}_p$-isotropic form; Otherwise, the answer is *false*.

`isK1f`(M): M, matrix $A_1(f)$ of a quadratic form f, over $\mathbb{Z}$, with determinant equal to a square.
Logical function. The answer is *true* if f is a K_1-form and *false* otherwise.

`isK2f`(M): M, matrix $A_1(f)$ of a quadratic form f, over $\mathbb{Z}$, with even determinant equal to a square.
Logical function. The answer is *true* if f is a K_2-form and *false* otherwise.

`isMaxOrder`(l): l list of 4 quaternions, $\mathbb{Z}$-basis of an order $\mathcal{O}$.
Logical function. The answer is *true* if $\mathcal{O}$ is a maximal order and *false* if it is not an order or if it is not maximal. With the optional argument `true`, if the answer is *false* the reason is shown.

`isNBasisOr`(l): l list of 4 quaternions, $\mathbb{Z}$-basis of an order $\mathcal{O}$.
Logical function. The answer is *true* if l is a normalized basis of the order $\mathcal{O}$ and *false* otherwise.

`isOrder`(l): l list of 4 quaternions.
Logical function. The answer is *true* if the list l defines an order and the answer is *false* otherwise. With the optional argument `true` at the end, it shows the reason why it is not an order, if this is the case. It is necessary to assign values to the parameters a and b in advance, with the instruction `defQuatAlg`.

`isPrimbfA`(p,N,ω): $p\equiv 3 \bmod 4$ prime number; $N\in\mathbb{N}$, $N\mid(p-1)/2$, square-free; ω quaternion of $H_A(p)$.
Logical function. The answer is *true* if the binary form $f_{\Phi(\omega)}$ is $\mathcal{O}_A(2p,N)$-primitive. The answer is *false* otherwise.

`isPrimbfB`(p,q,N,ω): p,q primes, $q\equiv 1 \bmod 4$ and $\left(\frac{p}{q}\right)=-1$; $N\in\mathbb{N}$ square-free, such that $\gcd(N,p)=1$ and $N\mid(q-1)/4$; ω quaternion of $H_B(p,q)$.
Logical function. The answer is *true* if the binary form $f_{\Phi(\omega)}$ is $\mathcal{O}_B(pq,N)$-primitive. The answer is *false* otherwise.

`isPrimf`(M): M, matrix $A_1(f)$ of a quadratic form f, over $\mathbb{Z}$.
Logical function. The answer is *true* if the quadratic form f is primitive and *false* otherwise.

`isRrepOrF`(d,m,R): $d\in\mathbb{Z}$, $d\neq k^2$ for all $k\in\mathbb{Z}$; $m\in\mathbb{N}$.

Logical function. The answer is *true* if the reduced binary normic form of the quadratic order of conductor m in the quadratic field $\mathbb{Q}(\sqrt{d})$ represents the value R. The answer is *false* otherwise.

`isSCMMOrFg`(D, N, d, m, R): $D \in \mathbb{N}$, product of an even number of different primes; $N \in \mathbb{N}$, $\gcd(D, N) = 1$; $d \in \mathbb{Z}$, $d \neq k^2$ for all $k \in \mathbb{Z}$; $m \in \mathbb{N}$; $R \in \mathbb{Z}^+$.
Logical function. The answer is *true* if the Shimura curve $X(D, N)$ has modified special complex multiplication points of value R by the quadratic order of conductor m in the quadratic field $\mathbb{Q}(\sqrt{d})$; The answer is *false* otherwise.

`isSCMOrF`(D, N, d, m): $D \in \mathbb{N}$, product of an even number of different primes; $N \in \mathbb{N}$, square-free, $\gcd(D, N) = 1$; $d \in \mathbb{Z}$, $d \neq k^2$ for all $k \in \mathbb{Z}$; $m \in \mathbb{N}$.
Logical function. The answer is *true* if the Shimura curve $X(D, N)$ has special complex multiplication points by the quadratic order of conductor m in the quadratic field $\mathbb{Q}(\sqrt{d})$; the answer is *false* otherwise.

`isSCMOrFg`(D, N, d, m): $D \in \mathbb{N}$, product of an even number of different primes; $N \in \mathbb{N}$, not necessarily square-free, $\gcd(D, N) = 1$; $d \in \mathbb{Z}$, $d \neq k^2$ for all $k \in \mathbb{Z}$; $m \in \mathbb{N}$.
Logical function. The answer is *true* if the Shimura curve $X(D, N)$ has special complex multiplication points by the quadratic order of conductor m in the quadratic field $\mathbb{Q}(\sqrt{d})$; the answer is *false* otherwise.

`kappa`(l, σ): l list of 4 quaternions, $\mathbb{Z}$-basis of an order $\mathcal{O}$; σ quaternion of $\mathrm{Nor}(\mathcal{O})$.
Gives the matrix $\kappa(\sigma) \in \mathrm{GL}(3, \mathbb{Z})$, cf. 4.35.

`leq`(z_1, z_2): $z_1, z_2 \in \mathbb{C}$.
Logical function. The answer is *true* if $\mathrm{Re}(z_1) \leq \mathrm{Re}(z_2)$. The answer is *false* otherwise.

`levf`(M): M, matrix $A_1(f)$ of a quadratic form f.
Gives the level $\mathrm{N}(f)$ of the quadratic form f.

`levnfH`(a, b): $a, b \in \mathbb{Z}$ defining the quaternion $\mathbb{Q}$-algebra $H = \left(\frac{a,b}{\mathbb{Q}}\right)$.
Gives the level of the quaternary and ternary normic forms, attached to H, $\mathrm{N}(n_{H,4}) = \mathrm{N}(n_{H,3})$.

`LevOr`(l): l list of 4 quaternions, basis of an Eichler order $\mathcal{O}$.
Gives the level $\mathrm{N}_{\mathcal{O}}$ of the order $\mathcal{O}$, assuming it is an Eichler order. Optional argument `true`.

`McbOr`(l_1, l_2): l_1, l_2, lists of 4 quaternions, $\mathbb{Z}$-bases of orders $\mathcal{O}_1$, $\mathcal{O}_2$, respectively.
Gives the matrix of change of basis from the basis l_1 to the basis l_2. Optional arguments `true`.

`Mcoor`(l): l list of quaternions.
Gives a matrix having as columns the coordinates of the quaternions in the list l with respect to the basis $\{1, i, j, ij\}$.

`multHom`(a, b, c, d): $a, b, c, d \in \mathbb{R}$, defining a transformation $\gamma = \left(\begin{smallmatrix} a & b \\ c & d \end{smallmatrix}\right) \in \mathrm{SL}(2, \mathbb{R})$.
Gives the multiplicator of the transformation γ.

`n1X1`(p): p prime number.
Gives the number $n_1(1, p)$ of accidental vertices of the fundamental domain for the curve $X(1, p) = X_0(p)$ given in Chapter 5.

`n2X1`(p): p prime number.
Gives the number $n_2(1, p)$ of elliptic vertices of order 2 of the fundamental domain for the curve $X(1, p) = X_0(p)$ given in Chapter 5.

`n3X1`(p): p prime number.

Gives the number $n_3(1,p)$ of elliptic vertices of order 3 of the fundamental domain for the curve $X(1,p) = X_0(p)$ given in Chapter 5.

`nBasisOr`(l): l list $[1, v_2, v_3, v_4]$, v_i quaternions, $\mathbb{Z}$-basis of an order $\mathcal{O}$.
Gives a normalized basis of the order $\mathcal{O}$. Optional argument `true`.

`nBasisZ2Or`(l): l list of 4 quaternions, $\mathbb{Z}$-basis of an order $\mathcal{O}$.
Gives as list a $\mathbb{Z}$-normalized basis of the order $\mathbb{Z}+2\mathcal{O}$. Optional argument `true`.

`neX`(D,N): $D \in \mathbb{N}$, product of an even number of different primes; $N \in \mathbb{N}$ such that $\gcd(D,N) = 1$.
Gives the number of vertices $n_e(D,N)$ of a fundamental domain for the Shimura curve $X(D,N)$ having all vertices elliptic, if it exists.

`nfH3`(a,b): $a, b \in \mathbb{Z}$ defining the quaternion $\mathbb{Q}$-algebra $H = \left(\frac{a,b}{\mathbb{Q}}\right)$.
Gives the matrix $A_1(n_{H,3})$ of the ternary normic form associated to the algebra H.

`nfH4`(a,b): $a, b \in \mathbb{Z}$ defining the quaternion $\mathbb{Q}$-algebra $H = \left(\frac{a,b}{\mathbb{Q}}\right)$.
Gives the matrix $A_1(n_{H,4})$ of the quaternary normic form associated to the algebra H.

`nfOr3`(l): l list of 4 quaternions, $\mathbb{Z}$-basis of an order $\mathcal{O}$.
Gives the matrix $A_1(n_{\mathcal{O},3})$ of the ternary normic form attached to the order $\mathcal{O}$ with respect to the basis l if this is normalized; otherwise, it computes first a normalized basis of $\mathcal{O}$, which is also shown in the output. Optional argument `true`.

`nfOr4`(l): l list of 4 quaternions, $\mathbb{Z}$-basis of an order $\mathcal{O}$.
Gives the matrix $A_1(n_{\mathcal{O},4})$ of the quaternary normic form attached to the order $\mathcal{O}$, with respect to the basis l if this is normalized; otherwise, it computes first a normalized basis of $\mathcal{O}$, which is also shown in the output. Optional argument `true`.

`ninfX1`(p): p prime number.
Gives the number $n_\infty(1,p)$ of parabolic vertices of the fundamental domain for the curve $X(1,p) = X_0(p)$ constructed in Chapter 5.

`nvX1`(p): p prime number.
Gives the total number of vertices $n(1,p)$ of the fundamental domain for the curve $X(1,p) = X_0(p)$ constructed in Chapter 5.

`OrM`(M): $M \in \mathrm{M}(4,\mathbb{Q})$.
Gives the list l of the columns of M. If M is the matrix of coordinates of an order $\mathcal{O}$, l is a basis of the order $\mathcal{O}$; it is thus the inverse instruction of `Mcoor`. With the optional argument `true`, it checks if l determines an order.

`pairEdX1`(p): p prime number.
Gives the pairs of identified edges of the fundamental domain for the curve $X(1,p) = X_0(p)$ given in Chapter 5.

`ParOr`(l): l list of 4 quaternions, $\mathbb{Z}$-basis of an order $\mathcal{O}$.
Gives the parity of the order $\mathcal{O}$: *0* if the order is even and *1* if the order is odd. Optional argument `true`.

`plotFDX1`(p): p prime number.
Gives the graphical representation of the fundamental domain for the modular curve $X(1,p) = X_0(p)$ constructed in Chapter 5.

`plotHypL`(z_1,z_2): $z_1, z_2 \in \mathbb{C}$, $\mathrm{Im}(z_1), \mathrm{Im}(z_2) \geq 0$.

Gives the graphical representation of the hyperbolic line through the two points z_1, z_2.

`plotHypPol(`l`)`: l a list $[z_1, \dots, z_n]$, $z_i \in \mathbb{C}$, $\mathrm{Im}(z_i) \geq 0$.
Gives the graphical representation of the hyperbolic polygon defined by the points of l.

`polH3(`a,b`)`: $a, b \in \mathbb{Z}$ defining the quaternion $\mathbb{Q}$-algebra $H = \left(\frac{a,b}{\mathbb{Q}}\right)$.
Gives the matrix $A_1\big(\mathrm{pol}(n_{H,3})\big)$ of the quadratic polar form of the ternary normic form of the algebra H.

`polH4(`a,b`)`: $a, b \in \mathbb{Z}$ defining the quaternion $\mathbb{Q}$-algebra $H = \left(\frac{a,b}{\mathbb{Q}}\right)$.
Gives the matrix $A_1\big(\mathrm{pol}(n_{H,4})\big)$ of the quadratic polar form of the quaternary normic form of the algebra H.

`polf(`M`)`: M, matrix $A_1(f)$ of a quadratic form f.
Gives the matrix $A_1\big(\mathrm{pol}(f)\big)$ of the quadratic polar form of the form f.

`prHom(`d,x,y,s`)`: $d \in \mathbb{Z}$, $d \neq k^2$ for all $k \in \mathbb{Z}$; $x, y \in \mathbb{Q}$ such that $x + y\sqrt{d}$ is the fundamental unit of $\mathbb{Q}(\sqrt{d})$; $s \in \mathbb{N}$.
Gives the matrix corresponding to the principal homothety, coming from the fundamental unit $(x + y\sqrt{d})^s$ in $\mathbb{Q}(\sqrt{d})$.

`Psip(`n,p`)`: $n \in \mathbb{N}$, p prime.
Gives the value of the arithmetical function $\psi_p(n) = p^{v_p(n)}(1 + 1/p)$.

`qbar(`ω`)`: ω quaternion.
Gives the conjugate $\overline{\omega}$ of ω.

`qcoeffs(`ω`)`: ω quaternion.
Gives a list formed by the coefficients of the quaternion ω; that is, a list with the coordinates of ω with respect to the basis $\{1, i, j, ij\}$.

`qconj(`ω,σ`)`: ω quaternion, σ invertible quaternion.
Gives the conjugate of ω by σ; that is, $\sigma^{-1}\omega\sigma$.

`qinv(`ω`)`: ω invertible quaternion.
Gives the inverse ω^{-1} of ω.

`qmul(`ω_1,ω_2`)`: ω_1, ω_2 quaternions.
Gives the product of the two quaternions, $\omega_1 \cdot \omega_2$. Note that the product can also be indicated by the symbol &q: ω_1 &q ω_2 gives also $\omega_1 \cdot \omega_2$.

`qnorm(`ω`)`: ω quaternion.
Gives the norm $\mathrm{n}(\omega)$ of ω.

`qtrace(`ω`)`: ω quaternion.
Gives the trace $\mathrm{tr}(\omega)$ of ω.

`radIC(`γ`)`: $\gamma \in \mathrm{SL}(2, \mathbb{R})$.
Gives the radius of the isometric circle attached to the transformation γ.

`rec1f(`M`)`: M, matrix $A_1(f)$ of a quadratic form f over $\mathbb{Z}$, with determinant equal to a square.
Gives the matrix $A_1\big(\mathrm{rec}_1(f)\big)$ of the 1-reciprocal form of the form f.

`rec1H3(`a,b`)`: $a, b \in \mathbb{Z}$ defining the quaternion $\mathbb{Q}$-algebra $H = \left(\frac{a,b}{\mathbb{Q}}\right)$.
Gives the matrix $A_1\big(\mathrm{rec}_1(n_{H,3})\big)$ of the 1-reciprocal form of the ternary normic form associated to the algebra H.

`rec1H4(`a,b`)`: $a, b \in \mathbb{Z}$ defining the quaternion $\mathbb{Q}$-algebra $H = \left(\frac{a,b}{\mathbb{Q}}\right)$.
Gives the matrix $A_1\big(\mathrm{rec}_1(n_{H,4})\big)$ of the 1-reciprocal form of the quaternary normic form associated to the algebra H.

`rec2f(`M`)`: M, matrix $A_1(f)$ of a quadratic form f over $\mathbb{Z}$, with even determinant equal to a square.
Gives the matrix $A_1\big(\mathrm{rec}_2(f)\big)$ of the 2-reciprocal form of the form f.

`rec2H4(`a,b`)`: $a,b \in \mathbb{Z}$ defining the quaternion $\mathbb{Q}$-algebra $H = \left(\frac{a,b}{\mathbb{Q}}\right)$.
Gives the matrix $A_1\big(\mathrm{rec}_2(n_{H,4})\big)$ of the 2-reciprocal form of the quaternary normic form associated to the algebra H.

`redbnfOrF(`d,m`)`: $d \in \mathbb{Z}$, $d \neq k^2$ for all $k \in \mathbb{Z}$. $m \in \mathbb{N}$.
Gives the reduced binary normic form of the quadratic order of conductor m in the quadratic field $\mathbb{Q}(\sqrt{d})$.

`relX1(`p`)`: p prime number.
Gives the relations of the group $\Gamma_0(p)/\pm\mathrm{Id}$ corresponding to the generators given by the instruction `geneX1(`p`)`.

`Sf(`M`)`: M, matrix $A_1(f)$ of a quadratic form f over $\mathbb{Z}$.
Gives the list of prime numbers p for which the quadratic form f is $\mathbb{Q}_p$-anisotropic.

`S1f(`M`)`: M, matrix $A_1(f)$ of a quadratic form f over $\mathbb{Z}$.
Gives a list of critical prime numbers for f.

`S2f(`M`)`: M, matrix $A_1(f)$ of a quadratic form f over $\mathbb{Z}$.
Gives the list of prime numbers p for which the Hasse–Witt invariant of the quadratic form f in $\mathbb{Q}_p$ is equal to -1.

`scm(`D,N`)`: $D \in \mathbb{N}$, product of an even number of different primes; $N \in \mathbb{N}$ such that $\gcd(D,N) = 1$.
Gives the number of special complex multiplication points of the Shimura curve $X(D,N)$.

`scmM(`D,N,R`)`: $D \in \mathbb{N}$, product of an even number of different primes; $N \in \mathbb{N}$ such that $\gcd(D,N) = 1$; $R \in \mathbb{N}$, $R \neq 0$.
Gives the number of modified special complex multiplication points of value R of the Shimura curve $X(D,N)$.

`SCMMOrF(`D,N,R`)`: $D \in \mathbb{N}$, product of an even number of different primes; $N \in \mathbb{N}$ such that $\gcd(D,N) = 1$; $R \in \mathbb{N}$, $R \neq 0$.
Gives the full list of pairs (d,m) such that the quadratic orders $\Lambda(d,m) \subseteq \mathbb{Q}(\sqrt{d})$ of conductor m give modified special complex multiplication points of value R for the Shimura curve $X(D,N)$.

`SCMOrF(`D,N`)`: $D \in \mathbb{N}$, product of an even number of different primes; $N \in \mathbb{N}$ such that $\gcd(D,N) = 1$.
Gives the full list of pairs (d,m) such that the quadratic orders $\Lambda(d,m) \subseteq \mathbb{Q}(\sqrt{d})$ of conductor m give special complex multiplication points for the Shimura curve $X(D,N)$.

`SCMOrFhn(`D,N`)`: $D \in \mathbb{N}$, product of an even number of different primes; $N \in \mathbb{N}$ such that $\gcd(D,N) = 1$.
Gives the full list of triples $[(d,m), h(d,m), \mathrm{cm}(D,N,d,m)]$ such that: the quadratic orders $\Lambda(d,m) \subseteq \mathbb{Q}(\sqrt{d})$ of conductor m give special complex multiplication points for the Shimura curve $X(D,N)$; $h(d,m)$ denotes the ideal class number of the order $\Lambda(d,m)$, and $\mathrm{cm}(D,N,d,m)$ the number of points of complex multiplication of $X(D,N)$ by the order $\Lambda(d,m)$.

`SCMOrFn(`D,N`)`: $D \in \mathbb{N}$, product of an even number of different primes; $N \in \mathbb{N}$ such that $\gcd(D,N) = 1$.

Gives the full list $[(d,m), \mathrm{cm}(D,N,d,m)]$ such that: the quadratic orders $\Lambda(d,m) \subseteq \mathbb{Q}(\sqrt{d})$ of conductor m give special complex multiplication points for the Shimura curve $X(D,N)$, and $\mathrm{cm}(D,N,d,m)$ denotes the number of points of complex multiplication of $X(D,N)$ by the order $\Lambda(d,m)$.

`Sf(`M`):` M, matrix $A_1(f)$ of a quadratic form f over $\mathbb{Z}$.
Gives the list of prime numbers p for which the quadratic form f is $\mathbb{Q}_p$-anisotropic.

`signf(`M`):` M, matrix $A_1(f)$ of a quadratic form f over $\mathbb{R}$.
Gives the signature of the quadratic form f.

`sqfr(`d`):` $d \in \mathbb{Z}$.
Gives the square-free part of d.

`symcS(`r, ε, k, s`):` $r \in \mathbb{R}^+$; ε fundamental unit of a real quadratic field $\mathbb{Q}(\sqrt{a})$; k equal to $\mathrm{n}(\varepsilon)$; $s \in \mathbb{N}$ such that ε^s is the fundamental unit of a quadratic order $\Lambda \subseteq \mathbb{Q}(\sqrt{a})$.
Gives the two radii corresponding to the hyperbolic strip symmetric with respect to the hyperbolic line of center 0 and radius r, fundamental domain for the group $\langle h \rangle$, where h is the principal homothety corresponding to the fundamental unit of the quadratic order Λ.

`symiS(`r, ε, k, s`):` $r \in \mathbb{R}^+$; ε fundamental unit of a real quadratic field $\mathbb{Q}(\sqrt{a})$; k equal to $\mathrm{n}(\varepsilon)$; $s \in \mathbb{N}$ such that ε^s is the fundamental unit of a quadratic order $\Lambda \subseteq \mathbb{Q}(\sqrt{a})$.
Gives the two radii corresponding to the hyperbolic strip of inferior radius r, fundamental domain for the group $\langle h \rangle$, where h is the principal homothety corresponding to the fundamental unit of the quadratic order Λ.

`symLIC(`γ`):` $\gamma \in \mathrm{SL}(2,\mathbb{R})$.
Gives the abscissa of the line L_γ such that the transformation γ is composition of the hyperbolic symmetry with respect to C_γ and the hyperbolic symmetry with respect to L_γ, where C_γ is the isometric circle attached to the transformation γ.

`symsS(`r, ε, k, s`):` $r \in \mathbb{R}^+$; ε fundamental unit of a real quadratic field $\mathbb{Q}(\sqrt{a})$; k equal to $\mathrm{n}(\varepsilon)$; $s \in \mathbb{N}$ such that ε^s is the fundamental unit of a quadratic order $\Lambda \subseteq \mathbb{Q}(\sqrt{a})$.
Gives the two radii corresponding to the hyperbolic strip of superior radius r, fundamental domain for the group $\langle h \rangle$, where h is the principal homothety corresponding to the fundamental unit of the quadratic order Λ.

`type(`ω`, purequaternion):` Logical function. The answer is *true* if ω is a pure quaternion and *false* otherwise.

`type(`ω`, quaternion):` Logical function. The answer is *true* if ω is a quaternion and *false* otherwise.

`typeH(`a, b`, T):` $a, b \in \mathbb{Z}$ defining the quaternion $\mathbb{Q}$-algebra $H = \left(\frac{a,b}{\mathbb{Q}}\right)$; `T`, one of the symbols `M`, `A` or `B`.
Logical function. The answer is *true* if the algebra H is nonramified, small ramified of type A or small ramified of type B, depending on whether the symbol `T` is `M`, `A` or `B`, respectively. The answer is *false* otherwise.

`typeHom(`a, b, c, d`, type):` $a, b, c, d \in \mathbb{R}$ defining a transformation $\gamma = \left(\begin{smallmatrix} a & b \\ c & d \end{smallmatrix}\right) \in \mathrm{SL}(2,\mathbb{R})$; `type`, one of the words `elliptic`, `hyperbolic` or `parabolic`.

Logical function. The answer is *true* if the transformation γ is elliptic, hyperbolic or parabolic, according to the word `type` is `elliptic`, `hyperbolic` or `parabolic`, respectively. The answer is *false* otherwise.

`valp(`n`, `p`):` $n \in \mathbb{Z}$, p prime.
Gives the p-adic valuation $v_p(n)$ of n.

`vaX1(`p`):` p prime.
Gives the list of accidental vertices of the fundamental domain for the curve $X(1,p) = X_0(p)$ given in Chapter 5.

`ve2X1(`p`):` p prime number.
Gives the list of elliptic vertices of order 2 of the fundamental domain for the curve $X(1,p) = X_0(p)$ given in Chapter 5.

`ve3X1(`p`):` p prime number.
Gives the list of elliptic vertices of order 3 of the fundamental domain for the curve $X(1,p) = X_0(p)$ given in Chapter 5.

`vintX1(`p`):` p prime number.
Gives the list of vertices obtained as intersection of isometric circles of the fundamental domain for the curve $X(1,p) = X_0(p)$ given in Chapter 5.

`volhX(`D, N`):` $D \in \mathbb{N}$, product of an even number of different primes; $N \in \mathbb{N}$ such that $\gcd(D,N) = 1$.
Gives the hyperbolic volume $V_h(D,N)$ of a polygon that is a fundamental domain for the Shimura curve $X(D,N)$.

`volhX1(`p`):` p prime number.
Gives the hyperbolic volume $V_h(1,p)$ of a polygon that is a fundamental domain for the curve $X(1,p) = X_0(p)$.

`volHypPol(`l`):` $l = [z_1, \dots, z_n]$, $z_1 \in \mathbb{C}$, $\mathrm{Im}(z_i) \geq 0$.
Gives the hyperbolic volume of the polygon determined by the n points of l, in terms of π.

`volX(`D, N`):` $D \in \mathbb{N}$, product of an even number of different primes; $N \in \mathbb{N}$ such that $\gcd(D,N) = 1$.
Gives the normalized hyperbolic volume $V(D,N)$ of a polygon that is fundamental domain for the Shimura curve $X(D,N)$.

`volX1(`p`):` p prime number.
Gives the normalized hyperbolic volume $V(1,p)$ of a hyperbolic polygon that is a fundamental domain for the curve $X(1,p) = X_0(p)$.

`vpX1(`p`):` p prime number.
Gives the list of parabolic vertices of the fundamental domain for the curve $X(1,p) = X_0(p)$ given in Chapter 5.

`vtcX1(`p`):` p prime number.
Gives the total list of the vertices of the fundamental domain for the curve $X(1,p) = X_0(p)$ given in Chapter 5, classified according to their type.

`vtoX1(`p`):` p prime number.
Gives the list total of the vertices of the fundamental domain for the curve $X(1,p) = X_0(p)$ given in Chapter 5, ordered according to the value of their real part.

APPENDIX A

Tables

A.1. Quaternion algebras

TABLE A.1. Representatives **H** of the isomorphy classes of the nonramified or small ramified quaternion $\mathbb{Q}$-algebras $H = \left(\frac{p,q}{\mathbb{Q}}\right)$, for p, $q \leq 55$ primes.

H	D_H	**H**
(2, 2)	1	(1, −1)
(2, 3)	6	(3, −1)
(2, 5)	10	(2, 5)
(2, 7)	1	(1, −1)
(2, 11)	22	(11, −1)
(2, 13)	26	(2, 13)
(2, 17)	1	(1, −1)
(2, 19)	38	(19, −1)
(2, 23)	1	(1, −1)
(2, 29)	58	(2, 29)
(2, 31)	1	(1, −1)
(2, 37)	74	(2, 37)
(2, 41)	1	(1, −1)
(2, 43)	86	(43, −1)
(2, 47)	1	(1, −1)
(2, 53)	106	(2, 53)
(3, 3)	6	(3, −1)
(3, 5)	15	(3, 5)
(3, 7)	14	(7, −1)
(3, 11)	6	(3, −1)
(3, 13)	1	(1, −1)
(3, 17)	51	(3, 17)
(3, 19)	38	(19, −1)
(3, 23)	6	(3, −1)
(3, 29)	87	(3, 29)
(3, 31)	62	(31, −1)
(3, 37)	1	(1, −1)
(3, 41)	123	(3, 41)
(3, 43)	86	(43, −1)
(3, 47)	6	(3, −1)
(3, 53)	159	(3, 53)
(5, 5)	1	(1, −1)
(5, 7)	35	(5, 7)
(5, 11)	1	(1, −1)
(5, 17)	85	(5, 17)
(5, 19)	1	(1, −1)
(5, 23)	115	(5, 23)
(5, 29)	1	(1, −1)
(5, 31)	1	(1, −1)
(5, 37)	185	(5, 37)
(5, 41)	1	(1, −1)
(5, 43)	215	(5, 43)
(5, 47)	235	(5, 47)
(5, 53)	265	(5, 53)
(7, 7)	14	(7, −1)
(7, 11)	22	(11, −1)
(7, 13)	91	(7, 13)
(7, 17)	119	(7, 17)
(7, 19)	14	(7, −1)
(7, 23)	46	(23, −1)
(7, 29)	1	(1, −1)
(7, 31)	14	(7, −1)
(7, 37)	1	(1, −1)
(7, 41)	287	(7, 41)
(7, 43)	86	(43, −1)
(7, 47)	14	(7, −1)
(7, 53)	1	(1, −1)
(11, 11)	22	(11, −1)
(11, 13)	143	(11, 13)
(11, 17)	187	(11, 17)

TABLE A.1 (continued)

H	D_H	$\mathbf{H}$
(11, 19)	22	(11, −1)
(11, 23)	46	(23, −1)
(11, 29)	319	(11, 29)
(11, 31)	62	(31, −1)
(11, 37)	1	(1, −1)
(11, 41)	451	(11, 41)
(11, 43)	22	(11, −1)
(11, 47)	94	(47, −1)
(11, 53)	1	(1, −1)
(13, 13)	1	(1, −1)
(13, 17)	1	(1, −1)
(13, 19)	247	(13, 19)
(13, 23)	1	(1, −1)
(13, 29)	1	(1, −1)
(13, 31)	403	(13, 31)
(13, 37)	481	(13, 37)
(13, 41)	533	(13, 41)
(13, 43)	1	(1, −1)
(13, 47)	611	(13, 47)
(13, 53)	1	(1, −1)
(17, 17)	1	(1, −1)
(17, 19)	1	(1, −1)
(17, 23)	391	(17, 23)
(17, 29)	493	(17, 29)
(17, 31)	527	(17, 31)
(17, 37)	629	(17, 37)
(17, 41)	697	(17, 41)
(17, 43)	1	(1, −1)
(17, 47)	1	(1, −1)
(17, 53)	1	(1, −1)
(19, 19)	38	(19, −1)
(19, 23)	46	(23, −1)
(19, 29)	551	(19, 29)
(19, 31)	38	(19, −1)
(19, 37)	703	(19, 37)
(19, 41)	779	(19, 41)
(19, 43)	86	(43, −1)

H	D_H	$\mathbf{H}$
(19, 47)	94	(47, −1)
(19, 53)	1007	(19, 53)
(23, 23)	46	(23, −1)
(23, 29)	1	(1, −1)
(23, 31)	62	(31, −1)
(23, 37)	851	(23, 37)
(23, 41)	1	(1, −1)
(23, 43)	46	(23, −1)
(23, 47)	94	(47, −1)
(23, 53)	1219	(23, 53)
(29, 29)	1	(1, −1)
(29, 31)	899	(29, 31)
(29, 37)	1073	(29, 37)
(29, 41)	1189	(29, 41)
(29, 43)	1247	(29, 43)
(29, 47)	1363	(29, 47)
(29, 53)	1	(1, −1)
(31, 31)	62	(31, −1)
(31, 37)	1147	(31, 37)
(31, 41)	1	(1, −1)
(31, 43)	62	(31, −1)
(31, 47)	94	(47, −1)
(31, 53)	1643	(31, 53)
(37, 37)	1	(1, −1)
(37, 41)	1	(1, −1)
(37, 43)	1591	(37, 43)
(37, 47)	1	(1, −1)
(37, 53)	1	(1, −1)
(41, 41)	1	(1, −1)
(41, 43)	1	(1, −1)
(41, 47)	1927	(41, 47)
(41, 53)	2173	(41, 53)
(43, 43)	86	(43, −1)
(43, 47)	94	(47, −1)
(43, 53)	1	(1, −1)
(47, 47)	94	(47, −1)
(47, 53)	1	(1, −1)

TABLE A.2. Representatives **H** of the isomorphy classes of the small ramified quaternion $\mathbb{Q}$-algebras of discriminant $D = p \cdot q < 250$, p, q primes.

D	$p \cdot q$	**H**
6	$2 \cdot 3$	$(3, -1)$
10	$2 \cdot 5$	$(2, 5)$
14	$2 \cdot 7$	$(7, -1)$
15	$3 \cdot 5$	$(3, 5)$
21	$3 \cdot 7$	$(21, -1)$
22	$2 \cdot 11$	$(11, -1)$
26	$2 \cdot 13$	$(2, 13)$
33	$3 \cdot 11$	$(33, -1)$
34	$2 \cdot 17$	$(34, -3)$
35	$5 \cdot 7$	$(5, 7)$
38	$2 \cdot 19$	$(19, -1)$
39	$3 \cdot 13$	$(39, -7)$
46	$2 \cdot 23$	$(23, -1)$
51	$3 \cdot 17$	$(3, 17)$
55	$5 \cdot 11$	$(55, -3)$
57	$3 \cdot 19$	$(57, -1)$
58	$2 \cdot 29$	$(2, 29)$
62	$2 \cdot 31$	$(31, -1)$
65	$5 \cdot 13$	$(5, 13)$
69	$3 \cdot 23$	$(69, -1)$
74	$2 \cdot 37$	$(2, 37)$
77	$7 \cdot 11$	$(77, -1)$
82	$2 \cdot 41$	$(82, -3)$
85	$5 \cdot 17$	$(5, 17)$
86	$2 \cdot 43$	$(43, -1)$
87	$3 \cdot 29$	$(3, 29)$
91	$7 \cdot 13$	$(7, 13)$
93	$3 \cdot 31$	$(93, -1)$
94	$2 \cdot 47$	$(47, -1)$
95	$5 \cdot 19$	$(95, -7)$
106	$2 \cdot 53$	$(2, 53)$
111	$3 \cdot 37$	$(111, -19)$
115	$5 \cdot 23$	$(5, 23)$
118	$2 \cdot 59$	$(59, -1)$
119	$7 \cdot 17$	$(7, 17)$
122	$2 \cdot 61$	$(2, 61)$
123	$3 \cdot 41$	$(3, 41)$

D	$p \cdot q$	**H**
129	$3 \cdot 43$	$(129, -1)$
133	$7 \cdot 19$	$(133, -1)$
134	$2 \cdot 67$	$(67, -1)$
141	$3 \cdot 47$	$(141, -1)$
142	$2 \cdot 71$	$(71, -1)$
143	$11 \cdot 13$	$(11, 13)$
145	$5 \cdot 29$	$(145, -3)$
146	$2 \cdot 73$	$(146, -5)$
155	$5 \cdot 31$	$(155, -7)$
158	$2 \cdot 79$	$(79, -1)$
159	$3 \cdot 53$	$(3, 53)$
161	$7 \cdot 23$	$(161, -1)$
166	$2 \cdot 83$	$(83, -1)$
177	$3 \cdot 59$	$(177, -1)$
178	$2 \cdot 89$	$(178, -3)$
183	$3 \cdot 61$	$(183, -7)$
185	$5 \cdot 37$	$(5, 37)$
187	$11 \cdot 17$	$(11, 17)$
194	$2 \cdot 97$	$(194, -5)$
201	$3 \cdot 67$	$(201, -1)$
202	$2 \cdot 101$	$(2, 101)$
203	$7 \cdot 29$	$(203, -2)$
205	$5 \cdot 41$	$(205, -3)$
206	$2 \cdot 103$	$(103, -1)$
209	$11 \cdot 19$	$(209, -1)$
213	$3 \cdot 71$	$(213, -1)$
214	$2 \cdot 107$	$(107, -1)$
215	$5 \cdot 43$	$(5, 43)$
217	$7 \cdot 31$	$(217, -1)$
218	$2 \cdot 109$	$(2, 109)$
219	$3 \cdot 73$	$(219, -7)$
221	$13 \cdot 17$	$(221, -5)$
226	$2 \cdot 113$	$(226, -3)$
235	$5 \cdot 47$	$(5, 47)$
237	$3 \cdot 79$	$(237, -1)$
247	$13 \cdot 19$	$(13, 19)$
249	$3 \cdot 83$	$(249, -1)$

TABLE A.3. Representatives $\mathbf{H}$ of the isomorphy classes of quaternion $\mathbb{Q}$-algebras of discriminant $D = p_1 \cdot p_2 \cdot p_3 \cdot p_4 < 1000$, p_1, p_2, p_3, p_4 different primes.

D	$p_1 \cdot p_2 \cdot p_3 \cdot p_4$	$\mathbf{H}$
210	$2 \cdot 3 \cdot 5 \cdot 7$	$(2 \cdot 3 \cdot 7, 5)$
330	$2 \cdot 3 \cdot 5 \cdot 11$	$(2, 3 \cdot 5 \cdot 11)$
390	$2 \cdot 3 \cdot 5 \cdot 13$	$(2 \cdot 3 \cdot 13, 5)$
462	$2 \cdot 3 \cdot 7 \cdot 11$	$(3 \cdot 7 \cdot 11, -1)$
510	$2 \cdot 3 \cdot 5 \cdot 17$	$(2 \cdot 3 \cdot 17, 5)$
546	$2 \cdot 3 \cdot 7 \cdot 13$	$(2 \cdot 13, 3 \cdot 7)$
570	$2 \cdot 3 \cdot 5 \cdot 19$	$(2, 3 \cdot 5 \cdot 19)$
690	$2 \cdot 3 \cdot 5 \cdot 23$	$(2 \cdot 3 \cdot 23, 5)$
714	$2 \cdot 3 \cdot 7 \cdot 17$	$(2 \cdot 7, 3 \cdot 7 \cdot 17)$
770	$2 \cdot 5 \cdot 7 \cdot 11$	$(2 \cdot 5, 11 \cdot 7)$
798	$2 \cdot 3 \cdot 7 \cdot 19$	$(3 \cdot 7 \cdot 19, -1)$
858	$2 \cdot 3 \cdot 11 \cdot 13$	$(2, 3 \cdot 11 \cdot 13)$
870	$2 \cdot 3 \cdot 5 \cdot 29$	$(2, 3 \cdot 5 \cdot 29)$
910	$2 \cdot 5 \cdot 7 \cdot 13$	$(2 \cdot 7 \cdot 13, 5)$
930	$2 \cdot 3 \cdot 5 \cdot 31$	$(3 \cdot 31, 2 \cdot 3 \cdot 5)$
966	$2 \cdot 3 \cdot 7 \cdot 23$	$(3 \cdot 7 \cdot 23, -1)$

TABLE A.4. Representatives $\mathcal{O}(6,N)$ of the conjugation classes of Eichler orders of level $N \leq 100$, in the quaternion algebra $H_{\mathbb{A}}(3)$.

N	$\mathcal{O}(6,N)$
1	$\mathbb{Z}[1,i,j,1/2+1/2i+1/2j+1/2ij]$
5	$\mathbb{Z}[1,5i,2i+j,1/2+3/2i+1/2j+1/2ij]$
7	$\mathbb{Z}[1,7i,i+j,1/2+5/2i+1/2j+1/2ij]$
11	$\mathbb{Z}[1,i,11j,1/2+1/2i+5/2j+1/2ij]$
13	$\mathbb{Z}[1,i,13j,1/2+1/2i+17/2j+1/2ij]$
17	$\mathbb{Z}[1,17i,i+j,1/2+7/2i+1/2j+1/2ij]$
19	$\mathbb{Z}[1,19i,14i+j,1/2+27/2i+1/2j+1/2ij]$
23	$\mathbb{Z}[1,23i,4i+j,1/2+5/2i+1/2j+1/2ij]$
25	$\mathbb{Z}[1,25i,3i+j,1/2+29/2i+1/2j+1/2ij]$
29	$\mathbb{Z}[1,29i,23i+j,1/2+45/2i+1/2j+1/2ij]$
31	$\mathbb{Z}[1,31i,i+j,1/2+9/2i+1/2j+1/2ij]$
35	$\mathbb{Z}[1,35i,20i+j,1/2+3/2i+1/2j+1/2ij]$
37	$\mathbb{Z}[1,i,37j,1/2+1/2i+59/2j+1/2ij]$
41	$\mathbb{Z}[1,41i,34i+j,1/2+67/2i+1/2j+1/2ij]$
43	$\mathbb{Z}[1,43i,2i+j,1/2+67/2i+1/2j+1/2ij]$
47	$\mathbb{Z}[1,i,47j,1/2+1/2i+35/2j+1/2ij]$
49	$\mathbb{Z}[1,49i,2i+j,1/2+35/2i+1/2j+1/2ij]$
53	$\mathbb{Z}[1,53i,6i+j,1/2+7/2i+1/2j+1/2ij]$
55	$\mathbb{Z}[1,55i,37i+j,1/2+43/2i+1/2j+1/2ij]$
59	$\mathbb{Z}[1,59i,10i+j,1/2+71/2i+1/2j+1/2ij]$
61	$\mathbb{Z}[1,61i,39i+j,1/2+71/2i+1/2j+1/2ij]$
65	$\mathbb{Z}[1,65i,45i+j,1/2+67/2i+1/2j+1/2ij]$
67	$\mathbb{Z}[1,67i,15i+j,1/2+17/2i+1/2j+1/2ij]$
71	$\mathbb{Z}[1,71i,i+j,1/2+13/2i+1/2j+1/2ij]$
73	$\mathbb{Z}[1,73i,5i+j,1/2+79/2i+1/2j+1/2ij]$
77	$\mathbb{Z}[1,77i,26i+j,1/2+31/2i+1/2j+1/2ij]$
79	$\mathbb{Z}[1,79i,49i+j,1/2+1/2i+1/2j+1/2ij]$
83	$\mathbb{Z}[1,83i,36i+j,1/2+97/2i+1/2j+1/2ij]$
85	$\mathbb{Z}[1,85i,67i+j,1/2+163/2i+1/2j+1/2ij]$
89	$\mathbb{Z}[1,89i,35i+j,1/2+129/2i+1/2j+1/2ij]$
91	$\mathbb{Z}[1,91i,47i+j,1/2+161/2i+1/2j+1/2ij]$
95	$\mathbb{Z}[1,95i,8i+j,1/2+9/2i+1/2j+1/2ij]$
97	$\mathbb{Z}[1,97i,25i+j,1/2+107/2i+1/2j+1/2ij]$

TABLE A.5. Representatives $\mathcal{O}(10, N)$ of the conjugation classes of Eichler orders of level $N \leq 85$, in the quaternion algebra $H_{\mathrm{B}}(2,5)$.

N	$\mathcal{O}(10, N)$
1	$\mathbb{Z}[1, i, 1/2 + 1/2j, 1/2i + 1/2ij]$
3	$\mathbb{Z}[1, 3i, 1/2 + 2i + 1/2j, 1/2i + 1/2ij]$
7	$\mathbb{Z}[1, i, 7/2 + 7/2j, 1/2 + 5/2i + 1/2j + 5/2ij]$
9	$\mathbb{Z}[1, 9i, 1/2 + 2i + 1/2j, 1/2i + 1/2ij]$
11	$\mathbb{Z}[1, 11i, 1/2 + 7i + 1/2j, 1/2i + 1/2ij]$
13	$\mathbb{Z}[1, 13i, 1/2 + 9i + 1/2j, 9/2i + 1/2ij]$
17	$\mathbb{Z}[1, i, 17/2 + 17/2j, 1/2 + 7i + 1/2j + 7ij]$
19	$\mathbb{Z}[1, 19i, 1/2 + 3i + 1/2j, 1/2i + 1/2ij]$
21	$\mathbb{Z}[1, 21i, 1/2 + 16i + 1/2j, 11/2i + 1/2ij]$
23	$\mathbb{Z}[1, 23i, 1/2 + 4i + 1/2j, 31/2i + 1/2ij]$
27	$\mathbb{Z}[1, 27i, 1/2 + 16i + 1/2j, 1/2i + 1/2ij]$
29	$\mathbb{Z}[1, 29i, 1/2 + i + 1/2j, 39/2i + 1/2ij]$
31	$\mathbb{Z}[1, i, 31/2 + 31/2j, 1/2 + 27/2i + 1/2j + 27/2ij]$
33	$\mathbb{Z}[1, 33i, 1/2 + 20i + 1/2j, 53/2i + 1/2ij]$
37	$\mathbb{Z}[1, 37i, 1/2 + 10i + 1/2j, 19/2i + 1/2ij]$
39	$\mathbb{Z}[1, 39i, 1/2 + 16i + 1/2j, 47/2i + 1/2ij]$
41	$\mathbb{Z}[1, 41i, 1/2 + 12i + 1/2j, 79/2i + 1/2ij]$
43	$\mathbb{Z}[1, 43i, 1/2 + 14i + 1/2j, 81/2i + 1/2ij]$
47	$\mathbb{Z}[1, 47i, 1/2 + 40i + 1/2j, 63/2i + 1/2ij]$
49	$\mathbb{Z}[1, i, 49/2 + 49/2j, 1/2 + 22i + 1/2j + 22ij]$
51	$\mathbb{Z}[1, 51i, 1/2 + i + 1/2j, 77/2i + 1/2ij]$
53	$\mathbb{Z}[1, 53i, 1/2 + 5i + 1/2j, 29/2i + 1/2ij]$
57	$\mathbb{Z}[1, 57i, 1/2 + 26i + 1/2j, 25/2i + 1/2ij]$
59	$\mathbb{Z}[1, 59i, 1/2 + 17i + 1/2j, 55/2i + 1/2ij]$
61	$\mathbb{Z}[1, 61i, 1/2 + 46i + 1/2j, 67/2i + 1/2ij]$
63	$\mathbb{Z}[1, 63i, 1/2 + 19i + 1/2j, 101/2i + 1/2ij]$
67	$\mathbb{Z}[1, 67i, 1/2 + 46i + 1/2j, 7/2i + 1/2ij]$
69	$\mathbb{Z}[1, 69i, 1/2 + 53i + 1/2j, 35/2i + 1/2ij]$
71	$\mathbb{Z}[1, 71i, 131/2i + 1/2ij, 1/2 + 11i + 1/2j]$
73	$\mathbb{Z}[1, 73i, 1/2 + 2i + 1/2j, 57/2i + 1/2ij]$
77	$\mathbb{Z}[1, 77i, 1/2 + 47i + 1/2j, 11/2i + 1/2ij]$
79	$\mathbb{Z}[1, 79i, 1/2 + 1/2j, 99/2i + 1/2ij]$
81	$\mathbb{Z}[1, 81i, 1/2 + 71i + 1/2j, 61/2i + 1/2ij]$
83	$\mathbb{Z}[1, 83i, 1/2 + 7i + 1/2j, 139/2i + 1/2ij]$

TABLE A.6. Representatives $\mathcal{O}(14, N)$ of the conjugation classes of Eichler orders of level $N \leq 80$, in the quaternion algebra $H_{\mathrm{A}}(7)$.

N	$\mathcal{O}(14, N)$
1	$\mathbb{Z}[1, i, j, 1/2 + 1/2i + 1/2j + 1/2ij]$
3	$\mathbb{Z}[1, i, 3j, 1/2 + 1/2i + 5/2j + 1/2ij]$
5	$\mathbb{Z}[1, 5i, i + j, 1/2 + 7/2i + 1/2j + 1/2ij]$
9	$\mathbb{Z}[1, 3i, 2i + 3j, 1/2 + 1/2i + 1/2j + 1/2ij]$
11	$\mathbb{Z}[1, 11i, 6i + j, 1/2 + 9/2i + 1/2j + 1/2ij]$
13	$\mathbb{Z}[1, 13i, 2i + j, 1/2 + 3/2i + 1/2j + 1/2ij]$
15	$\mathbb{Z}[1, 5i, 3i + 3j, 1/2 + 1/2i + 5/2j + 1/2ij]$
17	$\mathbb{Z}[1, 17i, j, 1/2 + 13/2i + 1/2j + 1/2ij]$
19	$\mathbb{Z}[1, 19i, 4i + j, 1/2 + 27/2i + 1/2j + 1/2ij]$
23	$\mathbb{Z}[1, 23i, i + j, 1/2 + 13/2i + 1/2j + 1/2ij]$
25	$\mathbb{Z}[1, 25i, 15i + j, 1/2 + 47/2i + 1/2j + 1/2ij]$
27	$\mathbb{Z}[1, i, 27j, 1/2 + 1/2i + 13/2j + 1/2ij]$
29	$\mathbb{Z}[1, 29i, 16i + j, 1/2 + 31/2i + 1/2j + 1/2ij]$
31	$\mathbb{Z}[1, i, 31j, 1/2 + 1/2i + 41/2j + 1/2ij]$
33	$\mathbb{Z}[1, 11i, 5i + 3j, 1/2 + 13/2i + 1/2j + 1/2ij]$
37	$\mathbb{Z}[1, 37i, 8i + j, 1/2 + 23/2i + 1/2j + 1/2ij]$
39	$\mathbb{Z}[1, 39i, 10i + j, 1/2 + 43/2i + 1/2j + 1/2ij]$
41	$\mathbb{Z}[1, 41i, 22i + j, 1/2 + 17/2i + 1/2j + 1/2ij]$
43	$\mathbb{Z}[1, 43i, 18i + j, 1/2 + 1/2i + 1/2j + 1/2ij]$
45	$\mathbb{Z}[1, 15i, 13i + 3j, 1/2 + 5/2i + 1/2j + 1/2ij]$
47	$\mathbb{Z}[1, 47i, 13i + j, 1/2 + 77/2i + 1/2j + 1/2ij]$
51	$\mathbb{Z}[1, 17i, 4i + 3j, 1/2 + 7/2i + 5/2j + 1/2ij]$
53	$\mathbb{Z}[1, 53i, i + j, 1/2 + 19/2i + 1/2j + 1/2ij]$
55	$\mathbb{Z}[1, 55i, 4i + j, 1/2 + 5/2i + 1/2j + 1/2ij]$
57	$\mathbb{Z}[1, 19i, 12i + 3j, 1/2 + 35/2i + 5/2j + 1/2ij]$
59	$\mathbb{Z}[1, 59i, 17i + j, 1/2 + 21/2i + 1/2j + 1/2ij]$
61	$\mathbb{Z}[1, 61i, 3i + j, 1/2 + 65/2i + 1/2j + 1/2ij]$
65	$\mathbb{Z}[1, 65i, 20i + j, 1/2 + 113/2i + 1/2j + 1/2ij]$
67	$\mathbb{Z}[1, 67i, 45i + j, 1/2 + 83/2i + 1/2j + 1/2ij]$
69	$\mathbb{Z}[1, 23i, 14i + 3j, 1/2 + 1/2i + 1/2j + 1/2ij]$
71	$\mathbb{Z}[1, 71i, 5i + j, 1/2 + 99/2i + 1/2j + 1/2ij]$
73	$\mathbb{Z}[1, 73i, 7i + j, 1/2 + 139/2i + 1/2j + 1/2ij]$
75	$\mathbb{Z}[1, 25i, 2i + 3j, 1/2 + 41/2i + 5/2j + 1/2ij]$
79	$\mathbb{Z}[1, 79i, 21i + j, 1/2 + 41/2i + 1/2j + 1/2ij]$

TABLE A.7. Representatives $\mathcal{O}(15, N)$ of the conjugation classes of Eichler orders of level $N \leq 60$, in the quaternion algebra $H_B(3, 5)$.

N	$\mathcal{O}(15, N)$
1	$\mathbb{Z}[1, i, 1/2 + 1/2j, 1/2i + 1/2ij]$
2	$\mathbb{Z}[1, i, j, 1/2 + 1/2i + 1/2j + 1/2ij]$
4	$\mathbb{Z}[1, 4i, 1/2 + i + 1/2j, 1/2i + 1/2ij]$
7	$\mathbb{Z}[1, 7i, 1/2 + 4i + 1/2j, 1/2i + 1/2ij]$
8	$\mathbb{Z}[1, 4i, 3i + j, 1/2 + 3/2i + 1/2j + 1/2ij]$
11	$\mathbb{Z}[1, i, 11/2 + 11/2j, 1/2 + 9/2i + 1/2j + 9/2ij]$
13	$\mathbb{Z}[1, 13i, 1/2 + 2i + 1/2j, 1/2i + 1/2ij]$
14	$\mathbb{Z}[1, 14i, 1/2 + 5i + 1/2j, 19/2i + 1/2ij]$
16	$\mathbb{Z}[1, 8i, 3i + j, 1/2 + 11/2i + 1/2j + 1/2ij]$
17	$\mathbb{Z}[1, 17i, 1/2 + i + 1/2j, 17/2i + 1/2ij]$
19	$\mathbb{Z}[1, 19i, 1/2 + 1/2j, 29/2i + 1/2ij]$
22	$\mathbb{Z}[1, 11i, 6i + j, 1/2 + 1/2i + 1/2j + 1/2ij]$
23	$\mathbb{Z}[1, 23i, 1/2 + 16i + 1/2j, 31/2i + 1/2ij]$
26	$\mathbb{Z}[1, i, 13j, 1/2 + 1/2i + 17/2j + 1/2ij]$
28	$\mathbb{Z}[1, 28i, 1/2 + 3i + 1/2j, 1/2i + 1/2ij]$
29	$\mathbb{Z}[1, 29i, 1/2 + 12i + 1/2j, 15/2i + 1/2ij]$
31	$\mathbb{Z}[1, 31i, 1/2 + 17i + 1/2j, 1/2i + 1/2ij]$
32	$\mathbb{Z}[1, 16i, 11i + j, 1/2 + 15/2i + 1/2j + 1/2ij]$
34	$\mathbb{Z}[1, 34i, 1/2 + 7i + 1/2j, 41/2i + 1/2ij]$
37	$\mathbb{Z}[1, 37i, 1/2 + 27i + 1/2j, 61/2i + 1/2ij]$
38	$\mathbb{Z}[1, 38i, 1/2 + i + 1/2j, 51/2i + 1/2ij]$
41	$\mathbb{Z}[1, 41i, 1/2 + 38i + 1/2j, 61/2i + 1/2ij]$
43	$\mathbb{Z}[1, 43i, 1/2 + 9i + 1/2j, 69/2i + 1/2ij]$
44	$\mathbb{Z}[1, 2i, i + 11j, 1/2 + 3/2i + 5/2j + 1/2ij]$
46	$\mathbb{Z}[1, 23i, 13i + j, 1/2 + 25/2i + 1/2j + 1/2ij]$
47	$\mathbb{Z}[1, i, 47/2 + 47/2j, 1/2 + 43/2i + 1/2j + 43/2ij]$
49	$\mathbb{Z}[1, 49i, 1/2 + 45i + 1/2j, 1/2i + 1/2ij]$
52	$\mathbb{Z}[1, 26i, 7i + j, 1/2 + 49/2i + 1/2j + 1/2ij]$
53	$\mathbb{Z}[1, 53i, 1/2 + 41i + 1/2j, 39/2i + 1/2ij]$
56	$\mathbb{Z}[1, 28i, 13i + j, 1/2 + 5/2i + 1/2j + 1/2ij]$
58	$\mathbb{Z}[1, 58i, 1/2 + 47i + 1/2j, 93/2i + 1/2ij]$
59	$\mathbb{Z}[1, 59i, 1/2 + 10i + 1/2j, 5/2i + 1/2ij]$

A.2. Shimura curves

TABLE A.8. Constants attached to the Shimura curves $X(D,1)$, $D < 200$, corresponding to small ramified cases.

D	V	e_2	e_3	g
6	1/3	2	2	0
10	2/3	0	4	0
14	1	2	0	1
15	4/3	0	2	1
21	2	4	0	1
22	5/3	2	4	0
26	2	0	0	2
33	10/3	4	2	1
34	8/3	0	4	1
35	4	0	0	3
38	3	2	0	2
39	4	0	0	3
46	11/3	2	4	1
51	16/3	0	2	3
55	20/3	0	4	3
57	6	4	0	3
58	14/3	0	4	2
62	5	2	0	3
65	8	0	0	5
69	22/3	4	2	3
74	6	0	0	4
77	10	4	0	5
82	20/3	0	4	3
85	32/3	0	4	5
86	7	2	0	4
87	28/3	0	2	5
91	12	0	0	7
93	10	4	0	5

D	V	e_2	e_3	g
94	23/3	2	4	3
95	12	0	0	7
106	26/3	0	4	4
111	12	0	0	7
115	44/3	0	4	7
118	29/3	2	4	4
119	16	0	0	9
122	10	0	0	6
123	40/3	0	2	7
129	14	4	0	7
133	18	4	0	9
134	11	2	0	6
141	46/3	4	2	7
142	35/3	2	4	5
143	20	0	0	11
145	56/3	0	4	9
146	12	0	0	7
155	20	0	0	11
158	13	2	0	7
159	52/3	0	2	9
161	22	4	0	11
166	41/3	2	4	6
177	58/3	4	2	9
178	44/3	0	4	7
183	20	0	0	11
185	24	0	0	13
187	80/3	0	4	13
194	16	0	0	9

TABLE A.9. Constants attached to the Shimura curves $X(6, N)$, $N \leq 200$, corresponding to small ramified cases of type A.

N	V	e_2	e_3	g
1	1	2	2	0
5	2	4	0	1
7	8/3	0	4	1
11	4	0	0	3
13	14/3	4	4	1
17	6	4	0	3
19	20/3	0	4	3
23	8	0	0	5
25	10	4	0	5
29	10	4	0	5
31	32/3	0	4	5
35	16	0	0	9
37	38/3	4	4	5
41	14	4	0	7
43	44/3	0	4	7
47	16	0	0	9
49	56/3	0	4	9
53	18	4	0	9
55	24	0	0	13
59	20	0	0	11
61	62/3	4	4	9
65	28	8	0	13
67	68/3	0	4	11
71	24	0	0	13
73	74/3	4	4	11
77	32	0	0	17
79	80/3	0	4	13
83	28	0	0	15
85	36	8	0	17
89	30	4	0	15
91	112/3	0	8	17
95	40	0	0	21
97	98/3	4	4	15
101	34	4	0	17
103	104/3	0	4	17
107	36	0	0	19
109	110/3	4	4	17
113	38	4	0	19
115	48	0	0	25
119	48	0	0	25
121	44	0	0	23
125	50	4	0	25
127	128/3	0	4	21
131	44	0	0	23
133	160/3	0	8	25
137	46	4	0	23
139	140/3	0	4	23
143	56	0	0	29
145	60	8	0	29
149	50	4	0	25
151	152/3	0	4	25
155	64	0	0	33
157	158/3	4	4	25
161	64	0	0	33
163	164/3	0	4	27
167	56	0	0	29
169	182/3	4	4	29
173	58	4	0	29
175	80	0	0	41
179	60	0	0	31
181	182/3	4	4	29
185	76	8	0	37
187	72	0	0	37
191	64	0	0	33
193	194/3	4	4	31
197	66	4	0	33
199	200/3	0	4	33

TABLE A.10. Constants attached to the Shimura curves $X(10, N)$, $N \leq 165$, corresponding to small ramified cases of type B.

N	V	e_2	e_3	g
1	2/3	0	4	0
3	8/3	0	4	1
7	16/3	0	8	1
9	8	0	0	5
11	8	0	0	5
13	28/3	0	8	3
17	12	0	0	7
19	40/3	0	8	5
21	63/3	0	8	9
23	16	0	0	9
27	24	0	0	13
29	20	0	0	11
31	64/3	0	8	9
33	32	0	0	17
37	76/3	0	8	11
39	112/3	0	8	17
41	28	0	0	15
43	88/3	0	8	13
47	32	0	0	17
49	112/3	0	8	17
51	48	0	0	25
53	36	0	0	19
57	160/3	0	8	25
59	40	0	0	21
61	124/3	0	8	19
63	64	0	0	33
67	136/3	0	8	21
69	64	0	0	33
71	48	0	0	25
73	148/3	0	8	23
77	64	0	0	33
79	160/3	0	8	25
81	72	0	0	37

N	V	e_2	e_3	g
83	56	0	0	29
87	80	0	0	41
89	60	0	0	31
91	224/3	0	16	33
93	256/3	0	8	41
97	196/3	0	8	31
99	96	0	0	49
101	68	0	0	35
103	208/3	0	8	33
107	72	0	0	37
109	220/3	0	8	35
111	304/3	0	8	49
113	76	0	0	39
117	112	0	0	57
119	96	0	0	49
121	88	0	0	45
123	112	0	0	57
127	256/3	0	8	41
129	352/3	0	8	57
131	88	0	0	45
133	320/3	0	16	49
137	92	0	0	47
139	280/3	0	8	45
141	128	0	0	65
143	112	0	0	57
147	448/3	0	8	73
149	100	0	0	51
151	304/3	0	8	49
153	144	0	0	73
157	316/3	0	8	51
159	144	0	0	73
161	128	0	0	65
163	328/3	0	8	53

TABLE A.11. Constants attached to the Shimura curves $X(14, N)$, $N \leq 160$, corresponding to small ramified case of type A.

N	V	e_2	e_3	g
1	1	2	0	1
3	4	0	0	3
5	6	4	0	3
9	12	0	0	7
11	12	0	0	7
13	14	4	0	7
15	24	0	0	13
17	18	4	0	9
19	20	0	0	11
23	24	0	0	13
25	30	4	0	15
27	36	0	0	19
29	30	4	0	15
31	32	0	0	17
33	48	0	0	25
37	38	4	0	19
39	56	0	0	29
41	42	4	0	21
43	44	0	0	23
45	72	0	0	37
47	48	0	0	25
51	72	0	0	37
53	54	4	0	27
55	72	0	0	37
57	80	0	0	41
59	60	0	0	31
61	62	4	0	31
65	84	8	0	41
67	68	0	0	35
69	96	0	0	49
71	72	0	0	37
73	74	4	0	37
75	120	0	0	61
79	80	0	0	41
81	108	0	0	55

N	V	e_2	e_3	g
83	84	0	0	43
85	108	8	0	53
87	120	0	0	61
89	90	4	0	45
93	128	0	0	65
95	120	0	0	61
97	98	4	0	49
99	144	0	0	73
101	102	4	0	51
103	104	0	0	53
107	108	0	0	55
109	110	4	0	55
111	152	0	0	77
113	114	4	0	57
115	144	0	0	73
117	168	0	0	85
121	132	0	0	67
123	168	0	0	85
125	150	4	0	75
127	128	0	0	65
129	176	0	0	89
131	132	0	0	67
135	216	0	0	109
137	138	4	0	69
139	140	0	0	71
141	192	0	0	97
143	168	0	0	85
145	180	8	0	89
149	150	4	0	75
151	152	0	0	77
153	216	0	0	109
155	192	0	0	97
157	158	4	0	79
159	216	0	0	109

TABLE A.12. Constants attached to the Shimura curves $X(15, N)$, $N \leq 125$, corresponding to small ramified cases of type B.

N	V	e_2	e_3	g
1	4/3	0	2	1
2	4	0	0	3
4	8	0	0	5
7	32/3	0	4	5
8	16	0	0	9
11	16	0	0	9
13	56/3	0	4	9
14	32	0	0	17
16	32	0	0	17
17	24	0	0	13
19	80/3	0	4	13
22	48	0	0	25
23	32	0	0	17
26	56	0	0	29
28	64	0	0	33
29	40	0	0	21
31	128/3	0	4	21
32	64	0	0	33
34	72	0	0	37
37	152/3	0	4	25
38	80	0	0	41
41	56	0	0	29
43	176/3	0	4	29
44	96	0	0	49
46	96	0	0	49
47	64	0	0	33
49	224/3	0	4	37
52	112	0	0	57
53	72	0	0	37
56	128	0	0	65
58	120	0	0	61
59	80	0	0	41
61	248/3	0	4	41
62	128	0	0	65
64	128	0	0	65
67	272/3	0	4	45
68	144	0	0	73
71	96	0	0	49
73	296/3	0	4	49
74	152	0	0	77
76	160	0	0	81
77	128	0	0	65
79	320/3	0	4	53
82	168	0	0	85
83	112	0	0	57
86	176	0	0	89
88	192	0	0	97
89	120	0	0	61
91	448/3	0	8	73
92	192	0	0	97
94	192	0	0	97
97	392/3	0	4	65
98	224	0	0	113
101	136	0	0	69
103	416/3	0	4	69
104	224	0	0	113
106	216	0	0	109
107	144	0	0	73
109	440/3	0	4	73
112	256	0	0	129
113	152	0	0	77
116	240	0	0	121
118	240	0	0	121
119	192	0	0	97
121	176	0	0	89
122	248	0	0	125
124	256	0	0	129

TABLE A.13. Constants attached to the Shimura curves $X(D,1)$, $D = p_1p_2p_3p_4 < 1000$.

D	V	e_2	e_3	g
210	8	0	0	5
330	40/3	0	8	5
390	16	0	0	9
462	20	8	0	9
510	64/3	0	8	9
546	24	0	0	13
570	24	0	0	13
690	88/3	0	8	13
714	32	0	0	17
770	40	0	0	21
798	36	8	0	17
858	40	0	0	21
870	112/3	0	8	17
910	48	0	0	25
930	40	0	0	21
966	44	8	0	21

TABLE A.14. Equations of Shimura curves $X(D,N)$, $D > 1$, according to [**Kur79, Jor81, Mic81a**].

D	N	g	Equation of $X(D,N)$
6	1	0	$x^2 + y^2 + 3 = 0$
10	1	0	$x^2 + y^2 + 2 = 0$
22	1	0	$x^2 + y^2 + 11 = 0$
14	1	1	$(x^2 - 13)^2 + 7^3 + 2y^2 = 0$
46	1	1	$(x^2 - 45)^2 + 23 + 2y^2 = 0$
15	1	1	$(x^2 + 243)(x^2 + 3) + 3y^2 = 0$
21	1	1	$x^4 - 658x^2 + 7^6 + 7y^2 = 0$
33	1	1	$x^4 + 30x^2 + 3^8 + 3y^2 = 0$

TABLE A.15. Full list of the hyperelliptic Shimura curves $X(D,N)$, $D>1$, and their hyperelliptic involution, according to [**Ogg83**].

D	N	$g(D,N)$	w
26	1	2	w_{26}
35	1	3	w_{35}
38	1	2	w_{38}
39	1	3	w_{39}
51	1	3	w_{51}
55	1	3	w_{55}
57	1	3	w_{19}
58	1	2	w_{29}
62	1	3	w_{62}
69	1	3	w_{69}
74	1	4	w_{74}
82	1	3	w_{41}
86	1	4	w_{86}
87	1	5	w_{87}
93	1	5	w_{31}
94	1	3	w_{94}
95	1	7	w_{95}
111	1	7	w_{111}
119	1	9	w_{119}
134	1	6	w_{134}
146	1	7	w_{146}
159	1	9	w_{159}
194	1	9	w_{194}
206	1	9	w_{206}

D	N	$g(D,N)$	w
6	11	3	w_{66}
6	17	3	w_{34}
6	19	3	w_{114}
6	29	5	w_{174}
6	31	5	w_{186}
6	37	5	w_{222}
10	11	5	w_{110}
10	13	3	w_{65}
10	19	5	w_{38}
10	23	9	w_{230}
14	3	3	w_{14}
14	5	3	w_{14}
15	2	3	w_{15}
15	4	5	w_{15}
21	2	3	w_7
22	3	3	w_{66}
22	5	5	w_{110}
26	3	5	w_{26}
39	2	7	w_{39}

A.3. Quaternion algebras and quadratic forms

TABLE A.16. Ternary quadratic forms attached to definite or indefinite quaternion $\mathbb{Q}$-algebras, of discriminant $D \leq 60$.

D	H	$f = n_{H,3}$	$\mathrm{d}_1(f)$	$\mathrm{N}(f)$	$S(f)$
1	$(1,-1)$	$-X^2+Y^2-Z^2$	-1	4	$\varnothing$
2	$(-1,-1)$	$X^2+Y^2+Z^2$	1	4	$\{2\}$
3	$(-3,-1)$	$3X^2+Y^2+3Z^2$	3	12	$\{3\}$
5	$(-5,-2)$	$5X^2+2Y^2+10Z^2$	10	40	$\{5\}$
6	$(3,-1)$	$-3X^2+Y^2-3Z^2$	-3	12	$\{2,3\}$
7	$(-7,-1)$	$7X^2+Y^2+7Z^2$	7	28	$\{7\}$
10	$(2,5)$	$-2X^2-5Y^2+10Z^2$	-10	40	$\{2,5\}$
11	$(-11,-1)$	$11X^2+Y^2+11Z^2$	11	44	$\{11\}$
13	$(-13,-2)$	$13X^2+2Y^2+26Z^2$	26	52	$\{13\}$
14	$(7,-1)$	$-7X^2+Y^2+7Z^2$	-7	28	$\{2,7\}$
15	$(3,5)$	$-3X^2-5Y^2+15Z^2$	-15	60	$\{3,5\}$
17	$(-17,-3)$	$17X^2+3Y^2+357Z^2$	357	1428	$\{17\}$
19	$(-19,-1)$	$19X^2+Y^2+19Z^2$	19	76	$\{19\}$
21	$(21,-1)$	$-21X^2+Y^2-21Z^2$	-21	82	$\{3,7\}$
22	$(11,-1)$	$-11X^2+Y^2-11Z^2$	-11	44	$\{2,11\}$
23	$(-23,-1)$	$23X^2+Y^2+23Z^2$	23	92	$\{23\}$
26	$(13,2)$	$-13X^2-2Y^2+26Z^2$	-26	104	$\{2,13\}$
29	$(-29,-2)$	$29X^2+2Y^2+58Z^2$	58	232	$\{29\}$
30	$(-10,-3)$	$10X^2+3Y^2+30Z^2$	30	120	$\{2,3,5\}$
31	$(-31,-1)$	$31X^2+Y^2+31Z^2$	31	124	$\{31\}$
33	$(33,-1)$	$-33X^2+Y^2-33Z^2$	-33	132	$\{3,11\}$
34	$(34,-3)$	$-34X^2+3Y^2-102Z^2$	-102	408	$\{2,17\}$
35	$(7,5)$	$-7X^2-5Y^2+35Z^2$	-35	140	$\{5,7\}$
37	$(-37,-2)$	$37X^2+2Y^2+74Z^2$	74	296	$\{37\}$
38	$(19,-1)$	$-19X^2+Y^2-19Z^2$	-19	76	$\{2,19\}$
39	$(39,-7)$	$-39X^2+7Y^2-273Z^2$	-273	1092	$\{3,13\}$
41	$(-41,-3)$	$41X^2+3Y^2+123Z^2$	123	492	$\{41\}$
42	$(-42,-1)$	$42X^2+Y^2+42Z^2$	42	168	$\{2,3,7\}$
43	$(-43,-1)$	$43X^2+Y^2+43Z^2$	43	172	$\{43\}$
46	$(23,-1)$	$-23X^2+Y^2-23Z^2$	-23	92	$\{2,23\}$
47	$(-47,-1)$	$47X^2+Y^2+47Z^2$	47	188	$\{47\}$
51	$(3,17)$	$-3X^2-17Y^2+51Z^2$	-51	204	$\{3,17\}$
53	$(-53,-2)$	$53X^2+2Y^2+106Z^2$	106	424	$\{53\}$
55	$(-55,-13)$	$55X^2+13Y^2+715Z^2$	-715	2860	$\{5,11\}$
57	$(57,-1)$	$-57X^2+Y^2-57Z^2$	-57	228	$\{3,19\}$
58	$(29,2)$	$-29X^2-2Y^2+58Z^2$	-58	232	$\{2,29\}$
59	$(-59,-1)$	$59X^2+Y^2+59Z^2$	59	236	$\{59\}$

TABLE A.17. Representatives of the $\mathbb{Z}$-equivalence classes of quaternary and ternary normic forms attached to the Eichler orders $\mathcal{O}(6,N)$, for $N \leq 20$.

N			
	$n_{\mathcal{O}(6,N),4}$		
	$\det_1(n_{\mathcal{O}(6,N),4})$	$d_1(n_{\mathcal{O}(6,N),4})$	$N(n_{\mathcal{O}(6,N),4})$
	$n_{\mathcal{O}(6,N),3}$		
	$\det_1(n_{\mathcal{O}(6,N),3})$	$d_1(n_{\mathcal{O}(6,N),3})$	$N(n_{\mathcal{O}(6,N),3})$
1	$X^2 - 3Y^2 + Z^2 - T^2 + XT - 3YT + ZT$		
	$9/4$	$-3/2$	6
	$-5X^2 + 6XY - 2XZ - 3Y^2 + Z^2$		
	9	-3	12
5	$X^2 + XT - 75Y^2 - 60YZ - 45YT - 11Z^2 - 17ZT - 7T^2$		
	$225/4$	$-15/2$	30
	$-29X^2 + 90XY + 34XZ - 75Y^2 - 60YZ - 11Z^2$		
	225	-15	60
7	$X^2 + XT - 147Y^2 - 42YZ - 105YT - 2Z^2 - 14ZT - 19T^2$		
	$441/4$	$-21/2$	42
	$-77X^2 + 210XY + 28XZ - 147Y^2 - 42YZ - 2Z^2$		
	441	-21	84
11	$X^2 + XT - 3Y^2 - 3YT + 121Z^2 + 55ZT + 5T^2$		
	$1089/4$	$-33/2$	66
	$19X^2 + 6XY - 110XZ - 3Y^2 + 121Z^2$		
	1089	-33	132
13	$X^2 + XT - 3Y^2 - 3YT + 169Z^2 + 221ZT + 71T^2$		
	$1521/4$	$-39/2$	78
	$283X^2 + 6XY - 442XZ - 3Y^2 + 169Z^2$		
	1521	-39	156
17	$X^2 + XT - 867Y^2 - 102YZ - 357YT - 2Z^2 - 20ZT - 37T^2$		
	$2601/4$	$-51/2$	102
	$-149X^2 + 714XY + 40XZ - 867Y^2 - 102YZ - 2Z^2$		
	2601	-51	204
19	$X^2 + XT - 1083Y^2 - 1596YZ - 1539YT - 587Z^2 - 1133ZT - 547T^2$		
	$3249/4$	$-57/2$	114
	$-2189X^2 + 3078XY + 2266XZ - 1083Y^2 - 1596YZ - 587Z^2$		
	3249	-57	228

TABLE A.18. Representatives of the $\mathbb{Z}$-equivalence classes of quaternary and ternary normic forms attached to the Eichler orders $\mathcal{O}(10, N)$, for $N \leq 17$.

N			
	$n_{\mathcal{O}(10,N),4}$		
	$\det_1(n_{\mathcal{O}(10,N),4})$	$\mathrm{d}_1(n_{\mathcal{O}(10,N),4})$	$N(n_{\mathcal{O}(10,N),4})$
	$n_{\mathcal{O}(10,N),3}$		
	$\det_1(n_{\mathcal{O}(10,N),3})$	$\mathrm{d}_1(n_{\mathcal{O}(10,N),3})$	$N(n_{\mathcal{O}(10,N),3})$
1	$X^2 + XT - 2Y^2 - 2YZ + 2Z^2 - T^2$		
	$25/4$	$-5/2$	10
	$-5X^2 - 2Y^2 - 2YZ + 2Z^2$		
	25	-5	20
3	$X^2 + XT - 18Y^2 - 6YZ - 24YT + 2Z^2 - 4ZT - 9T^2$		
	$225/4$	$-15/2$	30
	$-37X^2 + 48XY + 8XZ - 18Y^2 - 6YZ + 2Z^2$		
	225	-15	60
7	$X^2 + XT - 2Y^2 + 2YZ - 2YT - 18Z^2 - 34ZT - 9T^2$		
	$1225/4$	$-35/2$	70
	$-37X^2 + 4XY + 68XZ - 2Y^2 + 2YZ - 18Z^2$		
	1225	-35	140
9	$X^2 + XT - 162Y^2 - 18YZ - 72YT + 2Z^2 - 4ZT - 9T^2$		
	$2025/4$	$-45/2$	90
	$-37X^2 + 144XY + 8XZ - 162Y^2 - 18YZ + 2Z^2$		
	2025	-45	180
11	$X^2 + XT - 242Y^2 - 22YZ - 308YT + 2Z^2 - 14ZT - 99T^2$		
	$3025/4$	$-55/2$	110
	$-397X^2 + 616XY + 28XZ - 242Y^2 - 22YZ + 2Z^2$		
	3025	-55	220
13	$X^2 + XT - 338Y^2 - 234YZ - 468YT - 38Z^2 - 162ZT - 163T^2$		
	$4225/4$	$-65/2$	130
	$-653X^2 + 936XY + 324XZ - 338Y^2 - 234YZ - 38Z^2$		
	4225	-65	260
17	$X^2 + XT - 2Y^2 + 2YZ - 2YT - 43Z^2 - 169ZT - 149T^2$		
	$7225/4$	$-85/2$	170
	$-597X^2 + 4XY + 338XZ - 2Y^2 + 2YZ - 43Z^2$		
	7225	-85	340

TABLE A.19. Representatives of the $\mathbb{Z}$-equivalence classes of quaternary and ternary normic forms attached to the Eichler orders $\mathcal{O}(14, N)$, for $N \leq 15$.

N			
	$n_{\mathcal{O}(14,N),4}$		
	$\det_1(n_{\mathcal{O}(14,N),4})$	$\mathrm{d}_1(n_{\mathcal{O}(14,N),4})$	$N(n_{\mathcal{O}(14,N),4})$
	$n_{\mathcal{O}(14,N),3}$		
	$\det_1(n_{\mathcal{O}(14,N),3})$	$\mathrm{d}_1(n_{\mathcal{O}(14,N),3})$	$N(n_{\mathcal{O}(14,N),3})$
1	$X^2 + XT - 7Y^2 - 7YT + Z^2 + ZT - 3T^2$		
	$49/4$	$-7/2$	14
	$-13X^2 + 14XY - 2XZ - 7Y^2 + Z^2$		
	49	-7	28
3	$X^2 + XT - 7Y^2 - 7YT + 9Z^2 + 15ZT + 3T^2$		
	$441/4$	$-21/2$	42
	$11X^2 + 14XY - 30XZ - 7Y^2 + 9Z^2$		
	441	-21	84
5	$X^2 + XT - 175Y^2 - 70YZ - 245YT - 6Z^2 - 48ZT - 87T^2$		
	$1225/4$	$-35/2$	70
	$-349X^2 + 490XY + 96XZ - 175Y^2 - 70YZ - 6Z^2$		
	1225	-35	140
9	$X^2 + XT - 63Y^2 - 84YZ - 21YT - 19Z^2 - 11ZT - 3T^2$		
	$3969/4$	$-63/2$	126
	$-13X^2 + 42XY + 22XZ - 63Y^2 - 84YZ - 19Z^2$		
	3969	-63	252
11	$X^2 + XT - 847Y^2 - 924YZ - 693YT - 251Z^2 - 377ZT - 143T^2$		
	$5929/4$	$-77/2$	154
	$-573X^2 + 1386XY + 754XZ - 847Y^2 - 924YZ - 251Z^2$		
	5929	-77	308
13	$X^2 + XT - 1183Y^2 - 364YZ - 273YT - 27Z^2 - 41ZT - 17T^2$		
	$8281/4$	$-91/2$	182
	$-69X^2 + 546XY + 82XZ - 1183Y^2 - 364YZ - 27Z^2$		
	8281	-91	364
15	$X^2 + XT - 175Y^2 - 210YZ - 35YT - 54Z^2 - 6ZT + 3T^2$		
	$11025/4$	$-105/2$	210
	$11X^2 + 70XY + 12XZ - 175Y^2 - 210YZ - 54Z^2$		
	11025	-105	420

TABLE A.20. Representatives of the $\mathbb{Z}$-equivalence classes of quaternary and ternary normic forms attached to the Eichler orders $\mathcal{O}(15, N)$, for $N \leq 13$.

N	$n_{\mathcal{O}(15,N),4}$		
	$\det_1(n_{\mathcal{O}(15,N),4})$	$\mathrm{d}_1(n_{\mathcal{O}(15,N),4})$	$N(n_{\mathcal{O}(15,N),4})$
	$n_{\mathcal{O}(15,N),3}$		
	$\det_1(n_{\mathcal{O}(15,N),3})$	$\mathrm{d}_1(n_{\mathcal{O}(15,N),3})$	$N(n_{\mathcal{O}(15,N),3})$
1	$X^2 + XT - 3Y^2 - 3YZ + 3Z^2 - T^2$		
	$225/16$	$-15/4$	15
	$-5X^2 - 3Y^2 - 3YZ + 3Z^2$		
	$225/4$	$-15/2$	60
2	$X^2 + XT - 3Y^2 - 3YT - 5Z^2 - 5ZT + 2T^2$		
	$225/4$	$-15/2$	30
	$7X^2 + 6XY + 10XZ - 3Y^2 - 5Z^2$		
	225	-15	60
4	$X^2 + XT - 48Y^2 - 12YZ - 24YT + 3Z^2 - 3ZT - 4T^2$		
	225	-15	60
	$-17X^2 + 48XY + 6XZ - 48Y^2 - 12YZ + 3Z^2$		
	900	-30	60
7	$X^2 + XT - 147Y^2 - 21YZ - 168YT + 3Z^2 - 12ZT - 49T^2$		
	$11025/16$	$-105/4$	105
	$-197X^2 + 336XY + 24XZ - 147Y^2 - 21YZ + 3Z^2$		
	$11025/4$	$-105/2$	420
8	$X^2 + XT - 48Y^2 - 72YZ - 36YT - 32Z^2 - 32ZT - 4T^2$		
	900	-30	120
	$-17X^2 + 72XY + 64XZ - 48Y^2 - 72YZ - 32Z^2$		
	3600	-60	120
11	$X^2 + XT - 3Y^2 + 3YZ - 3YT - 42Z^2 - 81ZT - 28T^2$		
	$27225/16$	$-165/4$	165
	$-113X^2 + 6XY + 162XZ - 3Y^2 + 3YZ - 42Z^2$		
	$27225/4$	$-165/2$	660
13	$X^2 + XT - 507Y^2 - 39YZ - 156YT + 3Z^2 - 6ZT - 13T^2$		
	$38025/16$	$-195/4$	195
	$-53X^2 + 312XY + 12XZ - 507Y^2 - 39YZ + 3Z^2$		
	$38025/4$	$-195/2$	780

A.4. Embeddings and quadratic forms

TABLE A.21. Ternary normic forms $n_{\mathbb{Z}+2\mathcal{O}(6,N),3}$ for the quaternion orders in $H_{\mathbb{A}}(3)$ given in Table A.4, for $N \leq 100$.

N	$n_{\mathbb{Z}+2\mathcal{O}(6,N),3}$
1	$-12X^2 - 12ZX + 4Y^2 + 4YZ - 5Z^2$
5	$-300X^2 - 240YX - 180ZX - 44Y^2 - 68YZ - 29Z^2$
7	$-588X^2 - 168YX - 420ZX - 8Y^2 - 56YZ - 77Z^2$
11	$-12X^2 - 12ZX + 484Y^2 + 220YZ + 19Z^2$
13	$-12X^2 - 12ZX + 676Y^2 + 884YZ + 283Z^2$
17	$-3468X^2 - 408YX - 1428ZX - 8Y^2 - 80YZ - 149Z^2$
19	$-4332X^2 - 6384YX - 6156ZX - 2348Y^2 - 4532YZ - 2189Z^2$
23	$-6348X^2 - 2208YX - 1380ZX - 188Y^2 - 236YZ - 77Z^2$
25	$-7500X^2 - 1800YX - 8700ZX - 104Y^2 - 1040YZ - 2525Z^2$
29	$-10092X^2 - 16008YX - 15660ZX - 6344Y^2 - 12416YZ - 6077Z^2$
31	$-11532X^2 - 744YX - 3348ZX - 8Y^2 - 104YZ - 245Z^2$
35	$-14700X^2 - 16800YX - 1260ZX - 4796Y^2 - 716YZ - 29Z^2$
37	$-12X^2 - 12ZX + 5476Y^2 + 8732YZ + 3475Z^2$
41	$-20172X^2 - 33456YX - 32964ZX - 13868Y^2 - 27332YZ - 13469Z^2$
43	$-22188X^2 - 2064YX - 34572ZX - 44Y^2 - 1604YZ - 13469Z^2$
47	$-12X^2 - 12ZX + 8836Y^2 + 6580YZ + 1219Z^2$
49	$-28812X^2 - 2352YX - 20580ZX - 44Y^2 - 836YZ - 3677Z^2$
53	$-33708X^2 - 7632YX - 4452ZX - 428Y^2 - 500YZ - 149Z^2$
55	$-36300X^2 - 48840YX - 28380ZX - 16424Y^2 - 19088YZ - 5549Z^2$
59	$-41772X^2 - 14160YX - 50268ZX - 1196Y^2 - 8516YZ - 15125Z^2$
61	$-44652X^2 - 57096YX - 51972ZX - 18248Y^2 - 33224YZ - 15125Z^2$
65	$-50700X^2 - 70200YX - 52260ZX - 24296Y^2 - 36176YZ - 13469Z^2$
67	$-53868X^2 - 24120YX - 13668ZX - 2696Y^2 - 3056YZ - 869Z^2$
71	$-60492X^2 - 1704YX - 11076ZX - 8Y^2 - 152YZ - 509Z^2$
73	$-63948X^2 - 8760YX - 69204ZX - 296Y^2 - 4736YZ - 18725Z^2$
77	$-71148X^2 - 48048YX - 28644ZX - 8108Y^2 - 9668YZ - 2885Z^2$
79	$-74892X^2 - 92904YX - 948ZX - 28808Y^2 - 584YZ - 5Z^2$
83	$-82668X^2 - 71712YX - 96612ZX - 15548Y^2 - 41900YZ - 28229Z^2$
85	$-86700X^2 - 136680YX - 166260ZX - 53864Y^2 - 131048YZ - 79709Z^2$
89	$-95052X^2 - 74760YX - 137772ZX - 14696Y^2 - 54176YZ - 49925Z^2$
91	$-99372X^2 - 102648YX - 175812ZX - 26504Y^2 - 90800YZ - 77765Z^2$
95	$-108300X^2 - 18240YX - 10260ZX - 764Y^2 - 860YZ - 245Z^2$
97	$-112908X^2 - 58200YX - 124548ZX - 7496Y^2 - 32096YZ - 34349Z^2$

TABLE A.22. Ternary normic forms $n_{\mathbb{Z}+2\mathcal{O}(10,N),3}$ for the quaternion orders in $H_{\mathrm{B}}(2,5)$ given in Table A.5, for $N \leq 85$.

N	$n_{\mathbb{Z}+2\mathcal{O}(10,N),3}$
1	$-8X^2 - 8YX + 8Y^2 - 5Z^2$
3	$-72X^2 - 24YX - 96ZX + 8Y^2 - 16YZ - 37Z^2$
7	$-8X^2 - 56XY - 40XZ + 392Y^2 + 560YZ + 195Z^2$
9	$-648X^2 - 72YX - 288ZX + 8Y^2 - 16YZ - 37Z^2$
11	$-968X^2 - 88YX - 1232ZX + 8Y^2 - 56YZ - 397Z^2$
13	$-1352X^2 - 936YX - 1872ZX - 152Y^2 - 648YZ - 653Z^2$
17	$-8X^2 - 136XY - 112XZ + 2312Y^2 + 3808YZ + 1563Z^2$
19	$-2888X^2 - 152YX - 912ZX + 8Y^2 - 24YZ - 77Z^2$
21	$-3528X^2 - 1848YX - 5376ZX - 232Y^2 - 1408YZ - 2053Z^2$
23	$-4232X^2 - 5704YX - 1472ZX - 1912Y^2 - 992YZ - 133Z^2$
27	$-5832X^2 - 216YX - 6912ZX + 8Y^2 - 128YZ - 2053Z^2$
29	$-6728X^2 - 9048YX - 464ZX - 3032Y^2 - 312YZ - 13Z^2$
31	$-8X^2 - 248XY - 216XZ + 7688Y^2 + 13392YZ + 5827Z^2$
33	$-8712X^2 - 13992YX - 10560ZX - 5608Y^2 - 8480YZ - 3205Z^2$
37	$-10952X^2 - 5624YX - 5920ZX - 712Y^2 - 1520YZ - 805Z^2$
39	$-12168X^2 - 14664YX - 9984ZX - 4408Y^2 - 6016YZ - 2053Z^2$
41	$-13448X^2 - 25912YX - 7872ZX - 12472Y^2 - 7584YZ - 1157Z^2$
43	$-14792X^2 - 27864YX - 9632ZX - 13112Y^2 - 9072YZ - 1573Z^2$
47	$-17672X^2 - 23688YX - 30080ZX - 7928Y^2 - 20160YZ - 12805Z^2$
49	$-8X^2 - 392YX - 352XZ + 19208Y^2 + 34496YZ + 15483Z^2$
51	$-20808X^2 - 31416YX - 816ZX - 11848Y^2 - 616YZ - 13Z^2$
53	$-22472X^2 - 12296YX - 4240ZX - 1672Y^2 - 1160YZ - 205Z^2$
57	$-25992X^2 - 11400YX - 23712ZX - 1240Y^2 - 5200YZ - 5413Z^2$
59	$-27848X^2 - 25960YX - 16048ZX - 6040Y^2 - 7480YZ - 2317Z^2$
61	$-29768X^2 - 32696YX - 44896ZX - 8968Y^2 - 24656YZ - 16933Z^2$
63	$-31752X^2 - 50904YX - 19152ZX - 20392Y^2 - 15352YZ - 2893Z^2$
67	$-35912X^2 - 3752YX - 49312ZX - 88Y^2 - 2576YZ - 16933Z^2$
69	$-38088X^2 - 19320YX - 58512ZX - 2440Y^2 - 14840YZ - 22477Z^2$
71	$-8X^2 - 392YX - 352ZX + 19208Y^2 + 34496YZ + 15483Z^2$
73	$-42632X^2 - 33288YX - 2336ZX - 6488Y^2 - 912YZ - 37Z^2$
77	$-47432X^2 - 6776YX - 57904ZX - 232Y^2 - 4136YZ - 17677Z^2$
79	$-49928X^2 - 62568YX - 19592Y^2 - 5Z^2$
81	$-52488X^2 - 39528YX - 92016ZX - 7432Y^2 - 34648YZ - 40333Z^2$
83	$-55112X^2 - 92296YX - 9296ZX - 38632Y^2 - 7784YZ - 397Z^2$

TABLE A.23. Ternary normic forms $n_{\mathbb{Z}+2\mathcal{O}(14,N),3}$ for the quaternion orders in $H_{\mathrm{A}}(7)$ given in Table A.6, for $N \leq 80$.

N	$n_{\mathbb{Z}+2\mathcal{O}(14,N),3}$
1	$-28X^2 - 28ZX + 4Y^2 + 4YZ - 13Z^2$
3	$-28X^2 - 28ZX + 36Y^2 + 60YZ + 11Z^2$
5	$-700X^2 - 280YX - 980ZX - 24Y^2 - 192YZ - 349Z^2$
9	$-252X^2 - 336YX - 84ZX - 76Y^2 - 44YZ - 13Z^2$
11	$-3388X^2 - 3696YX - 2772ZX - 1004Y^2 - 1508YZ - 573Z^2$
13	$-4732X^2 - 1456YX - 1092ZX - 108Y^2 - 164YZ - 69Z^2$
15	$-700X^2 - 840YX - 140ZX - 216Y^2 - 24YZ + 11Z^2$
17	$-8092X^2 - 6188ZX + 4Y^2 + 4YZ - 1189Z^2$
19	$-10108X^2 - 4256YX - 14364ZX - 444Y^2 - 3020YZ - 5109Z^2$
23	$-14812X^2 - 1288YX - 8372ZX - 24Y^2 - 360YZ - 1189Z^2$
25	$-17500X^2 - 21000YX - 32900ZX - 6296Y^2 - 19736YZ - 15469Z^2$
27	$-28X^2 - 28ZX + 2916Y^2 + 1404YZ + 155Z^2$
29	$-23548X^2 - 25984YX - 25172ZX - 7164Y^2 - 13884YZ - 6733Z^2$
31	$-28X^2 - 28ZX + 3844Y^2 + 5084YZ + 1667Z^2$
33	$-3388X^2 - 3080YX - 4004ZX - 664Y^2 - 1808YZ - 1189Z^2$
37	$-38332X^2 - 16576YX - 23828ZX - 1788Y^2 - 5148YZ - 3709Z^2$
39	$-42588X^2 - 21840YX - 46956ZX - 2796Y^2 - 12036YZ - 12949Z^2$
41	$-47068X^2 - 50512YX - 19516ZX - 13548Y^2 - 10468YZ - 2029Z^2$
43	$-51772X^2 - 43344YX - 1204ZX - 9068Y^2 - 500YZ - 13Z^2$
45	$-6300X^2 - 10920YX - 2100ZX - 4696Y^2 - 1808YZ - 181Z^2$
51	$-8092X^2 - 3808YX - 3332ZX - 412Y^2 - 724YZ - 325Z^2$
53	$-78652X^2 - 2968YX - 28196ZX - 24Y^2 - 528YZ - 2533Z^2$
55	$-84700X^2 - 12320YX - 7700ZX - 444Y^2 - 556YZ - 181Z^2$
57	$-10108X^2 - 12768YX - 18620ZX - 3996Y^2 - 11700YZ - 8557Z^2$
59	$-97468X^2 - 56168YX - 34692ZX - 8088Y^2 - 9992YZ - 3093Z^2$
61	$-104188X^2 - 10248YX - 111020ZX - 248Y^2 - 5456YZ - 29581Z^2$
65	$-118300X^2 - 72800YX - 205660ZX - 11196Y^2 - 63276YZ - 89389Z^2$
69	$-14812X^2 - 18032YX - 644ZX - 5452Y^2 - 380YZ - 13Z^2$
75	$-17500X^2 - 2800YX - 28700ZX - 76Y^2 - 2236YZ - 11749Z^2$
79	$-174748X^2 - 92904YX - 90692ZX - 12344Y^2 - 24104YZ - 11773Z^2$

TABLE A.24. Ternary normic forms $n_{\mathbb{Z}+2\mathcal{O}(15,N),3}$ for the quaternion orders in $H_{\mathrm{B}}(3,5)$ given in Table A.7, for $N \leq 60$.

N	$n_{\mathbb{Z}+2\mathcal{O}(15,N),3}$
1	$-12X^2 - 12YX + 12Y^2 - 5Z^2$
2	$-12X^2 - 12ZX - 20Y^2 - 20YZ + 7Z^2$
4	$-192X^2 - 48YX - 96ZX + 12Y^2 - 12YZ - 17Z^2$
7	$-588X^2 - 84YX - 672ZX + 12Y^2 - 48YZ - 197Z^2$
8	$-192X^2 - 288YX - 144ZX - 128Y^2 - 128YZ - 17Z^2$
11	$-12X^2 - 132YX - 108ZX + 1452Y^2 + 2376YZ + 967Z^2$
13	$-2028X^2 - 156YX - 624ZX + 12Y^2 - 24YZ - 53Z^2$
14	$-2352X^2 - 3192YX - 1680ZX - 1068Y^2 - 1140YZ - 305Z^2$
16	$-768X^2 - 576YX - 1056ZX - 128Y^2 - 416YZ - 353Z^2$
17	$-3468X^2 - 3468YX - 408ZX - 852Y^2 - 204YZ - 17Z^2$
19	$-4332X^2 - 6612YX - 2508Y^2 - 5Z^2$
22	$-1452X^2 - 1584YX - 132ZX - 452Y^2 - 92YZ + 7Z^2$
23	$-6348X^2 - 8556YX - 8832ZX - 2868Y^2 - 5952YZ - 3077Z^2$
26	$-12X^2 - 12ZX - 3380Y^2 - 4420YZ - 1433Z^2$
28	$-9408X^2 - 336YX - 2016ZX + 12Y^2 - 36YZ - 113Z^2$
29	$-10092X^2 - 5220YX - 8352ZX - 660Y^2 - 2160YZ - 1733Z^2$
31	$-11532X^2 - 372YX - 12648ZX + 12Y^2 - 204YZ - 3473Z^2$
32	$-3072X^2 - 4224YX - 2880ZX - 1472Y^2 - 2000YZ - 665Z^2$
34	$-13872X^2 - 16728YX - 5712ZX - 5028Y^2 - 3444YZ - 593Z^2$
37	$-16428X^2 - 27084YX - 23976ZX - 11148Y^2 - 19764YZ - 8753Z^2$
38	$-17328X^2 - 23256YX - 912ZX - 7788Y^2 - 612YZ - 17Z^2$
41	$-20172X^2 - 30012YX - 37392ZX - 11148Y^2 - 27816YZ - 17333Z^2$
43	$-22188X^2 - 35604YX - 9288ZX - 14268Y^2 - 7452YZ - 977Z^2$
44	$-48X^2 - 48YX - 72ZX - 2432Y^2 - 1136YZ - 137Z^2$
46	$-6348X^2 - 7176YX - 6900ZX - 2048Y^2 - 3920YZ - 1865Z^2$
47	$-12X^2 - 564YX - 516ZX + 26508Y^2 + 48504YZ + 22183Z^2$
49	$-28812X^2 - 588YX - 52920ZX + 12Y^2 - 540YZ - 24305Z^2$
52	$-8112X^2 - 4368YX - 15288ZX - 608Y^2 - 4136YZ - 7193Z^2$
53	$-33708X^2 - 24804YX - 52152ZX - 4548Y^2 - 19188YZ - 20177Z^2$
56	$-9408X^2 - 8736YX - 1680ZX - 2048Y^2 - 800YZ - 65Z^2$
58	$-40368X^2 - 64728YX - 65424ZX - 25932Y^2 - 52452YZ - 26513Z^2$
59	$-41772X^2 - 3540YX - 14160ZX - 60Y^2 - 600YZ - 1205Z^2$

TABLE A.25. Generic binary form $f_{\mathcal{O}(6,N)}$ for the quaternion orders in $H_{\mathrm{A}}(3)$ given in Table A.4, for $N \leq 100$.

N	$f_{\mathcal{O}(6,N)}$
1	$(-z+x-x\sqrt{3}, -2(-x+y)\sqrt{3}, -z+x+x\sqrt{3})$
5	$(-z+x-x\sqrt{3}, -2(5y-3x+2z)\sqrt{3}, -z+x+x\sqrt{3})$
7	$(-z+x-x\sqrt{3}, -2(7y-5x+z)\sqrt{3}, -z+x+x\sqrt{3})$
11	$(-11z+5x-x\sqrt{3}, -2(-x+y)\sqrt{3}, -11z+5x+x\sqrt{3})$
13	$(-13z+17x-x\sqrt{3}, -2(-x+y)\sqrt{3}, -13z+17x+x\sqrt{3})$
17	$(-z+x-x\sqrt{3}, -2(17y-7x+z)\sqrt{3}, -z+x+x\sqrt{3})$
19	$(-z+x-x\sqrt{3}, -2(19y-27x+14z)\sqrt{3}, -z+x+x\sqrt{3})$
23	$(-z+x-x\sqrt{3}, -2(23y-5x+4z)\sqrt{3}, -z+x+x\sqrt{3})$
25	$(-z+x-x\sqrt{3}, -2(25y-29x+3z)\sqrt{3}, -z+x+x\sqrt{3})$
29	$(-z+x-x\sqrt{3}, -2(29y-45x+23z)\sqrt{3}, -z+x+x\sqrt{3})$
31	$(-z+x-x\sqrt{3}, -2(31y-9x+z)\sqrt{3}, -z+x+x\sqrt{3})$
35	$(-z+x-x\sqrt{3}, -2(35y-3x+20z)\sqrt{3}, -z+x+x\sqrt{3})$
37	$(-37z+59x-x\sqrt{3}, -2(-x+y)\sqrt{3}, -37z+59x+x\sqrt{3})$
41	$(-z+x-x\sqrt{3}, -2(41y-67x+34z)\sqrt{3}, -z+x+x\sqrt{3})$
43	$(-z+x-x\sqrt{3}, -2(43y-67x+2z)\sqrt{3}, -z+x+x\sqrt{3})$
47	$(-47z+35x-x\sqrt{3}, -2(-x+y)\sqrt{3}, -47z+35x+x\sqrt{3})$
49	$(-z+x-x\sqrt{3}, -2(49y-35x+2z)\sqrt{3}, -z+x+x\sqrt{3})$
53	$(-z+x-x\sqrt{3}, -2(53y-7x+6z)\sqrt{3}, -z+x+x\sqrt{3})$
55	$(-z+x-x\sqrt{3}, -2(55y-43x+37z)\sqrt{3}, -z+x+x\sqrt{3})$
59	$(-z+x-x\sqrt{3}, -2(59y-71x+10z)\sqrt{3}, -z+x+x\sqrt{3})$
61	$(-z+x-x\sqrt{3}, -2(61y-71x+39z)\sqrt{3}, -z+x+x\sqrt{3})$
65	$(-z+x-x\sqrt{3}, -2(65y-67x+45z)\sqrt{3}, -z+x+x\sqrt{3})$
67	$(-z+x-x\sqrt{3}, -2(67y-17x+15z)\sqrt{3}, -z+x+x\sqrt{3})$
71	$(-z+x-x\sqrt{3}, -2(71y-13x+z)\sqrt{3}, -z+x+x\sqrt{3})$
73	$(-z+x-x\sqrt{3}, -2(73y-79x+5z)\sqrt{3}, -z+x+x\sqrt{3})$
77	$(-z+x-x\sqrt{3}, -2(77y-31x+26z)\sqrt{3}, -z+x+x\sqrt{3})$
79	$(-z+x-x\sqrt{3}, -2(79y-x+49z)\sqrt{3}, -z+x+x\sqrt{3})$
83	$(-z+x-x\sqrt{3}, -2(83y-97x+36z)\sqrt{3}, -z+x+x\sqrt{3})$
85	$(-z+x-x\sqrt{3}, -2(85y-163x+67z)\sqrt{3}, -z+x+x\sqrt{3})$
89	$(-z+x-x\sqrt{3}, -2(89y+35z-129x)\sqrt{3}, -z+x+x\sqrt{3})$
91	$(-z+x-x\sqrt{3}, -2(91y-161x+47z)\sqrt{3}, -z+x+x\sqrt{3})$
95	$(-z+x-x\sqrt{3}, -2(95y-9x+8z)\sqrt{3}, -z+x+x\sqrt{3})$
97	$(-z+x-x\sqrt{3}, -2(97y-107x+25z)\sqrt{3}, -z+x+x\sqrt{3})$

TABLE A.26. Generic binary form $f_{\mathcal{O}(10,N)}$ for the quaternion orders in $H_{\mathrm{B}}(2,5)$ given in Table A.5, for $N \leq 80$.

N	$f_{\mathcal{O}(10,N)}$
1	$(-5x - 5/2t\sqrt{2}, -2(y + 1/2t)\sqrt{2}, x - 1/2t\sqrt{2})$
3	$(-5x - 5/2t\sqrt{2}, -2(3y + 1/2t - 4x)\sqrt{2}, x - 1/2t\sqrt{2})$
7	$(-5x - 5(7/2z - 5x)\sqrt{2}, -2(y + 7/2z - 5x)\sqrt{2}, x - (7/2z - 5x)\sqrt{2})$
9	$(-5x - 5/2t\sqrt{2}, -2(9y + 1/2t - 4x)\sqrt{2}, x - 1/2t\sqrt{2})$
11	$(-5x - 5/2t\sqrt{2}, -2(11y + 1/2t - 14x)\sqrt{2}, x - 1/2t\sqrt{2})$
13	$(-5x - 5/2t\sqrt{2}, -2(13y + 9/2t - 18x)\sqrt{2}, x - 1/2t\sqrt{2})$
17	$(-5x - 5(17/2z - 14x)\sqrt{2}, -2(y + 17/2z - 14x)\sqrt{2}, x - (17/2z - 14x)\sqrt{2})$
19	$(-5x - 5/2t\sqrt{2}, -2(19y + 1/2t - 6x)\sqrt{2}, x - 1/2t\sqrt{2})$
21	$(-5x - 5/2t\sqrt{2}, -2(21y + 11/2t - 32x)\sqrt{2}, x - 1/2t\sqrt{2})$
23	$(-5x - 5/2t\sqrt{2}, -2(23y + 31/2t - 8x)\sqrt{2}, x - 1/2t\sqrt{2})$
27	$(-5x - 5/2t\sqrt{2}, -2(27y + 1/2t - 32x)\sqrt{2}, x - 1/2t\sqrt{2})$
29	$(-5x - 5/2t\sqrt{2}, -2(29y + 39/2t - 2x)\sqrt{2}, x - 1/2t\sqrt{2})$
31	$(-5x - 5(31/2z - 27x)\sqrt{2}, -2(y + 31/2z - 27x)\sqrt{2}, x - (31/2z - 27x)\sqrt{2})$
33	$(-5x - 5/2t\sqrt{2}, -2(33y + 53/2t - 40x)\sqrt{2}, x - 1/2t\sqrt{2})$
37	$(-5x - 5/2t\sqrt{2}, -2(37y + 19/2t - 20x)\sqrt{2}, x - 1/2t\sqrt{2})$
39	$(-5x - 5/2t\sqrt{2}, -2(39y + 47/2t - 32x)\sqrt{2}, x - 1/2t\sqrt{2})$
41	$(-5x - 5/2t\sqrt{2}, -2(41y + 79/2t - 24x)\sqrt{2}, x - 1/2t\sqrt{2})$
43	$(-5x - 5/2t\sqrt{2}, -2(43y + 81/2t - 28x)\sqrt{2}, x - 1/2t\sqrt{2})$
47	$(-5x - 5/2t\sqrt{2}, -2(47y + 63/2t - 80x)\sqrt{2}, x - 1/2t\sqrt{2})$
49	$(-5x - 5(49/2z - 44x)\sqrt{2}, -2(y + 49/2z - 44x)\sqrt{2}, x - (49/2z - 44x)\sqrt{2})$
51	$(-5x - 5/2t\sqrt{2}, -2(51y + 77/2t - 2x)\sqrt{2}, x - 1/2t\sqrt{2})$
53	$(-5x - 5/2t\sqrt{2}, -2(53y + 29/2t - 10x)\sqrt{2}, x - 1/2t\sqrt{2})$
57	$(-5x - 5/2t\sqrt{2}, -2(57y + 25/2t - 52x)\sqrt{2}, x - 1/2t\sqrt{2})$
59	$(-5x - 5/2t\sqrt{2}, -2(59y + 55/2t - 34x)\sqrt{2}, x - 1/2t\sqrt{2})$
61	$(-5x - 5/2t\sqrt{2}, -2(61y + 67/2t - 92x)\sqrt{2}, x - 1/2t\sqrt{2})$
63	$(-5x - 5/2t\sqrt{2}, -2(63y + 101/2t - 38x)\sqrt{2}, x - 1/2t\sqrt{2})$
67	$(-5x - 5/2t\sqrt{2}, -2(67y + 7/2t - 92x)\sqrt{2}, x - 1/2t\sqrt{2})$
69	$(-5x - 5/2t\sqrt{2}, -2(69y + 35/2t - 106x)\sqrt{2}, x - 1/2t\sqrt{2})$
71	$([-5x - 5/2z\sqrt{2}, -2(71y + 131/2z - 22x)\sqrt{2}, x - 1/2z\sqrt{2})$
73	$(-5x - 5/2t\sqrt{2}, -2(73y + 57/2t - 4x)\sqrt{2}, x - 1/2t\sqrt{2})$
77	$(-5x - 5/2t\sqrt{2}, -2(77y + 11/2t - 94x)\sqrt{2}, x - 1/2t\sqrt{2})$
79	$(-5x - 5/2t\sqrt{2}, -2(79y + 99/2t)\sqrt{2}, x - 1/2t\sqrt{2})$
81	$(-5x - 5/2t\sqrt{2}, -2(81y + 61/2t - 142x)\sqrt{2}, x - 1/2t\sqrt{2})$
83	$(-5x - 5/2t\sqrt{2}, -2(83y + 139/2t - 14x)\sqrt{2}, x - 1/2t\sqrt{2}$

TABLE A.27. Generic binary form $f_{\mathcal{O}(14,N)}$ for the quaternion orders in $H_{\mathrm{A}}(7)$ given in Table A.6, for $N \leq 80$.

N	$f_{\mathcal{O}(14,N)}$
1	$(-z+x-x\sqrt{7}, -2(y-x)\sqrt{7}, -z+x+x\sqrt{7})$
3	$(-3z+5x-x\sqrt{7}, -2(y-x)\sqrt{7}, -3z+5x+x\sqrt{7})$
5	$(-z+x-x\sqrt{7}, -2(5y+z-7x)\sqrt{7}, -z+x+x\sqrt{7})$
9	$(-3z+x-x\sqrt{7}, -2(3y+2z-x)\sqrt{7}, -3z+x+x\sqrt{7})$
11	$(-z+x-x\sqrt{7}, -2(11y+6z-9x)\sqrt{7}, -z+x+x\sqrt{7})$
13	$(-z+x-x\sqrt{7}, -2(13y+2z-3x)\sqrt{7}, -z+x+x\sqrt{7})$
15	$(-3z+5x-x\sqrt{7}, -2(5y+3z-x)\sqrt{7}, -3z+5x+x\sqrt{7})$
17	$(-z+x-x\sqrt{7}, -2(17y-13x)\sqrt{7}, -z+x+x\sqrt{7})$
19	$(-z+x-x\sqrt{7}, -2(19y+4z-27x)\sqrt{7}, -z+x+x\sqrt{7})$
23	$(-z+x-x\sqrt{7}, -2(23y+z-13x)\sqrt{7}, -z+x+x\sqrt{7})$
25	$(-z+x-x\sqrt{7}, -2(25y+15z-47x)\sqrt{7}, -z+x+x\sqrt{7})$
27	$(-27z+13x-x\sqrt{7}, -2(y-x)\sqrt{7}, -27z+13x+x\sqrt{7})$
29	$(-z+x-x\sqrt{7}, -2(29y+16z-31x)\sqrt{7}, -z+x+x\sqrt{7})$
31	$(-31z+41x-x\sqrt{7}, -2(y-x)\sqrt{7}, -31z+41x+x\sqrt{7})$
33	$(-3z+x-x\sqrt{7}, -2(11y+5z-13x)\sqrt{7}, -3z+x+x\sqrt{7})$
37	$(-z+x-x\sqrt{7}, -2(37y+8z-23x)\sqrt{7}, -z+x+x\sqrt{7})$
39	$(-z+x-x\sqrt{7}, -2(39y+10z-43x)\sqrt{7}, -z+x+x\sqrt{7})$
41	$(-z+x-x\sqrt{7}, -2(41y+22z-17x)\sqrt{7}, -z+x+x\sqrt{7})$
43	$(-z+x-x\sqrt{7}, -2(43y+18z-x)\sqrt{7}, -z+x+x\sqrt{7})$
45	$(-3z+x-x\sqrt{7}, -2(15y+13z-5x)\sqrt{7}, -3z+x+x\sqrt{7})$
51	$(-3z+5x-x\sqrt{7}, -2(17y+4z-7x)\sqrt{7}, -3z+5x+x\sqrt{7})$
53	$(-z+x-x\sqrt{7}, -2(53y+z-19x)\sqrt{7}, -z+x+x\sqrt{7})$
55	$(-z+x-x\sqrt{7}, -2(55y+4z-5x)\sqrt{7}, -z+x+x\sqrt{7})$
57	$(-3z+5x-x\sqrt{7}, -2(19y+12z-35x)\sqrt{7}, -3z+5x+x\sqrt{7})$
59	$(-z+x-x\sqrt{7}, -2(59y+17z-21x)\sqrt{7}, -z+x+x\sqrt{7})$
61	$(-z+x-x\sqrt{7}, -2(61y+3z-65x)\sqrt{7}, -z+x+x\sqrt{7})$
65	$(-z+x-x\sqrt{7}, -2(65y+20z-113x)\sqrt{7}, -z+x+x\sqrt{7})$
69	$(-3z+x-x\sqrt{7}, -2(23y+14z-x)\sqrt{7}, -3z+x+x\sqrt{7})$
75	$(-3z+5x-x\sqrt{7}, -2(25y+2z-41x)\sqrt{7}, -3z+5x+x\sqrt{7})$
79	$(-z+x-x\sqrt{7}, -2(79y+21z-41x)\sqrt{7}, -z+x+x\sqrt{7})$

TABLE A.28. Generic binary form $f_{\mathcal{O}(15,N)}$ for the quaternion orders in $H_{\mathrm{B}}(3,5)$ given in Table A.7, for $N \le 60$.

N	$f_{\mathcal{O}(15,N)}$
1	$(-5x - 5/2t\sqrt{3}, -2(y + 1/2t)\sqrt{3}, x - 1/2t\sqrt{3})$
2	$(5z - 5x + 5x\sqrt{3}, -2(y - x)\sqrt{3}, -z + x + x\sqrt{3})$
4	$(-5x - 5/2t\sqrt{3}, -2(1/2t - 2x + 4y)\sqrt{3}, x - 1/2t\sqrt{3})$
7	$(-5x - 5/2t\sqrt{3}, -2(1/2t - 8x + 7y)\sqrt{3}, x - 1/2t\sqrt{3})$
8	$(5z - 5x + 5x\sqrt{3}, -2(3z - 3x + 4y)\sqrt{3}, -z + x + x\sqrt{3})$
11	$(-5x - 5(11/2z - 9x)\sqrt{3}, -2(y + 11/2z - 9x)\sqrt{3}, x - (11/2z - 9x)\sqrt{3})$
13	$(-5x - 5/2t\sqrt{3}, -2(1/2t - 4x + 13y)\sqrt{3}, x - 1/2t\sqrt{3})$
14	$(-5x - 5/2t\sqrt{3}, -2(19/2t - 10x + 14y)\sqrt{3}, x - 1/2t\sqrt{3})$
16	$(5z - 5x + 5x\sqrt{3}, -2(3z - 11x + 8y)\sqrt{3}, -z + x + x\sqrt{3})$
17	$(-5x - 5/2t\sqrt{3}, -2(17/2t - 2x + 17y)\sqrt{3}, x - 1/2t\sqrt{3})$
19	$(-5x - 5/2t\sqrt{3}, -2(19y + 29/2t)\sqrt{3}, x - 1/2t\sqrt{3})$
22	$(5z - 5x + 5x\sqrt{3}, -2(6z - x + 11y)\sqrt{3}, -z + x + x\sqrt{3})$
23	$(-5x - 5/2t\sqrt{3}, -2(31/2t - 32x + 23y)\sqrt{3}, x - 1/2t\sqrt{3})$
26	$(65z - 85x + 5x\sqrt{3}, -2(y - x)\sqrt{3}, -13z + 17x + x\sqrt{3})$
28	$(-5x - 5/2t\sqrt{3}, -2(1/2t - 6x + 28y)\sqrt{3}, x - 1/2t\sqrt{3})$
29	$(-5x - 5/2t\sqrt{3}, -2(15/2t - 24x + 29y)\sqrt{3}, x - 1/2t\sqrt{3})$
31	$(-5x - 5/2t\sqrt{3}, -2(1/2t - 34x + 31y)\sqrt{3}, x - 1/2t\sqrt{3})$
32	$(5z - 5x + 5x\sqrt{3}, -2(11z - 15x + 16y)\sqrt{3}, -z + x + x\sqrt{3})$
34	$(-5x - 5/2t\sqrt{3}, -2(41/2t - 14x + 34y)\sqrt{3}, x - 1/2t\sqrt{3})$
37	$(-5x - 5/2t\sqrt{3}, -2(61/2t - 54x + 37y)\sqrt{3}, x - 1/2t\sqrt{3})$
38	$(-5x - 5/2t\sqrt{3}, -2(51/2t - 2x + 38y)\sqrt{3}, x - 1/2t\sqrt{3})$
41	$(-5x - 5/2t\sqrt{3}, -2(61/2t - 76x + 41y)\sqrt{3}, x - 1/2t\sqrt{3})$
43	$(-5x - 5/2t\sqrt{3}, -2(69/2t - 18x + 43y)\sqrt{3}, x - 1/2t\sqrt{3})$
44	$(-25x + 55z + 5x\sqrt{3}, -2(z - 3x + 2y)\sqrt{3}, 5x - 11z + x\sqrt{3})$
46	$(5z - 5x + 5x\sqrt{3}, -2(13z - 25x + 23y)\sqrt{3}, -z + x + x\sqrt{3})$
47	$(-5x - 5(47/2z - 43x)\sqrt{3}, -2(y + 47/2z - 43x)\sqrt{3}, x - (47/2z - 43x)\sqrt{3})$
49	$(-5x - 5/2t\sqrt{3}, -2(1/2t - 90x + 49y)\sqrt{3}, x - 1/2t\sqrt{3})$
52	$(5z - 5x + 5x\sqrt{3}, -2(7z - 49x + 26y)\sqrt{3}, -z + x + x\sqrt{3})$
53	$(-5x - 5/2t\sqrt{3}, -2(39/2t - 82x + 53y)\sqrt{3}, x - 1/2t\sqrt{3})$
56	$(5z - 5x + 5x\sqrt{3}, -2(13z - 5x + 28y)\sqrt{3}, -z + x + x\sqrt{3})$
58	$(-5x - 5/2t\sqrt{3}, -2(93/2t - 94x + 58y)\sqrt{3}, x - 1/2t\sqrt{3})$
59	$(-5x - 5/2t\sqrt{3}, -2(5/2t - 20x + 59y)\sqrt{3}, x - 1/2t\sqrt{3}$

TABLE A.29. Generic binary forms $f_{\mathbb{Z}+2\mathcal{O}(6,N)}$ for the orders in $H_{\mathrm{A}}(3)$ given in Table A.4, $N \leq 100$.

N	$f_{\mathbb{Z}+2\mathcal{O}(6,N)}$
1	$(-2z-t+t\sqrt{3}, -2(t+2y)\sqrt{3}, -2z-t-t\sqrt{3})$
5	$(-2z-t+t\sqrt{3}, -2(3t+10y+4z)\sqrt{3}, -2z-t-t\sqrt{3})$
7	$(-2z-t+t\sqrt{3}, -2(5t+14y+2z)\sqrt{3}, -2z-t-t\sqrt{3})$
11	$(-22z-5t+t\sqrt{3}, -2(t+2y)\sqrt{3}, -22z-5t-t\sqrt{3})$
13	$(-26z-17t+t\sqrt{3}, -2(t+2y)\sqrt{3}, -26z-17t-t\sqrt{3})$
17	$(-2z-t+t\sqrt{3}, -2(7t+34y+2z)\sqrt{3}, -2z-t-t\sqrt{3})$
19	$(-2z-t+t\sqrt{3}, -2(27t+38y+28z)\sqrt{3}, -2z-t-t\sqrt{3})$
23	$(-2z-t+t\sqrt{3}, -2(5t+46y+8z)\sqrt{3}, -2z-t-t\sqrt{3})$
25	$(-2z-t+t\sqrt{3}, -2(29t+50y+6z)\sqrt{3}, -2z-t-t\sqrt{3})$
29	$(-2z-t+t\sqrt{3}, -2(45t+58y+46z)\sqrt{3}, -2z-t-t\sqrt{3})$
31	$(-2z-t+t\sqrt{3}, -2(9t+62y+2z)\sqrt{3}, -2z-t-t\sqrt{3})$
35	$(-2z-t+t\sqrt{3}, -2(3t+70y+40z)\sqrt{3}, -2z-t-t\sqrt{3})$
37	$(-74z-59t+t\sqrt{3}, -2(t+2y)\sqrt{3}, -74z-59t-t\sqrt{3})$
41	$(-2z-t+t\sqrt{3}, -2(67t+82y+68z)\sqrt{3}, -2z-t-t\sqrt{3})$
43	$(-2z-t+t\sqrt{3}, -2(67t+86y+4z)\sqrt{3}, -2z-t-t\sqrt{3})$
47	$(-94z-35t+t\sqrt{3}, -2(t+2y)\sqrt{3}, -94z-35t-t\sqrt{3})$
49	$(-2z-t+t\sqrt{3}, -2(35t+98y+4z)\sqrt{3}, -2z-t-t\sqrt{3})$
53	$(-2z-t+t\sqrt{3}, -2(7t+106y+12z)\sqrt{3}, -2z-t-t\sqrt{3})$
55	$(-2z-t+t\sqrt{3}, -2(43t+110y+74z)\sqrt{3}, -2z-t-t\sqrt{3})$
59	$(-2z-t+t\sqrt{3}, -2(71t+118y+20z)\sqrt{3}, -2z-t-t\sqrt{3})$
61	$(-2z-t+t\sqrt{3}, -2(71t+122y+78z)\sqrt{3}, -2z-t-t\sqrt{3})$
65	$(-2z-t+t\sqrt{3}, -2(67t+130y+90z)\sqrt{3}, -2z-t-t\sqrt{3})$
67	$(-2z-t+t\sqrt{3}, -2(17t+134y+30z)\sqrt{3}, -2z-t-t\sqrt{3})$
71	$(-2z-t+t\sqrt{3}, -2(13t+142y+2z)\sqrt{3}, -2z-t-t\sqrt{3})$
73	$(-2z-t+t\sqrt{3}, -2(79t+146y+10z)\sqrt{3}, -2z-t-t\sqrt{3})$
77	$(-2z-t+t\sqrt{3}, -2(31t+154y+52z)\sqrt{3}, -2z-t-t\sqrt{3})$
79	$(-2z-t+t\sqrt{3}, -2(t+158y+98z)\sqrt{3}, -2z-t-t\sqrt{3})$
83	$(-2z-t+t\sqrt{3}, -2(97t+166y+72z)\sqrt{3}, -2z-t-t\sqrt{3})$
85	$(-2z-t+t\sqrt{3}, -2(163t+170y+134z)\sqrt{3}, -2z-t-t\sqrt{3})$
89	$(-2z-t+\sqrt{3}, -2(129t+178y+70z)\sqrt{3}, -2z-t-t\sqrt{3})$
91	$(-2z-t+t\sqrt{3}, -2(161t+182y+94z)\sqrt{3}, -2z-t-t\sqrt{3})$
95	$(-2z-t+t\sqrt{3}, -2(9t+190y+16z)\sqrt{3}, -2z-t-t\sqrt{3})$
97	$(-2z-t+t\sqrt{3}, -2(107t+194y+50z)\sqrt{3}, (-2z-t-t\sqrt{3})$

TABLE A.30. Generic binary forms $f_{\mathbb{Z}+2\mathcal{O}(10,N)}$ for the orders in $H_{\mathrm{B}}(2,5)$ given in Table A.5, $N \leq 85$.

N	$f_{\mathbb{Z}+2\mathcal{O}(10,N)}$
1	$(5t-5z\sqrt{2}, -2(2y+z)\sqrt{2}, -t-z\sqrt{2})$
3	$(5t-5z\sqrt{2}, -2(z+4t+6y)\sqrt{2}, -t-z\sqrt{2})$
7	$(5t-5(7z+5t)\sqrt{2}, -2(2y+7z+5t)\sqrt{2}, -t-(7z+5t)\sqrt{2})$
9	$(5t-5z\sqrt{2}, -2(z+4t+18y)\sqrt{2}, -t-z\sqrt{2})$
11	$(5t-5z\sqrt{2}, -2(z+14t+22y)\sqrt{2}, -t-z\sqrt{2})$
13	$(5t-5z\sqrt{2}, -2(9z+18t+26y)\sqrt{2}, -t-z\sqrt{2})$
17	$(5t-5(17z+14t)\sqrt{2}, -2(2y+17z+14t)\sqrt{2}, -t-(17z+14t)\sqrt{2})$
19	$(5t-5z\sqrt{2}, -2(z+6t+38y)\sqrt{2}, -t-z\sqrt{2})$
21	$(5t-5z\sqrt{2}, -2(11z+32t+42y)\sqrt{2}, -t-z\sqrt{2})$
23	$(5t-5z\sqrt{2}, -2(31z+8t+46y)\sqrt{2}, -t-z\sqrt{2})$
27	$(5t-5z\sqrt{2}, -2(z+32t+54y)\sqrt{2}, -t-z\sqrt{2})$
29	$(5t-5z\sqrt{2}, -2(39z+2t+58y)\sqrt{2}, -t-z\sqrt{2})$
31	$(5t-5(31z+27t)\sqrt{2}, -2(2y+31z+27t)\sqrt{2}, -t-(31z+27t)\sqrt{2})$
33	$(5t-5z\sqrt{2}, -2(53z+40t+66y)\sqrt{2}, -t-z\sqrt{2})$
37	$(5t-5z\sqrt{2}, -2(19z+20t+74y)\sqrt{2}, -t-z\sqrt{2})$
39	$(5t-5z\sqrt{2}, -2(47z+32t+78y)\sqrt{2}, -t-z\sqrt{2})$
41	$(5t-5z\sqrt{2}, -2(79z+24t+82y)\sqrt{2}, -t-z\sqrt{2})$
43	$(5t-5z\sqrt{2}, -2(81z+28t+86y)\sqrt{2}, -t-z\sqrt{2})$
47	$(5t-5z\sqrt{2}, -2(63z+80t+94y)\sqrt{2}, -t-z\sqrt{2})$
49	$(5t-5(49z+44t)\sqrt{2}, -2(2y+49z+44t)\sqrt{2}, -t-(49z+44t)\sqrt{2})$
51	$(5t-5z\sqrt{2}, -2(77z+2t+102y)\sqrt{2}, -t-z\sqrt{2})$
53	$(5t-5z\sqrt{2}, -2(29z+10t+106y)\sqrt{2}, -t-z\sqrt{2})$
57	$(5t-5z\sqrt{2}, -2(25z+52t+114y)\sqrt{2}, -t-z\sqrt{2})$
59	$(5t-5z\sqrt{2}, -2(55z+34t+118y)\sqrt{2}, -t-z\sqrt{2})$
61	$(5t-5z\sqrt{2}, -2(67z+92t+122y)\sqrt{2}, -t-z\sqrt{2})$
63	$(5t-5z\sqrt{2}, -2(101z+38t+126y)\sqrt{2}, -t-z\sqrt{2})$
67	$(5t-5z\sqrt{2}, -2(7z+92t+134y)\sqrt{2}, -t-z\sqrt{2})$
69	$(5t-5z\sqrt{2}, -2(35z+106t+138y)\sqrt{2}, -t-z\sqrt{2})$
71	$(5t-5z\sqrt{2}, -2(142y+131z+22t)\sqrt{2}, -t-z\sqrt{2})$
73	$(5t-5z\sqrt{2}, -2(57z+4t+146y)\sqrt{2}, -t-z\sqrt{2})$
77	$(5t-5z\sqrt{2}, -2(11z+94t+154y)\sqrt{2}, -t-z\sqrt{2})$
79	$(5t-5z\sqrt{2}, -2(158y+99z)\sqrt{2}, -t-z\sqrt{2})$
81	$(5t-5z\sqrt{2}, -2(61z+142t+162y)\sqrt{2}, -t-z\sqrt{2})$
83	$(5t-5z\sqrt{2}, -2(139z+14t+166y)\sqrt{2}, -t-z\sqrt{2})$

TABLE A.31. Generic binary forms $f_{\mathbb{Z}+2\mathcal{O}(14,N)}$ for the orders in $H_{\mathrm{A}}(7)$ given in Table A.6, $N \leq 80$.

N	$f_{\mathbb{Z}+2\mathcal{O}(14,N)}$
1	$(-2z-t+t\sqrt{7}, -2(2y+t)\sqrt{7}, -2z-t-t\sqrt{7})$
3	$(-6z-5t+t\sqrt{7}, -2(2y+t)\sqrt{7}, -6z-5t-t\sqrt{7})$
5	$(-2z-t+t\sqrt{7}, -2(10y+2z+7t)\sqrt{7}, -2z-t-t\sqrt{7})$
9	$(-6z-t+t\sqrt{7}, -2(6y+4z+t)\sqrt{7}, -6z-t-t\sqrt{7})$
11	$(-2z-t+t\sqrt{7}, -2(22y+12z+9t)\sqrt{7}, -2z-t-t\sqrt{7})$
13	$(-2z-t+t\sqrt{7}, -2(26y+4z+3t)\sqrt{7}, -2z-t-t\sqrt{7})$
15	$(-6z-5t+t\sqrt{7}, -2(10y+6z+t)\sqrt{7}, -6z-5t-t\sqrt{7})$
17	$(-2z-t+t\sqrt{7}, -2(34y+13t)\sqrt{7}, -2z-t-t\sqrt{7})$
19	$(-2z-t+t\sqrt{7}, -2(38y+8z+27t)\sqrt{7}, -2z-t-t\sqrt{7})$
23	$(-2z-t+t\sqrt{7}, -2(46y+2z+13t)\sqrt{7}, -2z-t-t\sqrt{7})$
25	$(-2z-t+t\sqrt{7}, -2(50y+30z+47t)\sqrt{7}, -2z-t-t\sqrt{7})$
27	$(-54z-13t+t\sqrt{7}, -2(2y+t)\sqrt{7}, -54z-13t-t\sqrt{7})$
29	$(-2z-t+t\sqrt{7}, -2(58y+32z+31t)\sqrt{7}, -2z-t-t\sqrt{7})$
31	$(-62z-41t+t\sqrt{7}, -2(2y+t)\sqrt{7}, -62z-41t-t\sqrt{7})$
33	$(-6z-t+t\sqrt{7}, -2(22y+10z+13t)\sqrt{7}, -6z-t-t\sqrt{7})$
37	$(-2z-t+t\sqrt{7}, -2(74y+16z+23t)\sqrt{7}, -2z-t-t\sqrt{7})$
39	$(-2z-t+t\sqrt{7}, -2(78y+20z+43t)\sqrt{7}, -2z-t-t\sqrt{7})$
41	$(-2z-t+t\sqrt{7}, -2(82y+44z+17t)\sqrt{7}, -2z-t-t\sqrt{7})$
43	$(-2z-t+t\sqrt{7}, -2(86y+36z+t)\sqrt{7}, -2z-t-t\sqrt{7})$
45	$(-6z-t+t\sqrt{7}, -2(30y+26z+5t)\sqrt{7}, -6z-t-t\sqrt{7})$
51	$(-6z-5t+t\sqrt{7}, -2(34y+8z+7t)\sqrt{7}, -6z-5t-t\sqrt{7})$
53	$(-2z-t+t\sqrt{7}, -2(106y+2z+19t)\sqrt{7}, -2z-t-t\sqrt{7})$
55	$(-2z-t+t\sqrt{7}, -2(110y+8z+5t)\sqrt{7}, -2z-t-t\sqrt{7})$
57	$(-6z-5t+t\sqrt{7}, -2(38y+24z+35t)\sqrt{7}, -6z-5t-t\sqrt{7})$
59	$(-2z-t+t\sqrt{7}, -2(118y+34z+21t)\sqrt{7}, -2z-t-t\sqrt{7})$
61	$(-2z-t+t\sqrt{7}, -2(122y+6z+65t)\sqrt{7}, -2z-t-t\sqrt{7})$
65	$(-2z-t+t\sqrt{7}, -2(130y+40z+113t)\sqrt{7}, -2z-t-t\sqrt{7})$
69	$(-6z-t+t\sqrt{7}, -2(46y+28z+t)\sqrt{7}, -6z-t-t\sqrt{7})$
75	$(-6z-5t+t\sqrt{7}, -2(50y+4z+41t)\sqrt{7}, -6z-5t-t\sqrt{7})$
79	$(-2z-t+t\sqrt{7}, -2(158y+42z+41t)\sqrt{7}, -2z-t-t\sqrt{7})$

TABLE A.32. Generic binary forms $f_{\mathbb{Z}+2\mathcal{O}(15,N)}$ for the orders in $H_{\mathrm{B}}(3,5)$ given in Table A.7, $N \leq 60$.

N	$f_{\mathbb{Z}+2\mathcal{O}(15,N)}$
1	$(5t - 5z\sqrt{3}, -2(2y+z)\sqrt{3}, -t - z\sqrt{3})$
2	$(10z + 5t - 5t\sqrt{3}, -2(2y+t)\sqrt{3}, -2z - t - t\sqrt{3})$
4	$(5t - 5z\sqrt{3}, -2(8y+z+2t)\sqrt{3}, -t - z\sqrt{3})$
7	$(5t - 5z\sqrt{3}, -2(14y+z+8t)\sqrt{3}, -t - z\sqrt{3})$
8	$(10z + 5t - 5t\sqrt{3}, -2(8y+6z+3t)\sqrt{3}, -2z - t - t\sqrt{3})$
11	$(5t - 5(11z+9t)\sqrt{3}, -2(2y+11z+9t)\sqrt{3}, -t - (11z+9t)\sqrt{3})$
13	$(5t - 5z\sqrt{3}, -2(26y+z+4t)\sqrt{3}, -t - z\sqrt{3})$
14	$(5t - 5z\sqrt{3}, -2(28y+19z+10t)\sqrt{3}, -t - z\sqrt{3})$
16	$(10z + 5t - 5t\sqrt{3}, -2(16y+6z+11t)\sqrt{3}, -2z - t - t\sqrt{3})$
19	$(5t - 5z\sqrt{3}, -2(38y+29z)\sqrt{3}, -t - z\sqrt{3})$
22	$(10z + 5t - 5t\sqrt{3}, -2(22y+12z+t)\sqrt{3}, -2z - t - t\sqrt{3})$
23	$(5t - 5z\sqrt{3}, -2(46y+31z+32t)\sqrt{3}, -t - z\sqrt{3})$
26	$(130z + 85t - 5t\sqrt{3}, -2(2y+t)\sqrt{3}, -26z - 17t - t\sqrt{3})$
28	$(5t - 5z\sqrt{3}, -2(56y+z+6t)\sqrt{3}, -t - z\sqrt{3})$
29	$(5t - 5z\sqrt{3}, -2(58y+15z+24t)\sqrt{3}, -t - z\sqrt{3})$
31	$(5t - 5z\sqrt{3}, -2(62y+z+34t)\sqrt{3}, -t - z\sqrt{3})$
32	$(10z + 5t - 5t\sqrt{3}, -2(32y+22z+15t)\sqrt{3}, -2z - t - t\sqrt{3})$
34	$(5t - 5z\sqrt{3}, -2(68y+41z+14t)\sqrt{3}, -t - z\sqrt{3})$
37	$(5t - 5z\sqrt{3}, -2(74y+61z+54t)\sqrt{3}, -t - z\sqrt{3})$
38	$(5t - 5z\sqrt{3}, -2(76y+51z+2t)\sqrt{3}, -t - z\sqrt{3})$
41	$(5t - 5z\sqrt{3}, -2(82y+61z+76t)\sqrt{3}, -t - z\sqrt{3})$
43	$(5t - 5z\sqrt{3}, -2(86y+69z+18t)\sqrt{3}, -t - z\sqrt{3})$
46	$(10z + 5t - 5t\sqrt{3}, -2(46y+26z+25t)\sqrt{3}, -2z - t - t\sqrt{3})$
47	$(5t - 5(47z+43t)\sqrt{3}, -2(2y+47z+43t)\sqrt{3}, -t - (47z+43t)\sqrt{3})$
49	$(5t - 5z\sqrt{3}, -2(98y+z+90t)\sqrt{3}, -t - z\sqrt{3})$
52	$(10z + 5t - 5t\sqrt{3}, -2(52y+14z+49t)\sqrt{3}, -2z - t - t\sqrt{3})$
58	$(5t - 5z\sqrt{3}, -2(116y+93z+94t)\sqrt{3}, -t - z\sqrt{3})$
59	$(5t - 5z\sqrt{3}, -2(118y+5z+20t)\sqrt{3}, -t - z\sqrt{3})$

TABLE A.33. Class numbers $h(1, N, d, m)$, $N \leq 10$, $|d| \leq 5$ square-free, $m \leq 5$.

N	d	m	h
1	−1	1	1
		2	1
		3	2
		4	2
		5	2
	2	1	1
		2	2
		3	4
		4	4
		5	6
	−2	1	1
		2	2
		3	2
		4	4
		5	6
	3	1	1
		2	2
		3	3
		4	4
		5	6
	−3	1	1
		2	1
		3	1
		4	2
		5	2
	5	1	1
		2	3
		3	4
		4	6
		5	5
	−5	1	2
		2	4
		3	4
		4	8
		5	10
2	−1	1	1
		2	2
		3	2
		4	4
		5	2
	2	1	1
		2	4
		3	4
		4	8
		5	6
	−2	1	1
		2	4
		3	2
		4	8
		5	6
	3	1	1
		2	4
		3	3
		4	8
		5	6
	−3	2	2
		4	4
	5	2	6
		4	12
	−5	1	2
		2	8
		3	4
		4	16
		5	10
3	−1	3	4
	2	3	8
	−2	1	2
		2	4
		3	4
		4	8
		5	12
	3	1	1
		2	2
		3	6
		4	4
		5	6
	−3	1	1
		2	1
		3	2
		4	2
		5	2
	5	3	8
	−5	1	4
		2	8
		3	8
		4	16
		5	20
4	−1	2	3
		4	6
	2	2	6
		4	12
	−2	2	6
		4	12
	3	2	6
		4	12
	−3	2	2
		4	6
	5	2	6
		4	18
	−5	2	12
		4	24
5	−1	1	2
		2	2
		3	4
		4	4
		5	4
	2	5	12
	−2	5	12
	3	5	12
	−3	5	4
	5	1	1
		2	3
		3	4
		4	6
		5	10
	−5	1	2
		2	4
		3	4
		4	8
		5	20
6	−1	3	4
	2	3	8
	−2	1	2
		2	8
		3	4
		4	16
		5	12
	3	1	1
		2	4
		3	6
		4	8
		5	6

TABLE A.33 (continued)

N	d	m	h
	−3	2	2
		4	4
	−5	1	4
		2	16
		3	8
		4	32
		5	20
7	2	1	2
		2	4
		3	8
		4	8
		5	12
	−3	1	2
		2	2
		3	2
		4	4
		5	4
	−5	1	4
		2	8
		3	8
		4	16
		5	20

N	d	m	h
8	−1	2	2
		4	6
	2	2	4
		4	12
	−2	2	4
		4	12
	3	2	4
		4	12
	−3	4	6
	5	4	18
	−5	2	8
		4	24
9	−1	3	6
	2	3	12
	−2	1	2
		2	4
		3	10
		4	8
		5	12
	3	3	12
	−3	3	4
	5	3	12

N	d	m	h
	−5	1	4
		2	8
		3	20
		4	16
		5	20
10	−1	1	2
		2	4
		3	4
		4	8
		5	4
	2	5	12
	−2	5	12
	3	5	12
	5	2	6
		4	12
	−5	1	2
		2	8
		3	4
		4	16
		5	20

TABLE A.34. Class numbers $h(6, N, d, m)$, $N \leq 20$, $|d| \leq 10$ square-free, $m \leq 5$.

N	d	m	h
1	−1	1	2
		5	4
	2	1	2
		5	12
	3	1	1
		5	6
	−3	1	2
		5	4
	5	1	4
		5	20
	6	1	1
		5	4
	−6	1	2
		5	8
	−10	1	4
		5	20
5	−1	1	4
		5	8
	2	5	24
	3	5	12
	−3	5	8
	5	1	4
		5	40
	6	1	2
		5	8
	−6	1	4
		5	16
	−10	1	4
		5	40
7	2	1	4
		5	24
	−3	1	4
		5	8
	−6	1	4
		5	16
	−10	1	8
		5	40
11	3	1	2
		5	12
	5	1	8
		5	40
	−6	1	4
		5	16
	−10	1	8
		5	40
13	−1	1	4
		5	8
	3	1	2
		5	12
	−3	1	4
		5	8
	−10	1	8
		5	40
17	−1	1	4
		5	8
	2	1	4
		5	24
19	−3	1	4
		5	8
	5	1	8
		5	40
	6	1	2
		5	8
	−10	1	8
		5	40

TABLE A.35. Class numbers $h(10, N, d, m)$, $N \leq 20$, $|d| \leq 10$ square-free, $m \leq 5$.

N	d	m	h
1	2	1	2
		3	8
	−2	1	2
		3	4
	3	1	2
		3	6
	−3	1	4
		3	4
	5	1	2
		3	8
	−5	1	2
		3	4
	7	1	2
		3	4
	10	1	2
		3	4
	−10	1	2
		3	8
3	2	3	16
	−2	1	4
		3	8
	3	1	2
		3	12
	−3	1	4
		3	8
	5	3	16
	−5	1	4
		3	8
	7	1	4
		3	8
	10	1	4
		3	8
	−10	3	16
7	2	1	4
		3	16
	−3	1	8
		3	8
	−5	1	4
		3	8
	7	1	2
		3	4
	−10	1	4
		3	16
9	2	3	24
	−2	1	4
		3	20
	3	3	24
	−3	3	16
	5	3	24
	−5	1	4
		3	20
	7	1	4
		3	20
	10	1	4
		3	20
	−10	3	24
11	−2	1	4
		3	8
	3	1	4
		3	12
	5	1	4
		3	16
	−10	1	4
		3	16
13	3	1	4
		3	12
	−3	1	8
		3	8
	10	1	4
		3	8
	−10	1	4
		3	16
17	2	1	4
		3	16
	−2	1	4
		3	8
19	−2	1	4
		3	8
	−3	1	8
		3	8
	5	1	4
		3	16
	7	1	4
		3	8
	−10	1	4
		3	16

TABLE A.36. Class numbers $h(14, N, d, m)$, $N \leq 20$, $|d| \leq 10$ square-free, $m \leq 5$.

N	d	m	h
1	−1	1	2
		3	4
		5	4
	−2	1	2
		3	4
		5	12
	3	1	2
		3	6
		5	12
	5	1	4
		3	16
		5	20
	6	1	2
		3	6
		5	8
	7	1	1
		3	2
		5	6
	10	1	4
		3	8
		5	20
3	−1	3	8
	−2	1	4
		3	8
		5	24
	3	1	2
		3	12
		5	12
	5	3	32
	6	1	2
		3	12
		5	8
	7	1	2
		3	4
		5	12
	10	1	8
		3	16
		5	40
5	−1	1	4
		3	8
		5	8
	−2	5	24
	3	5	24
	5	1	4
		3	16
		5	40
	6	1	4
		3	12
		5	16
	7	5	12
	10	1	4
		3	8
		5	40
9	−1	3	12
	−2	1	4
		3	20
		5	24
	3	3	24
	5	3	48
	6	3	24
	7	1	2
		3	10
		5	12
	10	1	8
		3	40
		5	40
11	−2	1	4
		3	8
		5	24
	3	1	4
		3	12
		5	24
	5	1	8
		3	32
		5	40
13	−1	1	4
		3	8
		5	8
	3	1	4
		3	12
		5	24
	10	1	8
		3	16
		5	40
15	−1	3	16
	−2	5	48
	3	5	24
	5	3	32
	6	1	4
		3	24
		5	16
	7	5	24
	10	1	8
		3	16
		5	80
17	−1	1	4
		3	8
		5	8
	−2	1	4
		3	8
		5	24
19	−2	1	4
		3	8
		5	24
	5	1	8
		3	32
		5	40
	6	1	4
		3	12
		5	16
	7	1	2
		3	4
		5	12

TABLE A.37. Class numbers $h(15, N, d, m)$, $N \leq 20$, $|d| \leq 10$ square-free, $m \leq 5$.

N	d	m	h
1	2	1	4
		2	8
		4	16
	3	1	2
		2	4
		4	8
	−3	1	2
		2	2
		4	4
	5	1	2
		2	6
		4	12
	−7	1	4
		2	4
		4	8
	−10	1	4
		2	8
		4	16
2	2	1	4
		2	16
		4	32
	3	1	2
		2	8
		4	16
	−3	2	4
		4	8
	5	2	12
		4	24
	−7	1	8
		2	8
		4	16
	−10	1	4
		2	16
		4	32
4	2	2	24
		4	48
	3	2	12
		4	24
	−3	2	4
		4	12
	5	2	12
		4	36
	−7	1	8
		2	16
		4	24
	−10	2	24
		4	48
7	2	1	8
		2	16
		4	32
	−3	1	4
		2	4
		4	8
	−7	1	4
		2	4
		4	8
	−10	1	8
		2	16
		4	32
8	2	2	16
		4	48
	3	2	8
		4	24
	−3	4	12
	5	4	36
	−7	1	8
		2	8
		4	24
	−10	2	16
		4	48
11	3	1	4
		2	8
		4	16
	5	1	4
		2	12
		4	24
	−7	1	8
		2	8
		4	16
	−10	1	8
		2	16
		4	32
13	3	1	4
		2	8
		4	16
	−3	1	4
		2	4
		4	8
	−10	1	8
		2	16
		4	32
14	2	1	8
		2	32
		4	64
	−3	2	8
		4	16
	−7	1	8
		2	8
	−7	4	16
	−10	1	8
		2	32
		4	64
16	2	4	96
	3	4	48
	−3	4	16
	5	4	48
	−7	1	8
		2	8
		4	64
	−10	4	96
17	2	1	8
		2	16
		4	32
19	−3	1	4
		2	4
		4	8
	5	1	4
		2	12
		4	24
	−10	1	8
		2	16
		4	32

A.5. Complex multiplication points in Shimura curves

TABLE A.38. Quadratic orders $\Lambda(d,m)$ for which $X(1,N)$ has special complex multiplication points by $\Lambda(d,m)$, for $N \leq 25$.

N	(d,m)	h	cm
1	$(-1,1)$	1	1
	$(-3,1)$	1	1
	$\mathrm{scm}(1,1)=2$		
2	$(-1,1)$	1	1
	$(-2,1)$	1	1
	$(-7,1)$	1	2
	$\mathrm{scm}(1,2)=4$		
3	$(-2,1)$	1	2
	$(-3,1)$	1	1
	$(-3,2)$	1	1
	$(-11,1)$	1	2
	$\mathrm{scm}(1,3)=6$		
5	$(-1,1)$	1	2
	$(-1,2)$	1	2
	$(-5,1)$	2	2
	$(-11,1)$	1	2
	$(-19,1)$	1	2
	$\mathrm{scm}(1,5)=10$		
6	$(-2,1)$	1	2
	$(-5,1)$	2	4
	$(-6,1)$	2	2
	$(-15,1)$	2	4
	$(-23,1)$	3	12
	$\mathrm{scm}(1,6)=24$		
7	$(-3,1)$	1	2
	$(-3,2)$	1	2
	$(-3,3)$	1	2
	$(-6,1)$	2	4
	$(-7,1)$	1	1
	$(-7,2)$	1	1
	$(-19,1)$	1	2
	$\mathrm{scm}(1,7)=14$		
10	$(-1,1)$	1	2
	$(-1,3)$	2	4
	$(-6,1)$	2	4
	$(-10,1)$	2	2
	$(-15,1)$	2	4
	$(-31,1)$	3	12
	$(-39,1)$	4	16
	$\mathrm{scm}(1,10)=44$		
11	$(-2,1)$	1	2
	$(-7,1)$	1	2

N	(d,m)	h	cm
	$(-7,2)$	1	2
	$(-10,1)$	2	4
	$(-11,1)$	1	1
	$(-11,2)$	3	3
	$(-19,1)$	1	2
	$(-35,1)$	2	4
	$(-43,1)$	1	2
	$\mathrm{scm}(1,11)=22$		
13	$(-1,1)$	1	2
	$(-1,2)$	1	2
	$(-1,3)$	2	4
	$(-3,1)$	1	2
	$(-3,2)$	1	2
	$(-3,3)$	1	2
	$(-3,4)$	2	4
	$(-13,1)$	2	2
	$(-43,1)$	1	2
	$(-51,1)$	2	4
	$\mathrm{scm}(1,13)=26$		
14	$(-5,1)$	2	4
	$(-7,1)$	1	2
	$(-10,1)$	2	4
	$(-13,1)$	2	4
	$(-14,1)$	4	4
	$(-31,1)$	3	12
	$(-47,1)$	5	20
	$(-55,1)$	4	16
	$\mathrm{scm}(1,14)=66$		
15	$(-6,1)$	2	4
	$(-11,1)$	1	4
	$(-11,2)$	3	12
	$(-14,1)$	4	16
	$(-15,1)$	2	2
	$(-15,2)$	2	2
	$(-35,1)$	2	4
	$(-51,1)$	2	4
	$(-59,1)$	3	12
	$\mathrm{scm}(1,15)=60$		
17	$(-1,1)$	1	2
	$(-1,2)$	1	2
	$(-1,4)$	2	4
	$(-2,1)$	1	2

N	(d,m)	h	cm
	$(-2,2)$	2	4
	$(-13,1)$	2	4
	$(-17,1)$	4	4
	$(-19,1)$	1	2
	$(-43,1)$	1	2
	$(-59,1)$	3	6
	$(-67,1)$	1	2
	$\mathrm{scm}(1,17)=34$		
19	$(-2,1)$	1	2
	$(-2,3)$	2	4
	$(-3,1)$	1	2
	$(-3,2)$	1	2
	$(-3,3)$	1	2
	$(-3,5)$	2	4
	$(-10,1)$	2	4
	$(-15,1)$	2	4
	$(-15,2)$	2	4
	$(-19,1)$	1	1
	$(-19,2)$	3	3
	$(-51,1)$	2	4
	$(-67,1)$	1	2
	$\mathrm{scm}(1,19)=38$		
21	$(-3,1)$	1	2
	$(-3,2)$	1	2
	$(-3,4)$	2	4
	$(-3,5)$	2	4
	$(-5,1)$	2	8
	$(-5,2)$	4	16
	$(-17,1)$	4	16
	$(-21,1)$	4	4
	$(-35,1)$	2	4
	$(-59,1)$	3	12
	$(-83,1)$	3	12
	$\mathrm{scm}(1,21)=84$		
22	$(-2,1)$	1	2
	$(-2,3)$	2	4
	$(-6,1)$	2	4
	$(-7,1)$	1	4
	$(-7,3)$	4	16
	$(-13,1)$	2	4
	$(-21,1)$	4	8
	$(-22,1)$	2	2

TABLE A.38 (continued)

N	(d,m)	h	cm
	$(-39,1)$	4	16
	$(-79,1)$	5	20
	$(-87,1)$	6	24
	$\mathrm{scm}(1,22) = 104$		
23	$(-7,1)$	1	2
	$(-7,2)$	1	2

N	(d,m)	h	cm
	$(-11,1)$	1	2
	$(-14,1)$	4	8
	$(-19,1)$	1	2
	$(-19,2)$	3	6
	$(-22,1)$	2	4
	$(-23,1)$	3	3

N	(d,m)	h	cm
	$(-23,2)$	3	3
	$(-43,1)$	1	2
	$(-67,1)$	1	2
	$(-83,1)$	3	6
	$(-91,1)$	2	4
	$\mathrm{scm}(1,23) = 46$		

TABLE A.39. Quadratic orders $\Lambda(d,m)$ for which $X(6,N)$ has special complex multiplication points by $\Lambda(d,m)$, for $N \leq 25$.

N	(d,m)	h	cm
1	$(-6,1)$	2	2
	$\mathrm{scm}(6,1) = 2$		
5	$(-21,1)$	4	8
	$(-30,1)$	4	4
	$\mathrm{scm}(6,5) = 12$		
7	$(-6,1)$	2	4
	$(-33,1)$	4	8
	$(-42,1)$	4	4
	$\mathrm{scm}(6,7) = 16$		
11	$(-30,1)$	4	8
	$(-57,1)$	4	8

N	(d,m)	h	cm
	$(-66,1)$	8	8
	$\mathrm{scm}(6,11) = 24$		
13	$(-42,1)$	4	8
	$(-69,1)$	8	16
	$(-78,1)$	4	4
	$\mathrm{scm}(6,13) = 28$		
17	$(-21,1)$	4	8
	$(-66,1)$	8	16
	$(-93,1)$	4	8
	$(-102,1)$	4	4
	$\mathrm{scm}(6,17) = 36$		

N	(d,m)	h	cm
19	$(-33,1)$	4	8
	$(-78,1)$	4	8
	$(-105,1)$	8	16
	$(-114,1)$	8	8
	$\mathrm{scm}(6,19) = 40$		
23	$(-57,1)$	4	8
	$(-102,1)$	4	8
	$(-129,1)$	12	24
	$(-138,1)$	8	8
	$\mathrm{scm}(6,23) = 48$		

TABLE A.40. Quadratic orders $\Lambda(d,m)$ for which $X(10,N)$ has special complex multiplication points by $\Lambda(d,m)$, for $N \leq 25$.

N	(d,m)	h	cm
1	$(-10,1)$	2	2
	$\mathrm{scm}(10,1)=2$		
3	$(-5,1)$	2	4
	$(-30,1)$	4	4
	$\mathrm{scm}(10,3)=8$		
7	$(-5,1)$	2	4
	$(-5,3)$	4	8
	$(-70,1)$	4	4
	$\mathrm{scm}(10,7)=16$		
11	$(-10,1)$	2	4
	$(-85,1)$	4	8
	$(-110,1)$	12	12
	$\mathrm{scm}(10,11)=24$		
13	$(-30,1)$	4	8
	$(-105,1)$	8	16
	$(-130,1)$	4	4
	$\mathrm{scm}(10,13)=28$		
17	$(-70,1)$	4	8
	$(-145,1)$	8	16
	$(-170,1)$	12	12
	$\mathrm{scm}(10,17)=36$		
19	$(-10,1)$	2	4
	$(-10,3)$	8	16
	$(-165,1)$	8	16
	$(-190,1)$	4	4
	$\mathrm{scm}(10,19)=40$		
21	$(-110,1)$	12	48
	$(-185,1)$	16	64
	$(-210,1)$	8	8
	$\mathrm{scm}(10,21)=120$		
23	$(-5,1)$	2	4
	$(-130,1)$	4	8
	$(-205,1)$	8	16
	$(-230,1)$	20	20
	$\mathrm{scm}(10,23)=48$		

TABLE A.41. Quadratic orders $\Lambda(d,m)$ for which $X(15,N)$ has special complex multiplication points by $\Lambda(d,m)$, for $N \leq 22$.

N	(d,m)	h	cm
1	$(-15,1)$	2	2
	$(-15,2)$	2	2
	$\mathrm{scm}(15,1)=4$		
2	$(-30,1)$	4	4
	$\mathrm{scm}(15,2)=4$		
7	$(-105,1)$	8	8
	$(-195,1)$	4	8
	$\mathrm{scm}(15,7)=16$		
11	$(-165,1)$	8	8
	$(-435,1)$	4	8
	$\mathrm{scm}(15,11)=16$		
13	$(-195,1)$	4	4
	$(-195,2)$	12	12
	$(-555,1)$	4	8
	$\mathrm{scm}(15,13)=24$		
14	$(-210,1)$	8	8
	$(-615,1)$	20	80
	$\mathrm{scm}(15,14)=88$		
17	$(-30,1)$	4	8
	$(-255,1)$	12	12
	$(-255,2)$	12	12
	$(-795,1)$	4	8
	$\mathrm{scm}(15,17)=40$		
19	$(-15,1)$	2	4
	$(-15,2)$	2	4
	$(-15,4)$	4	8
	$(-285,1)$	16	16
	$(-915,1)$	8	16
	$\mathrm{scm}(15,19)=48$		
22	$(-105,1)$	8	16
	$(-330,1)$	8	8
	$(-1095,1)$	28	112
	$\mathrm{scm}(15,22)=136$		

APPENDIX B

Further Contributions to the Study of Shimura Curves

Shimura generalized his results on arithmetic Fuchsian groups and on curves to reductive groups acting on symmetric bounded domains, cf. [**Shi70a, Shi70b**]. This point of view was axiomatized by Deligne [**Del71, Del79**]. Shimura's work aroused the interest of many authors, who produced a succession of articles on the subject. Nowadays, the study of Shimura curves, or more generally that of Shimura varieties, has become a substantial focus of research. Shimura curves and Shimura varieties have played an outstanding role in the proof of many of the latest results in number theory. We shall outline some of these advances restricting ourselves to the framework of curves. Since modular curves are discussed in great detail in the literature, we concentrate on the non modular cases.

Let R denote the ring of integers of a number field K. Fix a finite place v of K and let K_v be the completion of K in v. Čerednik [**Čer76**] proved the existence of a v-adic uniformization of the Shimura curve by arithmetic discrete subgroups of $\mathrm{PGL}(2, K_v)$. Čerednik's proof of the existence of a model over the local ring R_v is indirect. By using the modular interpretation of Shimura curves, Drinfel′d [**Dri76**] obtained a more direct proof of Čerednik's theorem in the case $K = \mathbb{Q}$. These results have become basic for many later works. We note the existence of several papers devoted to their exposition, such as [**vdP89, BC91**]. Čerednik's theorem has been generalized by Varshavsky [**Var98**].

The theory of the modular curves $X_0(N)$ and $X_1(N)$ is studied and documented in the works of Deligne–Rapoport [**DR73**] and Katz–Mazur [**KM85**]. In this case, the primes p with good reduction are the primes $p \nmid N$. The corresponding results for non modular Shimura curves need to take into account a new parameter: the discriminant D of the quaternion algebra. For a Shimura curve $S_{\mathbb{Q}}$ associated to an Eichler order of level N in a quaternion $\mathbb{Q}$-algebra of discriminant D, the primes p of good reduction are those $p \nmid DN$. Results for the reduction of Shimura curves are due to Morita [**Mor81**] (for $K = \mathbb{Q}$) and Carayol [**Car83, Car86**] (for $K \neq \mathbb{Q}$). We also note the papers of Ihara [**Iha68a, Iha68b, Iha69**], Ogg [**Ogg85**] and the paper of Buzzard [**Buz97**].

Kurihara [**Kur79**] computed equations for some Shimura curves of genus ≤ 1. He combined the results of Shimura ([**Shi67, Shi70a, Shi75**]) and results for v-adic uniformizations. Jordan and Livné [**Jor81**] and Michon [**Mic81a**] computed three more equations in the case of genus ≤ 1. Later, Kurihara [**Kur94**] presented partial results towards a possible computation of more equations.

All the hyperelliptic Shimura curves are known when $K = \mathbb{Q}$ and $\mathcal{O}$ is an Eichler order. Ishii [**Ish75**] determined the hyperelliptic curves corresponding to maximal orders with some restrictions. Michon [**Mic81b**] solved the case of maximal order, using results of Ogg [**Ogg74**] in the modular case. Ogg [**Ogg83, Ogg85**] completed the list of hyperelliptic Shimura curves for $K = \mathbb{Q}$; he studied the involutions w of Shimura curves $S_{\mathbb{Q}}$, in the case of Eichler order of level N, and described the connected components of the set of real points of the curve $S_{\mathbb{Q}}/\langle w\rangle$.

Jordan and Livné [**Jor84, JL85**] gave necessary and sufficient criteria for the existence of points of $S_{\mathbb{Q}}$ in the local fields $\mathbb{Q}_p$, for the case of maximal order. Ogg [**Ogg85**] dealt with the case of Eichler order and characterized the existence of points of $S_{\mathbb{Q}}$ and $S_{\mathbb{Q}}/\langle w\rangle$ in the local fields $\mathbb{Q}_p$. For the primes with good reduction, the problem is limited to the existence of points of the reduced curve over finite fields, which is determined by the trace formula of Eichler–Selberg; for the primes of bad reduction, it is necessary to use uniformization results by [**Dri76, Kur79**]. In [**Jor86**] there are results on $S_{\mathbb{Q}}(F)$, where F is an imaginary quadratic field. Jordan and Livné [**JL87**] dealt with the existence of F-rational divisors and classes of divisors of fixed degree d, in the case F is a finite extension of $\mathbb{Q}_p$. Kamienny [**Kam90**] obtained results on the finiteness of the number of rational points of some Shimura curves $S_{\mathbb{Q}}$ over quadratic fields. Recently, Jordan and Livné [**JL99**] studied local Diophantine properties of the quotients $S_{\mathbb{Q}}/\langle w\rangle$ for certain families of Shimura curves $S_{\mathbb{Q}}$. In the work of Elkies [**Elk98**], some explicit points in quotients $S_{\mathbb{Q}}/\langle w\rangle$ of Shimura curves associated to a maximal order in the quaternion algebras of discriminant $D = 6, 10, 14, 15$ are computed.

Among later works on Shimura curves we also note [**Lin93, Bes95, HM95, Ji98, Rot02**]. For further references cf. [**AAB**].

APPENDIX C

Applications of Shimura Curves

In recent years, Shimura curves have become key tools in the study of arithmetical problems and in the proof of important results.

K. Ribet [**Rib80**] showed that Shimura curves can be used in an innovative way. From the results of Eichler [**Eic58**] and Shimizu [**Shz65**] for the trace formula, Ribet proved the existence of an isogeny between the new part of the Jacobian of the modular curve $X_0(N)$, with N equal to a product of an even number of different primes, and the Jacobian of the Shimura curve associated to a maximal order in a quaternion algebra of discriminant N.

Shimura curves, and in particular their Jacobians, play a strategic role in the study of the Galois representations associated to modular forms. Jordan and Livné [**JL86**] studied the reduction of the Néron model of the Jacobian of the Shimura curve $S_{\mathbb{Q}}$, in the case of maximal order, for the primes of bad reduction and determined the group of connected components, using the descriptions of the regular model of the curve over $\mathbb{Z}_p$, given by Drinfeld and Kurihara, and results by Raynaud [**Ray70**] and Deligne–Rapoport [**DR73**]. Later, Ribet [**Rib90, Rib89**] interrelated the reductions of the Néron model of the Jacobians of a Shimura curve and a modular curve to prove the particular case of the conjecture (3.2.4?) of Serre [**Ser87**] known as *epsilon conjecture*. Using this, it is proved that the Shimura–Taniyama–Weil conjecture in the semistable case implies Fermat's Last Theorem. In [**JL89**] results on the cohomology of some Shimura curves are proved, in an attempt to generalize Ribet's results to modular forms of weight greater than 2; the authors use results by Faltings and Jordan [**FJ95**]. In later generalizations of the work of Ribet, cf. [**Rib91, BLR91, MR91**], other results for Shimura curves are stated. In [**Rib94**], Ribet gave a description of them.

Among the authors that used Shimura curves in their approach to Serre's conjecture, we should mention Diamond [**Dia97a**]. Diamond and Taylor [**DT94**] stated results for Shimura curves analogous to the known results for modular curves. To do this they used the cohomology of the canonical models, the reduction and the modular interpretation of Shimura curves. In [**Dia97a**], looking for a generalization of results by [**Rib90**], Diamond used Shimura curves analogous to the modular curves coming from the groups $\Gamma_0(N)\cap\Gamma_1(N')$. Further results have been obtained by Ribet and Takahashi [**RT97**], Diamond [**Dia97b**], etc.

As a consequence of these studies, Shimura curves intervene in the version of the Shimura–Taniyama–Weil conjecture proved by Wiles [**Wil95**] and Taylor–Wiles [**TW95**]. In [**DDT97, CSS97**] an exposition of this topic can be found. Thus, Shimura curves play a dual role in the proof of Fermat's Last Theorem.

Bibliography

[AAB] M. Alsina, A. Arenas, and P. Bayer (eds.), *Corbes de Shimura i aplicacions*, Publ. Universitat de Barcelona, (to appear).

[Alb34] A. A. Albert, *Integral domains of rational generalized quaternion algebras*, Bull. Amer. Math. Soc. **40** (1934), 164–176.

[Als97] M. Alsina, *Fundamental domains of the upper half-plane by the action of matrix groups*, Meeting on Matrix Analysis and Applications (Dept. Mat. Apl. I, Fac. Informática y Estadística, Univ. de Sevilla), 1997, pp. 10–17.

[Als00] ______, *Dominios fundamentales modulares*, Rev. R. Acad. Cienc. Exactas Fís. Nat. (Esp.) **94** (2000), no. 3, 309–322.

[Are87] A. Arenas, *On integral representations by quadratic forms*, Linear and Multilinear Algebra **22** (1987), no. 2, 149–160.

[BC91] J.-F. Boutot and H. Carayol, *Uniformisation p-adique des courbes de Shimura: les théorèmes de Čerednik et de Drinfel'd*, Courbes modulaires et courbes de Shimura (Orsay 1987/1988), Astérisque, vol. 196-197, Soc. Math. France, Paris, 1991, pp. 45–158.

[Bes95] A. Besser, *CM cycles over Shimura curves*, J. Algebraic Geom. **4** (1995), no. 4, 659–691.

[BLR91] N. Boston, H. W. Lenstra, and K. A. Ribet, *Quotients of group rings arising from two-dimensional representations*, C. R. Acad. Sci. Paris Sér. I Math. **312** (1991), no. 4, 323–328.

[Bra24] H. Brandt, *Der Kompositionsbegriff bei den quaternären quadratischen Formen*, Math. Ann. **91** (1924), 300–315.

[Bra28] ______, *Idealtheorie in Quaternionenalgebren*, Math. Ann. **99** (1928), 1–29.

[Bra43] ______, *Zur Zahlentheorie der Quaternionen*, Jber. Deutsch. Math. Verein. **53** (1943), 23–57.

[Brz80] J. Brzeziński, *Arithmetical quadratic surfaces of genus* 0. I, Math. Scand. **46** (1980), no. 2, 183–208.

[Brz82] ______, *A characterization of Gorenstein orders in quaternion algebras*, Math. Scand. **50** (1982), no. 1, 19–24.

[Brz83] ______, *On orders in quaternion algebras*, Comm. Algebra **11** (1983), no. 5, 501–522.

[Brz90] ______, *On automorphisms of quaternion orders*, J. Reine Angew. Math. **403** (1990), 166–186.

[Brz95] ______, *Definite quaternion orders of class number one*, J. Théor. Nombres Bordeaux **7** (1995), no. 1, 93–96.

[BS66] Z. I. Borevich and I. R. Shafarevich, *Number theory*, Academic Press, New York, 1966.

[Buz97] K. Buzzard, *Integral models of certain Shimura curves*, Duke Math. J. **87** (1997), 591–612.

[Car83] H. Carayol, *Sur la mauvaise réduction des courbes de Shimura*, C. R. Acad. Sc. Paris Sér. I Math. **296** (1983), no. 13, 557–560.

[Car86] ______, *Sur la mauvaise réduction des courbes de Shimura*, Compositio Math. **59** (1986), no. 2, 151–230.

[Čer76] I. V. Čerednik, *Uniformization of algebraic curves by discrete arithmetic subgroups of* $\mathrm{PGL}_2(k_w)$ *with compact quotients*, Math. USSR-Sb. **29** (1976), no. 1, 55–78.

[Coh95] H. Cohen, *A course in computational algebraic number theory*, Grad. Texts in Math., vol. 138, Springer, Berlin, 1995.

[CSS97] G. Cornell, J. H. Silverman, and G. Stevens (eds.), *Modular forms and Fermat's last theorem*, Boston, MA, 1995, Springer, New York, 1997.

[DDT97] H. Darmon, F. Diamond, and R. Taylor, *Fermat's last theorem*, Elliptic Curves, Modular Forms & Fermat's Last Theorem (Hong Kong, 1993), Internat. Press, Cambridge, MA, 1997, pp. 2–140.

[Del71] P. Deligne, *Travaux de Shimura*, Séminaire Bourbaki, 23e année 1970-1971, Lecture Notes in Math., vol. 244, Springer, Berlin, 1971, pp. 123–165.

[Del79] ______, *Variétés de Shimura: interprétation modulaire, et techniques de construction de modèles canoniques*, Automorphic Forms, Representations and L-Functions (Corvallis, OR, 1977), Proc. Sympos. Pure Math., vol. 33, Part 2, Amer. Math. Soc., Providence, RI, 1979, pp. 247–290.

[Deu68] M. Deuring, *Algebren*, Ergeb. Math. Grenzgeb., vol. 41, Springer-Verlag, Berlin–New York, 1968.

[Dia97a] F. Diamond, *The refined conjecture of Serre*, Elliptic Curves, Modular Forms, & Fermat's Last Theorem (Hong Kong, 1993), Ser. Number Theory, vol. 1, Internat. Press, Cambridge, MA, 1997, pp. 22–37.

[Dia97b] ______, *The Taylor–Wiles construction and multiplicity one*, Invent. Math. **128** (1997), no. 2, 379–391.

[DR73] P. Deligne and M. Rapoport, *Les schémas de modules de courbes elliptiques*, Modular functions of one variable. II (Antwerp, 1972), Lecture Notes in Math., vol. 349, Springer, Berlin, 1973, pp. 143–316.

[Dri76] V. G. Drinfel′d, *Coverings of p-adic symmetric regions*, Funct. Anal. Appl. **10** (1976), 107–115.

[DT94] F. Diamond and R. Taylor, *Nonoptimal levels of mod l modular representations*, Invent. Math. **115** (1994), no. 3, 435–462.

[Eic37] M. Eichler, *Bestimmung der Idealklassenzahl in gewissen normalen einfachen Algebren*, J. Reine Angew. Math. **176** (1937), 192–202.

[Eic38] ______, *Über die Idealklassenzahl hyperkomplexer Systeme*, Math. Z. **43** (1938), 481–494.

[Eic55a] ______, *Über die Darstellbarkeit von Modulformen durch Thetareihen*, J. Reine Angew. Math. **195** (1955), 156–171.

[Eic55b] ______, *Zur Zahlentheorie der Quaternionen-Algebren*, J. Reine Angew. Math. **195** (1955), 127–151.

[Eic58] ______, *Quadratische Formen und Modulfunktionen*, Acta Arith. **4** (1958), 217–239.

[Elk98] N. D. Elkies, *Shimura curve computations*, Algorithmic Number Theory (Portland, OR, 1998), Lecture Notes in Comput. Sci., vol. 1423, Springer, Berlin, 1998, pp. 1–49.

[FJ95] G. Faltings and B. W. Jordan, *Crystalline cohomology and* $\mathrm{GL}(2,\mathbb{Q})$, Israel J. Math. **90** (1995), 1–66.

[FK97] R. Fricke and F. Klein, *Vorlesungen über die Theorie der automorphen Funktionen*, Leipzig, 1897.

[For51] L. R. Ford, *Automorphic functions*, 2nd ed., Chelsea, 1951.

[Gau01] C. F. Gauss, *Disquisitiones arithmeticae*, Gerh. Fleischer, 1801; *Disquisicions aritmètiques*, Societat Catalana de Matemàtiques, 1996 (Catalan translation by G. Pascual).

[Gér79] P. Gérardin, *Algèbres de quaternions*, Variétés de Shimura et fonctions L, Publ. Math. Univ. Paris VII, vol. 6, 1979, pp. 145–155.

[God87] R. Godement, *Cours d'algèbre*, Collect. Enseign. Sci., vol. 5, Hermann, Paris, 1987.

[Has33] H. Hasse, *Die Struktur der R. Brauerschen algebrenklassengruppe über einem algebraischen Zahlkörper*, Math. Ann. **107** (1933), 731–760.

[Hij74] H. Hijikata, *Explicit formula of the traces of Hecke operators for* $\Gamma_0(N)$, J. Math. Soc. Japan **26** (1974), 56–82.

[HM95] K. Hashimoto and N. Murabayashi, *Shimura curves as intersections of Humbert surfaces and defining equations of QM-curves of genus two*, Tohoku Math. J. (2) **47** (1995), no. 2, 271–296.

[Iha68a] Y. Ihara, *The congruence monodromy problem*, J. Math. Soc. Japan **20** (1968), 107–121.

[Iha68b] ______, *On congruence monodromy problems.* I, Lecture Notes, vol. 1, Dept. of Mathematics, University of Tokyo, Tokyo, 1968.

[Iha69] ______, *On congruence monodromy problems.* II, Lecture Notes, vol. 2, Dept. of Mathematics, University of Tokyo, Tokyo, 1969.

[Ish75] N. Ishii, *An application of the Fricke formula for quaternion groups*, Math. Japon. **20** (1975), 171–177.

[Ji98] S. Ji, *Analogs of* $\Delta(z)$ *for triangular Shimura curves*, Acta Arith. **84** (1998), no. 2, 97–108.

[JL85] B. W. Jordan and R. A. Livné, *Local Diophantine properties of Shimura curves*, Math. Ann. **270** (1985), no. 2, 235–248.

[JL86] ———, *On the Néron model of Jacobians of Shimura curves*, Compositio Math. **60** (1986), no. 2, 227–236.

[JL87] ———, *Divisor classes on Shimura curves rational over local fields*, J. Reine Angew. Math. **378** (1987), 46–52.

[JL89] ———, *Conjecture "epsilon" for weight $k > 2$*, Bull. Amer. Math. Soc. (N.S.) **21** (1989), no. 1, 51–56.

[JL99] ———, *On Atkin–Lehner quotients of Shimura curves*, Bull. London Math. Soc. **31** (1999), no. 6, 681–685.

[Jon67] B. W. Jones, *The arithmetic theory of quadratic forms*, The Mathematical Association of America, 1967.

[Jor81] B. W. Jordan, *On the Diophantine arithmetic of Shimura curves*, Ph.D. thesis, Harvard University, 1981.

[Jor84] ———, *p-adic points on Shimura curves*, Séminaire de théorie des nombres de Paris 1982–1983 (M.-J. Bertin and C. Goldstein, eds.), Progr. Math., vol. 51, Birkhäuser, Boston, MA, 1984, pp. 135–142.

[Jor86] ———, *Points on Shimura curves rational over number fields*, J. Reine Angew. Math. **371** (1986), 92–114.

[Kam90] S. Kamienny, *Points on Shimura curves over fields of even degree*, Math. Ann. **286** (1990), no. 4, 731–734.

[KM85] N. M. Katz and B. Mazur, *Arithmetic moduli of elliptic curves*, Ann. of Math. Stud., vol. 108, Princeton Univ. Press, Princeton, NJ, 1985.

[Kur79] A. Kurihara, *On some examples of equations defining Shimura curves and the Mumford uniformization*, J. Fac. Sci. Tokyo Sect. IA Math. **25** (1979), no. 3, 277–300.

[Kur94] ———, *On p-adic Poincaré series and Shimura curves*, Internat. J. Math. **5** (1994), no. 5, 747–763.

[Lat36] C. G. Latimer, *The quadratic subfields of a generalized quaternion algebra*, Duke Math. J. **2** (1936), 681–684.

[Lat37] ———, *The classis of integral sets in a quaternion algebra*, Duke Math. J. **3** (1937), 237–247.

[Leh64] J. Lehner, *Discontinuous groups and automorphic functions*, Math. Surveys, vol. 8, Amer. Math. Soc., Providence, RI, 1964.

[Leh92] J. L. Lehman, *Levels of positive definite ternary quadratic forms*, Math. Comp. **58** (1992), no. 197, 399–417.

[Lin93] S. Ling, *Shimura subgroups of Jacobians of Shimura curves*, Proc. Amer. Math. Soc. **118** (1993), no. 2, 385–390.

[Mic81a] J.-F. Michon, *Courbes de Shimura de genre* 1, Séminaire Delange–Pisot–Poitou, 1981.

[Mic81b] ———, *Courbes de Shimura hyperelliptiques*, Bull. Soc. Math. France **109** (1981), no. 2, 217–225.

[Mor81] Y. Morita, *Reduction modulo $\mathfrak{P}$ of Shimura curves*, Hokkaido Math. J. **10** (1981), no. 2, 209–238.

[MR91] B. Mazur and K. A. Ribet, *Two-dimensional representations in the arithmetic of modular curves*, Courbes modulaires et courbes de Shimura (Orsay, 1987/1988), Astérisque, vol. 196-197, Soc. Math. France, Paris, 1991, pp. 215–255.

[Ogg69] A. P. Ogg, *Modular forms and Dirichlet series*, W. A. Benjamin, Inc., New York–Amsterdam, 1969.

[Ogg74] ———, *Hyperelliptic modular curves*, Bull. Soc. Math. France **102** (1974), 449–462.

[Ogg83] ———, *Real points on Shimura curves*, Arithmetic and Geometry. I, Progr. Math., vol. 35, Birkhäuser, Boston, MA, 1983, pp. 277–303.

[Ogg85] ———, *Mauvaise réduction des courbes de Shimura*, Séminaire de théorie des nombres (Paris, 1983-1984), Progr. Math., vol. 59, Birkhäuser, Boston, MA, 1985, pp. 199–217.

[Piz73] A. Pizer, *Type numbers of Eichler orders*, J. Reine Angew. Math. **264** (1973), 76–102.

[Piz76a] ———, *On the arithmetic of quaternion algebras*, Acta Arith. **31** (1976), no. 1, 61–89.

[Piz76b] ———, *On the arithmetic of quaternion algebras*. II, J. Math. Soc. Japan **28** (1976), no. 4, 676–688.

[Piz80] ———, *An algorithm for computing modular forms on $\Gamma_0(N)^*$*, J. Algebra **64** (1980), no. 2, 340–390.

[Poi87] H. Poincaré, *Les fonctions fuchsiennes et l'arithmétique*, J. Math. **4** (1887), 405–464.

[Ray70] M. Raynaud, *Spécialisation du foncteur de Picard*, Inst. Hautes Études Sci. Publ. Math. **38** (1970), 27–76.

[Rei75] I. Reiner, *Maximal orders*, London Math. Soc. Monogr., vol. 5, Academic Press, London–New York, 1975.

[Rib80] K. A. Ribet, *Sur les variétés abéliennes à multiplications réelles*, C. R. Acad. Sci. Paris Sér. A-B **291** (1980), no. 2, 121–123.

[Rib89] ———, *Bimodules and abelian surfaces*, Algebraic Number Theory, Adv. Stud. Pure Math., vol. 17, Academic Press, Boston, MA, 1989, pp. 359–407.

[Rib90] ———, *On modular representations of* $\mathrm{Gal}(\overline{\mathbb{Q}}/\mathbb{Q})$ *arising from modular forms*, Invent. Math. **100** (1990), no. 2, 431–476.

[Rib91] ———, *Lowering the levels of modular representations without multiplicity one*, Internat. Math. Res. Notices (1991), no. 2, 15–19.

[Rib94] ———, *Report on* mod l *representations of* $\mathrm{Gal}(\overline{\mathbb{Q}}/\mathbb{Q})$, Motives (Seattle, WA, 1991), Proc. Sympos. Pure Math., vol. 55, Part 2, Amer. Math. Soc., Providence, RI, 1994, pp. 639–676.

[Rot02] V. Rotger, *Abelian varieties with quaternionic multiplication and their moduli*, Ph.D. thesis, Universitat de Barcelona, 2002.

[RT97] K. A. Ribet and S. Takahashi, *Parametrizations of elliptic curves by Shimura curves and by classical modular curves*, Proc. Nat. Acad. Sci. U.S.A. **94** (1997), no. 21, 11110–11114.

[Ser73] J-P. Serre, *A course in arithmetic*, Grad. Texts in Math., vol. 7, Springer-Verlag, New York–Heidelberg, 1973.

[Ser87] ———, *Sur les représentations modulaires de degré* 2 *de* $\mathrm{Gal}(\overline{\mathbb{Q}}/\mathbb{Q})$, Duke Math. J. **54** (1987), no. 1, 179–230.

[Shi59] G. Shimura, *On the theory of automorphic functions*, Ann. of Math. (2) **70** (1959), 101–144.

[Shi67] ———, *Construction of class fields and zeta functions of algebraic curves*, Ann. of Math. (2) **85** (1967), 58–159.

[Shi70a] ———, *On canonical models of arithmetic quotients of bounded symetric domains.* I, Ann. of Math. (2) **91** (1970), 144–222.

[Shi70b] ———, *On canonical models of arithmetic quotients of bounded symetric domains.* II, Ann. of Math. (2) **92** (1970), 528–549.

[Shi71] ———, *Introduction to the arithmetic theory of automorphic functions*, Publ. Math. Soc. Japan, vol. 11, Iwanami Shoten, Tokyo; Princeton Univ. Press, Princeton, NJ, 1971.

[Shi75] ———, *On the real points of arithmetic quotient of a bounded symetric domain*, Math. Ann. **215** (1975), 135–164.

[Shz65] H. Shimizu, *On zeta functions of quaternion algebras*, Ann. of Math. (2) **81** (1965), 166–193.

[Sie35] C. L. Siegel, *Über die analytische Theorie der quadratischen Formen*, Ann. of Math. (2) **36** (1935), 527–606.

[Sie44] ———, *On the theory of indefinite quadratic forms*, Ann. of Math. (2) **45** (1944), 577–622.

[Sie71] ———, *Topics in complex function theory.* II. *Automorphic functions and abelian integrals*, Wiley-Interscience, New York, 1971.

[TW95] R. Taylor and A. Wiles, *Ring-theoretic properties of certain Hecke algebras*, Ann. of Math.(2) **141** (1995), no. 3, 553–572.

[Var98] Y. Varshavsky, *p-adic uniformization of unitary Shimura varieties*, Inst. Hautes Études Sci. Publ. Math. **87** (1998), 57–119.

[vdP89] M. van der Put, *Les courbes de Shimura*, Sém. Théorie Nombres Bordeaux (2) **1** (1989), no. 1, 89–102.

[Vig80] M.-F. Vignéras, *Arithmétique des algèbres de quaternions*, Lecture Notes in Math., vol. 800, Springer, Berlin, 1980.

[Wil95] A. Wiles, *Modular elliptic curves and Fermat's last theorem*, Ann. of Math. (2) **141** (1995), no. 3, 443–551.

[Zag81] D. B. Zagier, *Zetafunktionen und quadratische Körper*, Hochschultex, Springer, Berlin–New York, 1981.

Index

Titles in This Series

22 **Montserrat Alsina and Pilar Bayer,** Quaternion orders, quadratic forms, and Shimura curves, 2004

21 **Andrei Tyurin,** Quantization, classical and quantum field theory and theta functions, 2003

20 **Joel Feldman, Horst Knörrer, and Eugene Trubowitz,** Riemann surfaces of infinite genus, 2003

19 **L. Lafforgue,** Chirurgie des grassmanniennes, 2003

18 **G. Lusztig,** Hecke algebras with unequal parameters, 2003

17 **Michael Barr,** Acyclic models, 2002

16 **Joel Feldman, Horst Knörrer, and Eugene Trubowitz,** Fermionic functional integrals and the renormalization group, 2002

15 **José I. Burgos Gil,** The regulators of Beilinson and Borel, 2002

14 **Eyal Z. Goren,** Lectures on Hilbert modular varieties and modular forms, 2002

13 **Michael Baake and Robert V. Moody, Editors,** Directions in mathematical quasicrystals, 2000

12 **Masayoshi Miyanishi,** Open algebraic surfaces, 2001

11 **Spencer J. Bloch,** Higher regulators, algebraic K-theory, and zeta functions of elliptic curves, 2000

10 **James D. Lewis,** A survey of the Hodge conjecture, Second Edition, 1999

9 **Yves Meyer,** Wavelets, vibrations and scaling, 1998

8 **Ioannis Karatzas,** Lectures on the mathematics of finance, 1996

7 **John Milton,** Dynamics of small neural populations, 1996

6 **Eugene B. Dynkin,** An introduction to branching measure-valued processes, 1994

5 **Andrew Bruckner,** Differentiation of real functions, 1994

4 **David Ruelle,** Dynamical zeta functions for piecewise monotone maps of the interval, 1994

3 **V. Kumar Murty,** Introduction to Abelian varieties, 1993

2 **M. Ya. Antimirov, A. A. Kolyshkin, and Rémi Vaillancourt,** Applied integral transforms, 1993

1 **D. V. Voiculescu, K. J. Dykema, and A. Nica,** Free random variables, 1992